HUMAN BIOLOGY

HUMAN BIOLOGY

DAVID WRIGHT

Heinemann Educational,
a division of Heinemann Educational Books Ltd
Halley Court, Jordan Hill, Oxford OX2 8EJ

OXFORD LONDON EDINBURGH MADRID
ATHENS BOLOGNA PARIS MELBOURNE SYDNEY
AUCKLAND SINGAPORE TOKYO IBADAN NAIROBI
HARARE GABORONE PORTSMOUTH NH (USA)

ISBN 0 435 59960 7

First published 1989
93 94 95 11 10 9 8 7 6 5 4

Designed by The Pen and Ink Book Company Ltd,
Huntingdon, Cambridgeshire

Printed in Great Britain by The Bath Press, Avon

Acknowledgements

Associated Press, p.232, p.233 (right); Associated Sports Photography, p.101; Atomic Energy of Canada Ltd, 14.2 Fig 6; Barnabys, p.100, 13.4 Fig 1, 13.14 Fig 1, 14.3 Fig 2; Biophoto Associates, 1.1 Fig 1, 4.4 Fig 1, 13.1 Fig 1, 13.4 Fig 1; Camera Press, p.48, 11.4 Fig 1; John Cleare/Mountain Camera, p.62; Chris Davies/Network p.74; Farmers Weekly, 14.1 Fig 5; Geoff Franklin/Network, p.194; Friends of the Earth, 14.4 Fig 2; Granada TV, p.170; Sally & Richard Greenhill, p.iii, p.126, 8.5 Figs 2, 4, 8.6 Fig 3, p.186, p.233 (left); Greenpeace, 14.5 Fig 3; Health Education Council, 12.1 Fig 2, 13.6 Fig 1; ICI, p.195; Imperial War Museum, 14.4 Fig 3; Institute of Environmental Health Officers, 13.12 Figs 2, 4; J Allan Cash, p.63, p.160 (top & bottom right), 11.5 Fig 3, 13.14 Fig 2, 14.3 Fig 1; Jenny Mathews/Format, 13.12 Fig 1; New Zealand High Commission, 14.2 Fig 7; Novo Industri A/S, p.3 (bottom); Oxford Scientific Films, 2.12 Fig 1, 13.3 Fig 2, 13.5 Fig 1 (bottom right); Brenda Prince/Format, 13.12 Fig 3, 14.4 Fig 5; Rex Features, p.3 (top), p.16; Science Photo Library, 5.3 Fig 1 (top & bottom), 8.3 Fig 3, 13.2 Fig 2 (4), 13.3 Fig 5, 13.4 Fig 1 (bottom left), 13.5 Fig 1 (5); Raissa Page/Format, p.160 (left), 11.4 Fig 4, 11.5 Fig 4; St Bartholomew's Hospital, 2.2 Figs 1, 2, 2.8 Fig 2, 6.2 Fig 2 (2), p.112, 11.5 Fig 2, 13.4 Fig 2 (3); John Sturrock/Network, 10.2 Fig 1; Topham, 14.5 Fig 5; Bob Watkins, 13.9 Figs 2, 3; WHO, 13.3 Fig 4.

Illustrated by Chris Etheridge
and Nancy Sutcliffe

Contents

Chapter 1 The organisation of the body

Enzymes beat washday blues 2
1.1 Cell structure 4
1.2 Cell division 6
1.3 Cells into bodies 8
1.4 Metabolism 10
1.5 Movement into cells 12
Questions 14

Chapter 2 Nutrition

New ways to feed the world 16
2.1 Food 18
2.2 Minerals and vitamins 20
2.3 Food sources 22
2.4 Recycling the nutrients 24
2.5 Growing food 26
2.6 Agriculture 28
2.7 A healthy diet 30
2.8 Too much or too little 32
2.9 The world's food problems 34
2.10 The digestive system 36
2.11 Digestion 38
2.12 Absorption and assimilation 40
2.13 The structure of the teeth 42
2.14 Care of teeth 44
Questions 46

Chapter 3 Transport

How to help save a life 48
3.1 The composition of blood 50
3.2 The circulation of the blood 52
3.3 The heart 54
3.4 Heart disease 56
3.5 The functions of the blood 58
Questions 60

Chapter 4 Respiration

Breathing in strange places 62
4.1 Energy 64
4.2 The air you breathe 66
4.3 Breathing 68
4.4 Smoking and lung diseases 70
Questions 72

Chapter 5 Sensitivity and co-ordination

How new drugs are developed 74
5.1 The senses 76
5.2 The eye 78
5.3 Focusing 80
5.4 The ear 82
5.5 The nervous system 84
5.6 The central nervous system 86
5.7 The nerves 88
5.8 Drugs and the nervous system 90
5.9 Social drugs 92
5.10 Mental illness 94
5.11 The endocrine system 96
Questions 98

Chapter 6 Support and locomotion

Muscle power 100
6.1 The skeleton 102
6.2 Bone 104
6.3 Joints 106
6.4 Muscles and movement 108
Questions 110

Chapter 7 Homeostasis

Genetic engineering solves the insulin problem 112
7.1 Temperature control 114
7.2 Overheating and overcooling 116
7.3 Excretion of waste materials 118
7.4 Excretion and osmoregulation 120
7.5 Blood sugar regulation 122
Questions 124

Chapter 8 Reproduction

Infertile couples helped to have babies 126
8.1 Kinds of reproduction 128
8.2 Getting the gametes together 130
8.3 Development of the foetus 132
8.4 Birth 134
8.5 Antenatal care 136
8.6 Drugs and diet 138
Questions 140

Chapter 9 Growth and development

New parts for old 142
9.1 The newborn baby 144
9.2 The child 146
9.3 The growing child 148
9.4 The adolescent 150
9.5 Ageing 152
9.6 Exercise and rest 154
9.7 Social diseases 156
Questions 158

Chapter 10 Growth of populations

People, people everywhere! 160
10.1 Overpopulation 162
10.2 Population control 164
10.3 Birth control 166
Questions 168

Chapter 11 Inheritance and variation

Scientists harness bacteria 170
11.1 Genes 172
11.2 Mendel and inheritance 174
11.3 Inheritance 176
11.4 Variation 178
11.5 Variation and evolution 180
11.6 Evolution 182
Questions 184

Chapter 12 Health

Health and social class 186
12.1 Influences on health 188
12.2 You and your health 190
Questions 192

Chapter 13 Disease

Bugs can be useful! 194
13.1 Organisms that cause disease 196
13.2 Bacteria 198
13.3 Protozoa 200
13.4 Fungi 202
13.5 Insects and worms 204
13.6 The spread of disease 206
13.7 Fighting disease 208
13.8 Safe water and sewage disposal 210
13.9 Safe food 212
13.10 Preserving food 214
13.11 Health care 216
13.12 Environmental health 218
13.13 Preventing disease 220
13.14 Antiseptics, disinfectants and antibiotics 222
13.15 The body fights back 224
13.16 Helping the body fight back 226
13.17 AIDS 228
Questions 230

Chapter 14 The environment

Recycling is becoming big business 232
14.1 Living with the environment 234
14.2 Management of resources 236
14.3 Pollution 238
14.4 Polluting the air 240
14.5 Polluting the water 242
14.6 Conservation 244
Questions 246

Index 248

To the reader

Human Biology has been specially written to cover **all** the topics you are likely to come across during a GCSE Human Biology examination course but it is also suitable for other courses at this level. For those of you who are not studying for such an examination, I hope it will be a worthwhile and stimulating read.

Each major topic is presented in double-page spreads, each spread containing one easily-managed unit of work. Most of the headings are in question form. Here I have tried to anticipate the questions you might ask. At the end of each unit are some more questions for *you* to answer to test your recall and understanding of the work in the unit.

One of the aims of all GCSE Human Biology courses is to link the subject to everyday life. With this in mind I have started each topic with a double-page spread relating to the personal, social, economic or technological applications of Biology. These themes in fact are spread throughout the whole book. Even the questions at the end of each topic use data which have been taken from real situations.

I hope that you enjoy using this book and that it helps you appreciate the wondrous nature of human life and the Earth on which we live. I also hope that it helps you achieve success in your examination.

David Wright

The start of a new life is a very special moment.

CHAPTER 1: THE ORGANISATION OF

Your body is made up of millions of cells. Inside each cell many chemical reactions are taking place. Some of these reactions would not normally happen at body temperature. Others would only take place very slowly.

To speed these reactions up, the cells produce various **enzymes**. Each enzyme speeds up one particular reaction. In this chapter you will learn more about the cells in your body, enzymes and the movement of chemicals between cells.

Enzymes beat washday blues

This section explains how enzymes are being used to speed up reactions outside the body.

Biological washing powders

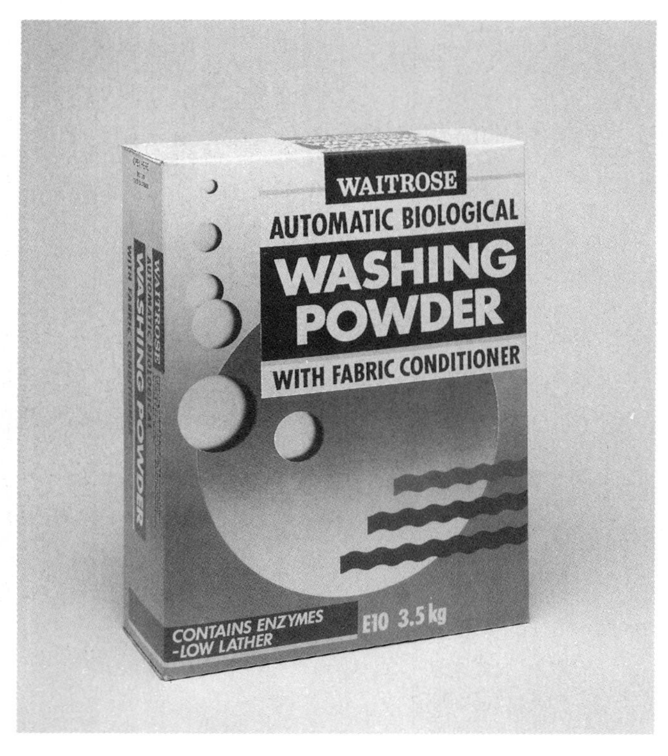

You have probably heard of 'biological' washing powders. This means that the powder contains enzymes. Although common today, they have only been available since the 1960s.

Some stains like blood are very difficult to remove. If you soak bloodstained material in ordinary washing powder and cold water, the blood will only come out very slowly, if at all. If you use hot water in a washing machine it will come out more quickly, but may still take several washes.

A biological powder will usually get rid of the blood in one wash, because the enzymes speed up the reaction. However, to get the best results, clothes should be soaked in the powder and water first. Suggested soaking times are:

Temperature (°C)	**Soaking time** (min)
20	960
30	480
40	120–240
50	60–120
60	15–30

Biological washing powders are especially good for certain materials that cannot be washed in hot water. They should not however be used on natural fibres such as wool. Can you think why?

Where do the enzymes come from?

Enzymes can only be made in living cells. Most are made by micro-organisms grown in large containers called **fermenters** (page 194).

The enzymes for use in washing powders are made like this. A common bacterium called *Bacillus subtilis* is fed on a mixture of potato starch, soy meal and minerals. As it grows, it produces the enzymes. These are removed by filtration and then dried. They look like a fine brown powder which can be added to the detergent.

Enzymes in other industries

Of the 2000 or so enzymes known, about 300 have found uses in industrial processes.

- **Baking** Enzymes are used to convert the starch in flour into sugar. It is this sugar that the yeast ferments, producing carbon dioxide to make the dough rise.

- **Brewing** Enzymes are again used to turn starch into sugar. Yeast ferments this sugar producing alcohol.
  Enzymes are also used to improve the clarity of beers and wines.
- **Dairying** Enzymes are used to separate milk into curd and whey. The curd is turned into cheese. More enzymes are added to speed up the ripening of this cheese.
- **Detergents** Enzymes are added to washing powders to help remove biological stains such as blood.
- **Leather** Enzymes are used to remove hairs from the skin and soften it.
- **Medical and pharmaceutical** Enzymes are used to produce high specification drugs and hormones such as insulin (page 113). They are also used to clean wounds and break down blood clots.
- **Syrups** Enzymes are used to produce very sweet fructose syrups from less sweet sugars. These are used to sweeten things like coca cola.
- **Textiles** Starch is added to fabrics as a strengthener during the weaving process. Enzymes are used to remove this. This removal process is called **desizing**.

Desizing of fabrics used to involve strong chemicals such as acids. Enzymes now do the job without damaging the fabric.

What are the advantages of using enzymes?

Enzymes are being used in many different processes nowadays because:

- They are very specific and therefore the operator has greater control over the process.
- They are only needed in very tiny amounts and are re-usable, so reducing costs.
- The product they produce is always the same quality.

1.1 Cell structure

What are you made from?

You may be surprised to know that most of your body is made from a jelly-like substance which looks like it hasn't quite set! This substance is called **protoplasm**. It contains about 80 per cent water, the rest being **proteins, fats, oils** and **sugars** (see page 18).

How do you stay in shape?

Although protoplasm looks like a runny jelly it does not move about freely in your body. It is packaged into small units called **cells**. This makes it easier to build into the correct shape. You can compare this to a house which is made from clay. In order to make the building of the house easier, the clay is packaged into small bricks. In a similar way cells are the building bricks in your body.

What does a cell look like?

Cells are very small – so small in fact that you need a **microscope** to see them. The cells in figure 1 opposite have been enlarged 500 times. The actual size of these cells was 1/100 cm.

The protoplasm of a cell has two distinct areas. The darker area in the photograph opposite is called the **nucleus**. This is the control centre of the cell. The remainder of the protoplasm is called the **cytoplasm**. The outside covering of the cell is called the **cell membrane**. This acts as a barrier controlling the entry and exit of substances into or out of the protoplasm.

Cell ultrastructure

The most powerful microscopes reveal complex features within the protoplasm. For example the cytoplasm contains many smaller structures, collectively called **organelles**. Some of these organelles, together with the jobs they do are shown in figure 2.

What are plants made from?

Plants are also made from cells but there are several important structural differences between plant cells and animal cells. Some of these are shown in figure 2.

How many cells does your body contain?

A human baby contains about 30 million million cells. An adult body contains in excess of 100 million million cells.

Are all these cells the same?

You have only to look at yourself to see that all cells are *not* the same, even though they are all made from protoplasm covered by a cell membrane. Cells are designed to do all sorts of different jobs. For example skin cells are designed for protection and muscle cells are designed for moving. Figure 3 shows some of the types of cells that make up your body.

Questions

1. A typical factory has an office where records are kept and decisions made, a production area and a boundary wall. Suggest why a cell might be compared to this.
2. Why do you think it took a long time to discover that a body was made from cells?
3. Draw a typical animal cell and label the nucleus, cytoplasm, cell membrane and a mitochondrion.
4. The actual size of the cell you have drawn would be about 1/100 cm. Work out the magnification of your drawing.
5. Your body will contain somewhere between 30 and 100 million million cells. These will not all be the same since cells are designed to do a particular job. With this in mind make a list of the many different types of cells your body is likely to contain.
6. Draw up a table of differences and similarities between plant and animal cells.

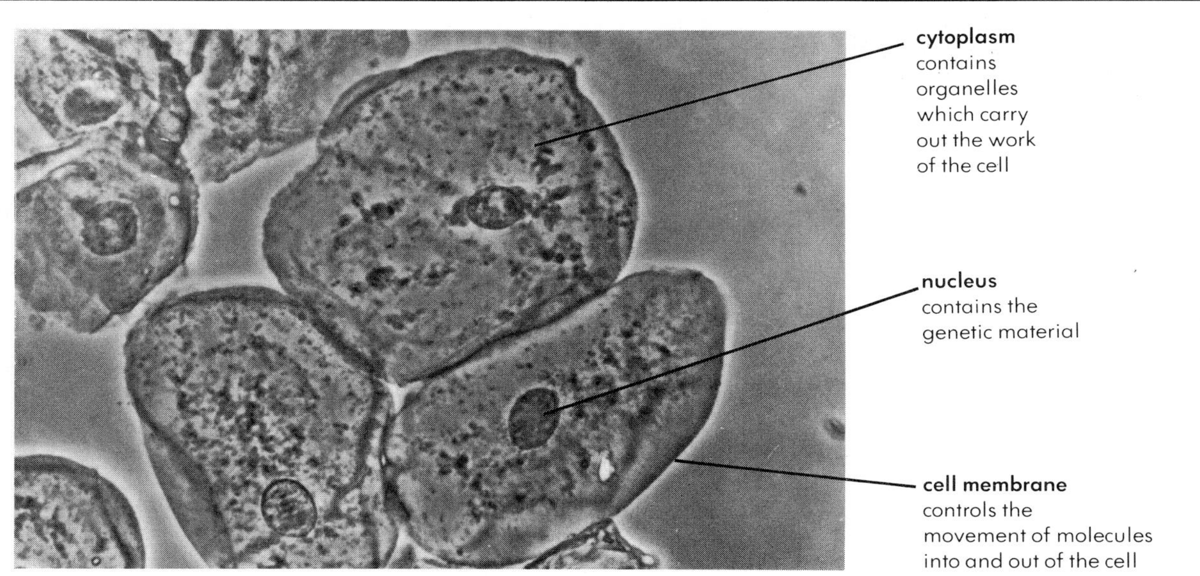

Figure 1 These human cheek cells have been enlarged 500 times

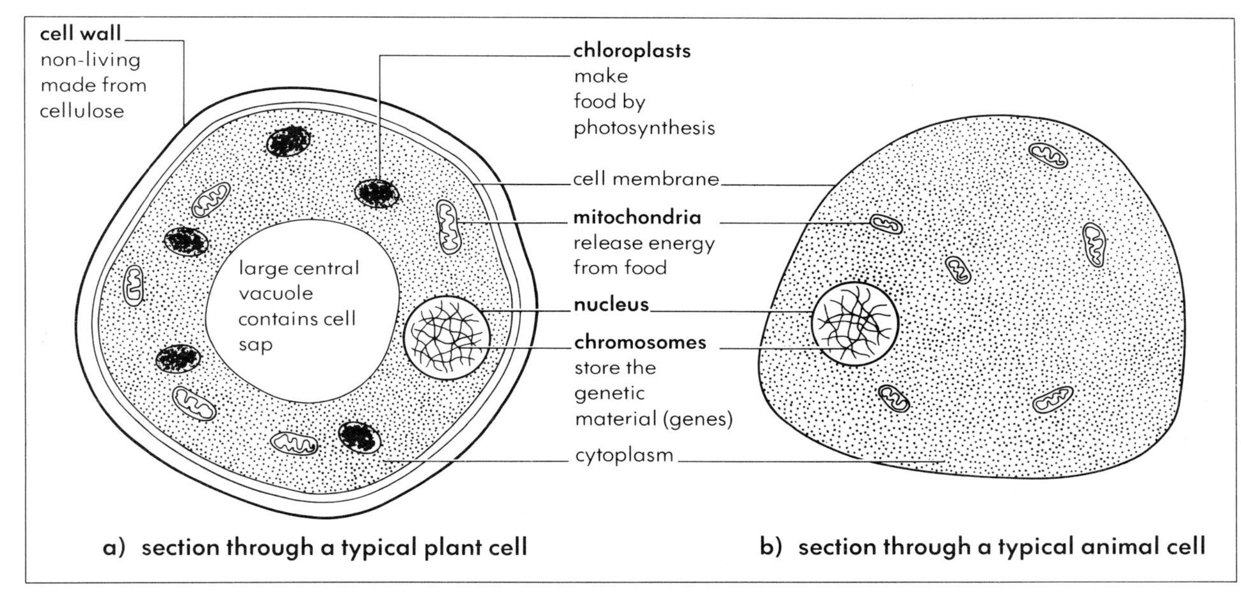

Figure 2 Plant and animal cells have many similarities and a few important differences

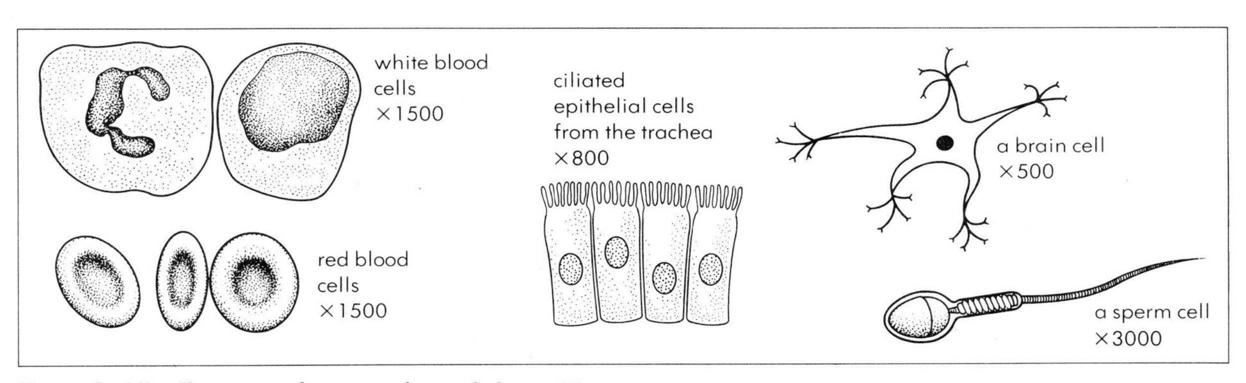

Figure 3 All cells are not the same size and shape. The design depends on the job they are required to do

1.2 Cell division

Cells and more cells

Your body makes more cells as it grows. When you're fully grown, your body goes on making new cells, to replace cells worn out and damaged from all the living you've been doing!

How are new cells made?

A cell uses some of the food it receives to make new protoplasm. The cell therefore gets bigger and when it reaches a certain size it divides into two smaller ones. These then repeat the process until all the cells the body needs have been produced. This is how **growth** occurs. This kind of cell division is called **mitosis** and is shown in figure 1. During mitosis, each new cell produced receives exactly the same **genetic information** (see page 172) as the original cell.

What is a cancer cell?

A great number of people die from some form of **cancer** every year in the UK (table 1). You may have wondered what cancer is and why it is so serious.

Sometimes a cell divides for no apparent reason and goes on dividing, forming a ball of cells. A cell which does this is called a **cancer** cell and the ball of cells it forms is called a **tumour**. There are two kinds of tumour: **benign** and **malignant tumours**. Benign tumours stop growing after a while and, unless they are in a vital organ, they seldom cause problems. Malignant tumours never stop growing and often take over important organs, stopping them working properly. Sometimes some of the cancerous cells break away and spread to other parts of the body, where they produce new tumours.

What causes cancer?

The exact cause or causes of cancer are unknown. A lot of research has been done into the links between some forms of cancer and smoking. Cancer can also be brought on by other harmful substances such as atomic radiation and asbestos dust. It is also thought that too much sunbathing can increase your chances of developing forms of skin cancer.

Can cancer be cured?

Some cancers, if discovered early enough, can be cured. Unfortunately one of the many problems with cancer is that it is often very difficult to detect in the early stages. However, once diagnosed, cancer drugs, radiation therapy and surgery are all used successfully to treat people with cancer. These treatments kill or remove the tumour cells or at least stop them from dividing and spreading. The success of the treatment depends partly on the type of cancer and how soon the cancerous cells are discovered. Unfortunately many people still die of cancer, despite research and advances in treatment.

Leukaemia

Leukaemia is a cancer of the white blood cells and the cells that make these. In 1984 it affected about 4000 people in the UK, resulting in the death of about 300 of them.

Questions

1. Describe with diagrams what happens during mitosis. Why are cells produced by mitosis always identical when fully grown?
2. Explain how a cancerous tumour is formed. Why are malignant tumours often so much more damaging to health than benign tumours?
3. Make a list, commonest first, of the main types of cancer that affected (a) men, and (b) women in the UK in 1982 (figure 2). Suggest reasons for the differences.
4. Draw a line graph of the deaths from all cancers during the period 1974 to 1984 (table 1). Does this show that the number of deaths is increasing or decreasing? Is this increase/decrease greater during the first five or the second five years of this period?
5. Name a type of cancer where the number of deaths is decreasing. Suggest possible reasons for this.

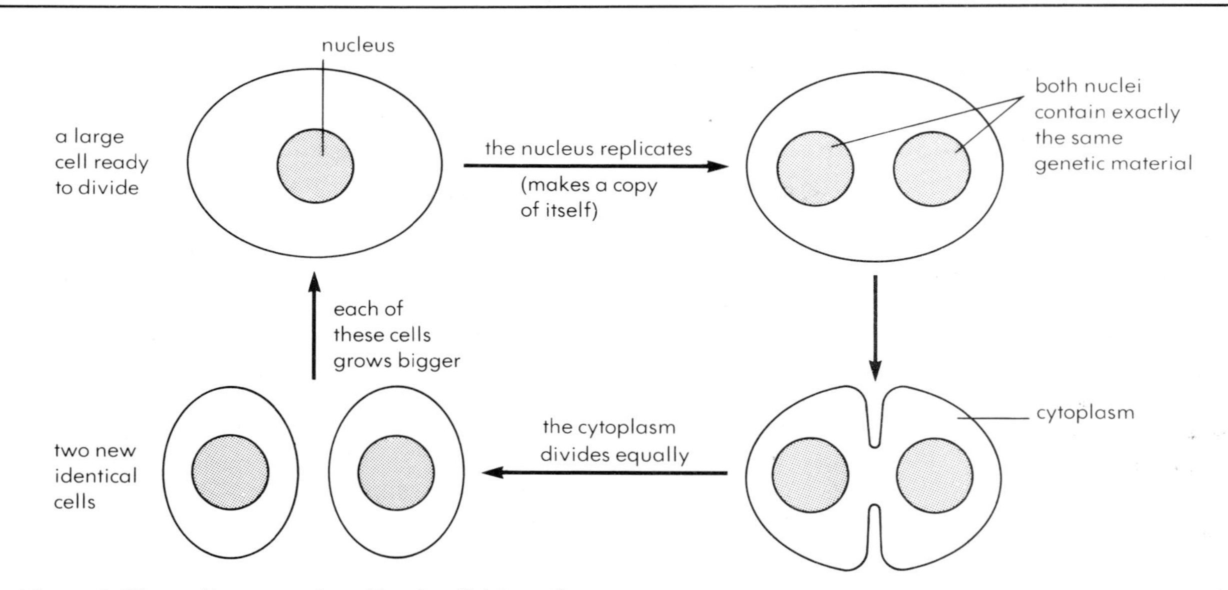

Figure 1 New cells are produced by the division of one cell into two. This is called mitosis

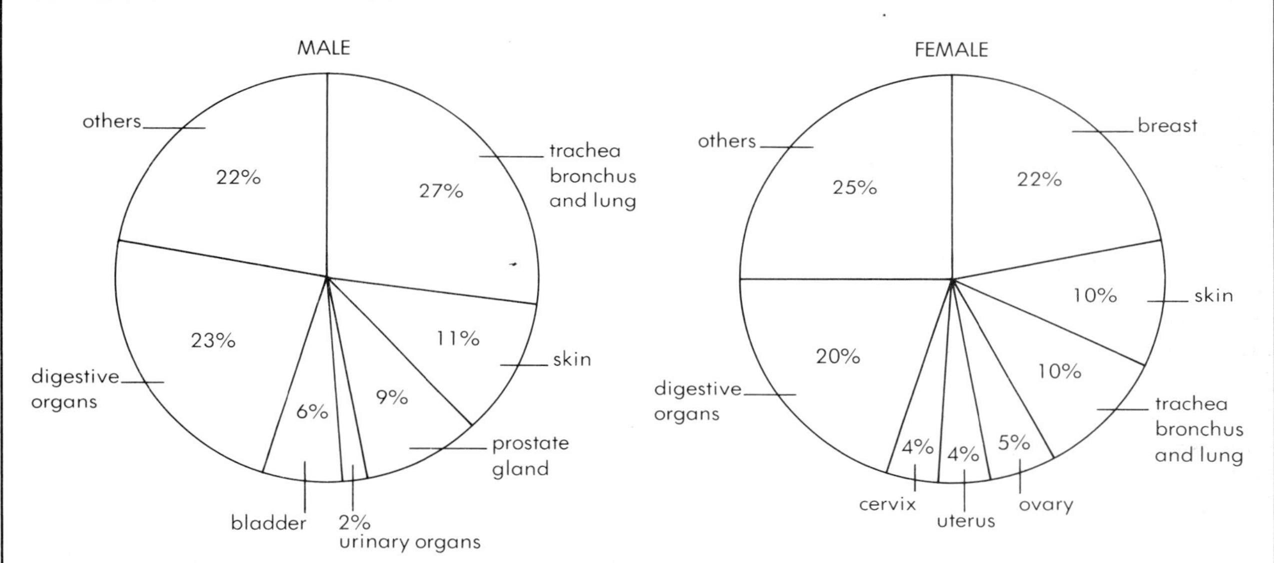

Figure 2 The most common cancers affecting men and women in 1982

Type of cancer / Year	1974	1979	1982	1984
Trachea, bronchus and lung	33 057	34 756	34 832	35 739
Skin	1 111	1 184	1 358	1 432
Cervix	2 068	2 087	1 932	1 899
Breast	11 319	12 091	12 405	13 405
Digestive organs	39 507	40 199	38 896	41 112
All	123 022	129 638	130 561	140 101

Table 1 The number of deaths in the UK from cancer is generally increasing

1.3 Cells into bodies

How are cells organised into a body?

There are many different types of cells in your body. These cells do not just float around any old how but are organised in a very fixed way into all the complex organs and systems of your body. This organisation can be understood in three stages:

1 Cells form tissues

As you know different kinds of cell in your body do different jobs. For example, all muscle cells help you to move and skin cells form a protective covering. Cells which do the *same* job in a body are grouped together into a **tissue**.

The main tissues in the human body are **muscle tissue** made from muscle cells; **nervous tissue** made from nerve cells and **lining (epithelial) tissue** made from skin cells.

2 Tissues form organs

How are the cells held together in these tissues in your body? The answer is with another tissue: **connective tissue** which is made from connective cells. You could say this is the cement.

Sometimes connective tissue may hold more than one tissue together, as in the heart. The heart is made from muscle tissue, nervous tissue and lining tissue, all held together by connective tissue. All these tissues work together to enable the heart to pump blood round the body.

When more than one tissue is present in a structure, that structure is called an **organ**. Figure 1 shows the three tissues in the small intestine. The positions of some of the main organs in your body are shown in figure 2.

3 Organs work in systems

For you to see anything, you need your eyes and brain, to hear anything, you need your ears and brain. For almost *any* action carried out by your body, several organs are used. When organs work together like this, they are grouped together and called an **organ system**.

The organ systems in the human body

- The **heart, blood vessels** and **blood** make up the **circulatory system**.
- **The lungs, windpipe** and **diaphragm** make up the **respiratory stystem**.
- The **muscles** of the body make up the **muscular system**.
- The **bones** of the body make up the **skeletal system**.
- The **kidneys, bladder, ureters** and **urethra** make up the **urinary system**.
- The **sex organs** make up the **reproductive system**.
- The **gullet, stomach, intestines** and **liver** make up the **digestive system**.
- The **pituitary, thyroid, adrenal** and other glands make up the **glandular (endocrine) system**.
- The **nerves, spinal cord** and **brain** make up the nervous system.
- The **eyes, ears, tongue, nose** and **skin** make up the **sensory system**.

Many of the organs have a function in more than one system, for example, the male urethra acts as a duct for both urine and sperm.

Questions

1 Complete the table:

Name of tissue	Cells in tissue	Function of tissue
Muscle Nervous Epithelial Connective		

2 Copy and fill in the 'organ trees'.

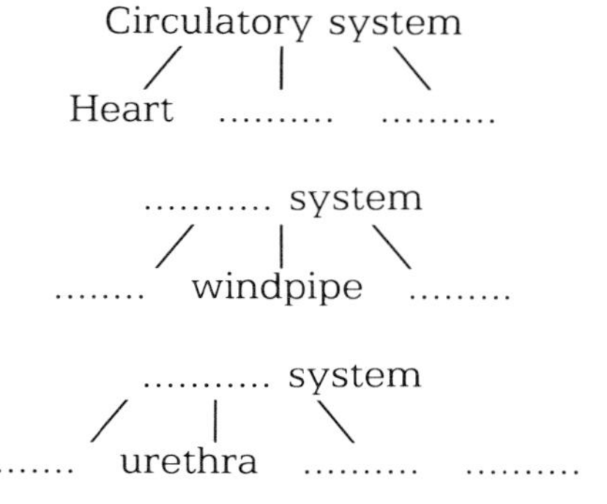

3 List the other organ systems found in your body and for each describe its function.

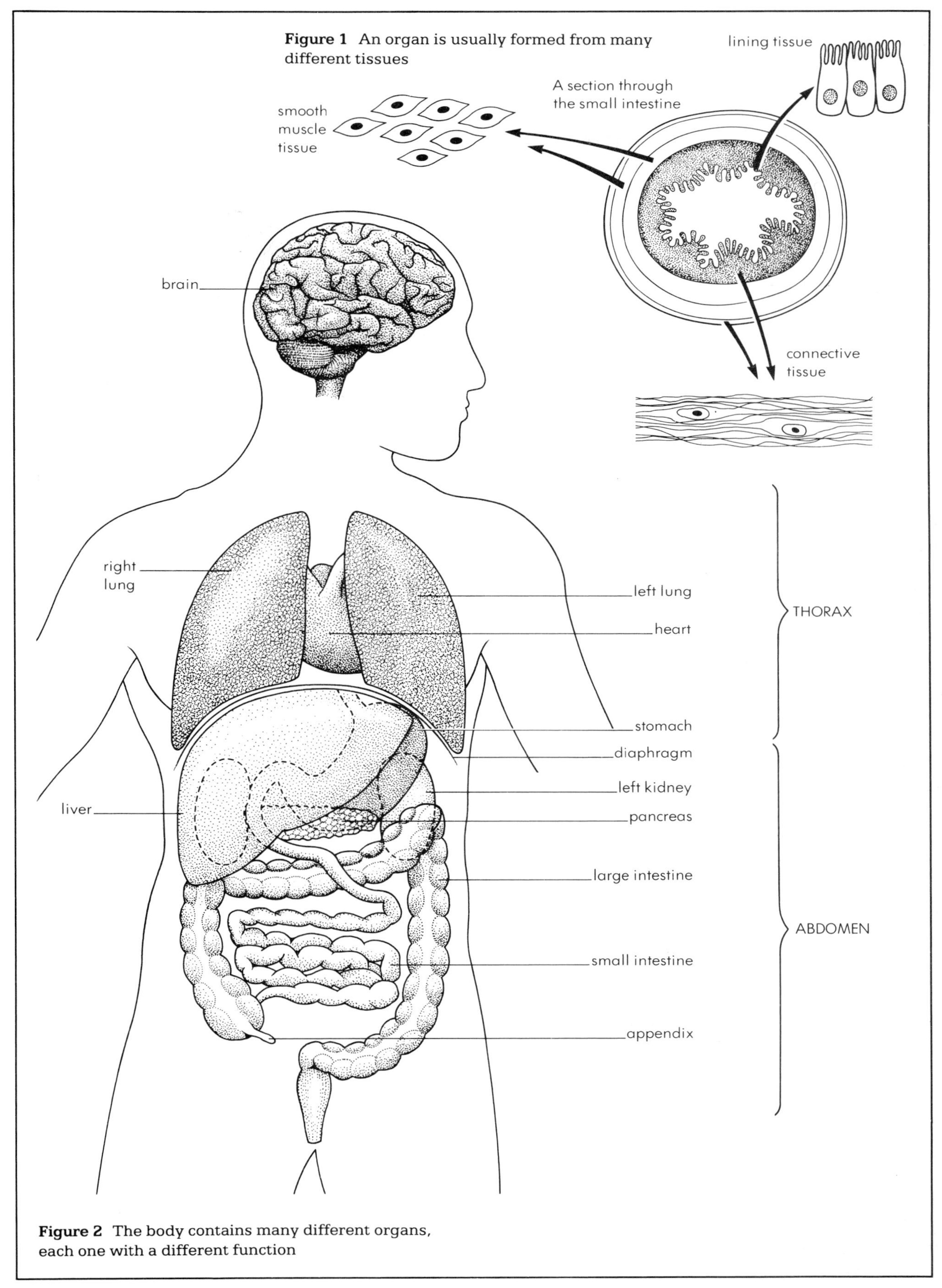

Figure 1 An organ is usually formed from many different tissues

Figure 2 The body contains many different organs, each one with a different function

1.4 Metabolism

What goes on inside cells?

Even though cells are tiny in size an awful lot goes on in each individual cell! Each cell is like a chemical factory. Within the protoplasm hundreds of chemical reactions are usually taking place. Some of these reactions make use of energy and raw materials to *build up* more complex substances, such as proteins. These reactions are examples of **anabolism**. Other chemical reactions *break down* complex substances which releases energy. These kind of reactions are examples of **catabolism**.

All the chemical activity taking place within the cells of an organism (anabolic and catabolic) is usually referred to as its **metabolism**. Figure 1 shows some examples of this.

Can a cell control its metabolism?

A cell controls its metabolism by making **enzymes**. Enzymes are very special chemicals produced by the cells to speed up their chemical reactions. Without these, metabolism would be too slow to carry on life.

An enzyme will only speed up one particular reaction. By producing the correct sequence of enzymes, a cell can therefore control its metabolism.

What kind of chemical is an enzyme?

All enzymes are proteins. They can only be produced by living cells. Some more of their important characteristics are:

- They are **affected by heat**. In general, high temperatures destroy enzymes. Cooler temperatures slow their action down (figure 2a). The enzymes in your body work best at temperatures around 37 °C (see page 114).
- They are **affected by pH**. Most enzymes have an optimum pH and will only work within a narrow range around this (see figure 2b). Enzymes can therefore be switched on and off by altering the pH, as happens in digestion (see page 38).
- Enzymes are very **specific**. Each enzyme catalyses one reaction only.
- Although enzymes speed up and in some cases even start reactions, they themselves remain unchanged and can therefore be reused. This means they are very efficient. They are also only needed in very tiny amounts.

Some important enzymes

Enzymes can either be produced to work inside or outside a cell. Those which work inside a cell are collectively known as **intracellular enzymes**. The enzymes involved in the digestion of food are all **extracellular**. These are produced by special glandular cells and secreted into the alimentary canal.

Digestive enzymes are of three types:

- **Carbohydrases**. These catalyse reactions involving carbohydrates, e.g. amylase, maltase.
- **Proteases**. These catalyse reactions involving proteins, e.g. pepsin, trypsin.
- **Lipases**. These catalyse reactions involving lipids, e.g. lipase.

Questions

1 Explain how the characteristics of enzymes are made use of by cells to control their metabolism.

2 a) What is the difference between intracellular and extracellular enzymes?
b) What are the three important groups of digestive enzymes?

3 Pepsin, trypsin and amylase are enzymes which are important in digestion. Digestion takes place in your mouth, stomach and intestines. The pH in these places is 6, 2 and 8 respectively. Suggest, with reasons, which of the three enzymes will probably be found in which of the regions (see figure 2b).

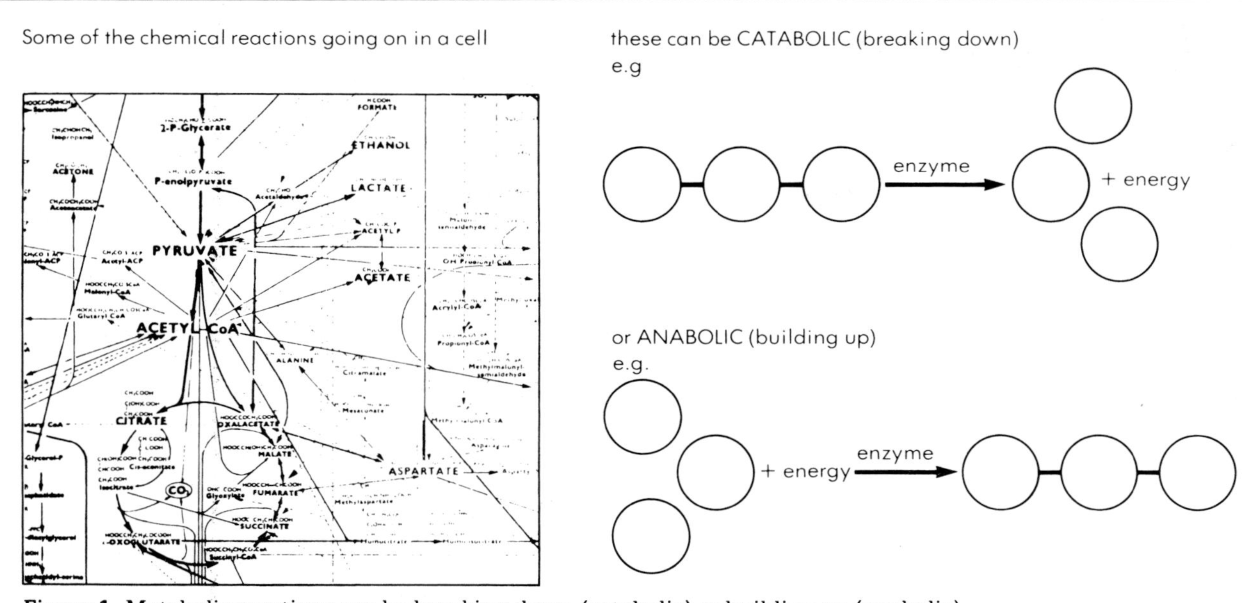

Figure 1 Metabolic reactions can be breaking down (catabolic) or building up (anabolic)

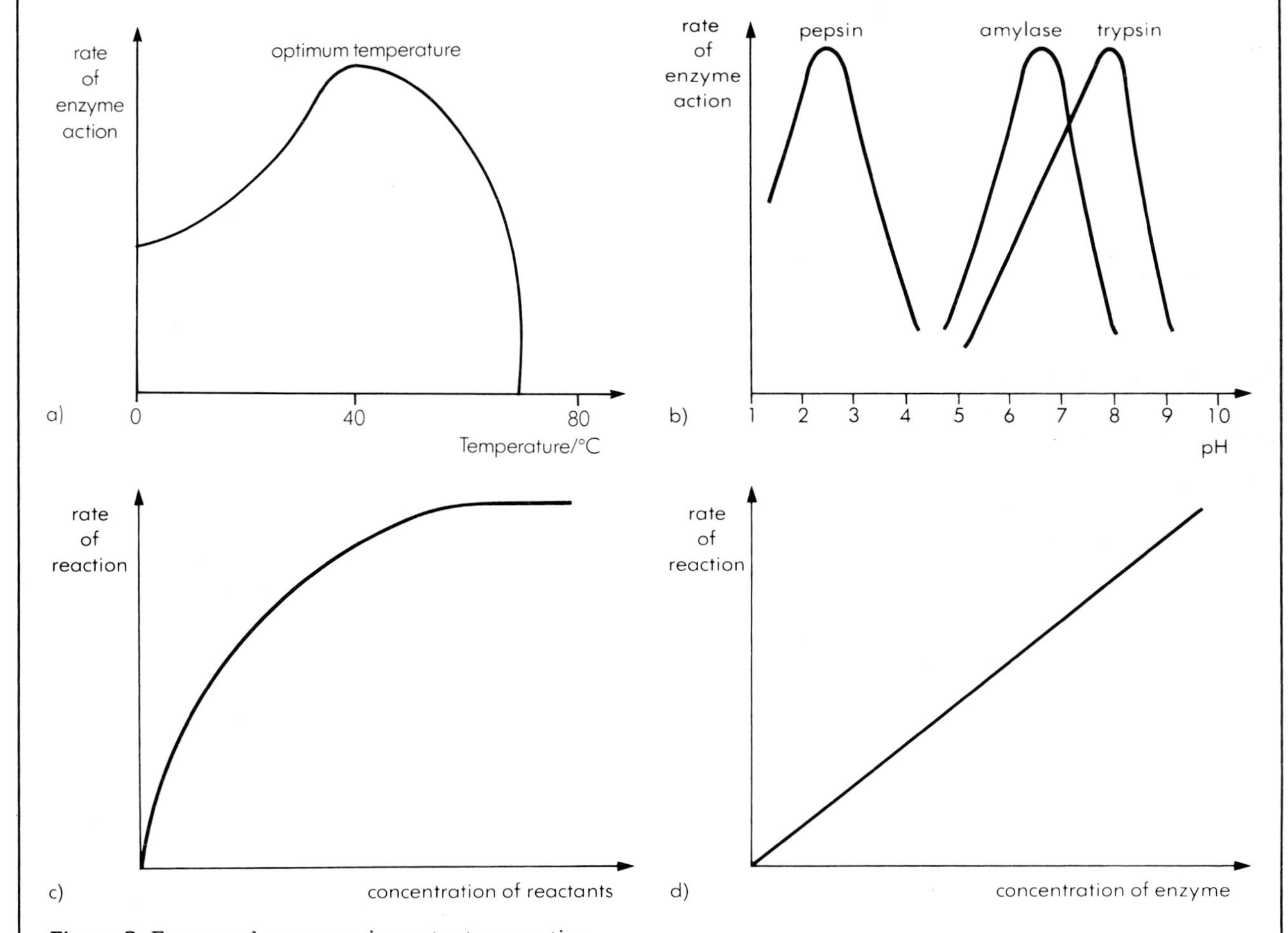

Figure 2 Enzymes have some important properties

a) Enzymes in your body work best at around 37°C

b) Most enzymes only work well within a narrow range of pH

c) For a given amount of enzyme the speed of the reaction is dependent on the concentration of the reactants

d) If there is more enzyme present the rate of the reaction increases up to a limit

1.5 Movement into cells

What do cells need?

The chemical reactions in cells (see page 10) take place in a watery solution. This is one reason why your body needs plenty of water. Your cells also need a constant supply of oxygen and other chemicals to carry on their work.

Where do these substances come from?

Most of the chemicals and water come from the food you eat. The oxygen comes from the air you breathe. All these important substances are transported to the cells by the blood.

How do these substances get into the cells?

Many of the substances simply **diffuse** from the blood into the cells. Others have to be **actively transported** into the cells.

Diffusion

Diffusion is the movement of **molecules** of a substance from an area where there are a lot of them to an area where there are very few, or none at all, figure 1. Diffusion will always occur when there are differences in concentration (a **concentration gradient**). Cells can usually manage to maintain the right concentrations of substances in their protoplasm so that diffusion will take place.

Active transport

There are some substances which are always in a higher concentration inside the cell than in the blood, yet the cell still manages to take them in. This is done by a process called **active transport**. Unlike diffusion, which does not require energy, active transport requires a great deal of energy. This energy is produced in the mitochondria and, therefore, cells which carry out a lot of active transport contain a lot of mitochondria.

What can enter a cell?

The cell membrane contains many small pores, through which only the smallest of molecules can pass. It therefore *selects* what can enter according to size. Such a membrane is said to be **selectively permeable**.

Osmosis – a special kind of diffusion

The diffusion of water molecules across a selectively permeable membrane is called **osmosis**, figure 2. Whenever a weak solution is separated from a strong solution by a selectively permeable membrane, the water molecules will diffuse from the weak solution into the strong solution.

Osmosis is very important to living cells. If an animal's body fluids suddenly become diluted, water starts to enter the cells by osmosis. The cells start to swell and may eventually burst. See figure 3.

Cells eat germs

Some of your body cells, for example white blood cells, can take in very large particles. They do this by a process called **phagocytosis**, which means 'cell eating'.

White blood cells use their ability to move to seek out foreign particles and engulf them. They then digest the particles by secreting enzymes onto them. Details are shown in figure 4.

Questions

1. Look carefully at figures 1 and 2. Fully explain the results.
2. When animal cells are put into a solution which is more dilute than their contents, they swell and may burst. Explain why.
3. How does the cell membrane control the entry and exit of substances into a cell? Use diagrams to show how a larger substance may enter a cell.
4. Which of the following processes uses energy? Diffusion, active transport, osmosis.
5. Use the index to find out by what process the following get in or out of cells:
   a) water,
   b) oxygen,
   c) carbon dioxide,
   d) glucose.

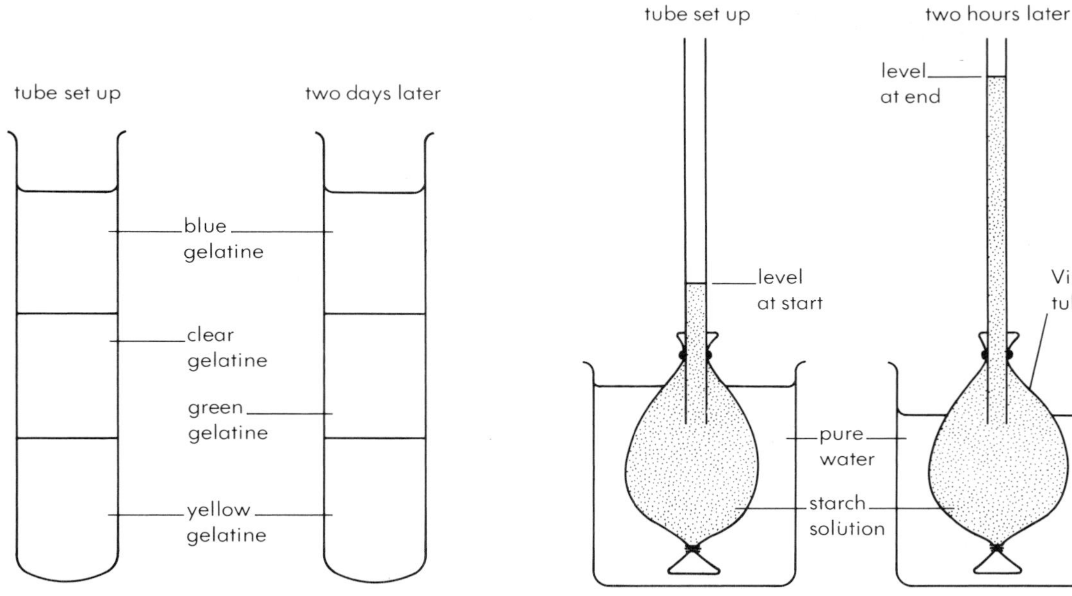

Figure 1 Diffusion occurs when there is a difference in concentration

Figure 2 Diffusion of water across a selectively permeable membrane is called osmosis

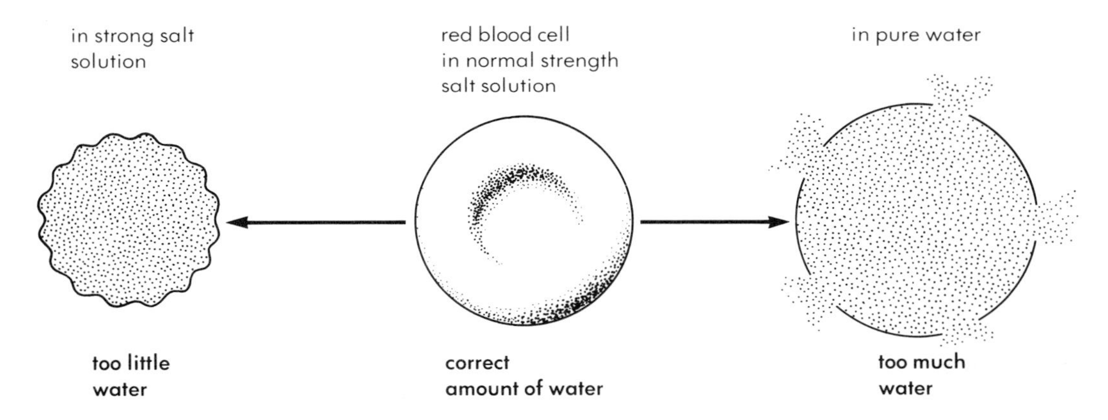

Figure 3 Animal cells take in water and lose water by osmosis. This can sometimes result in their destruction

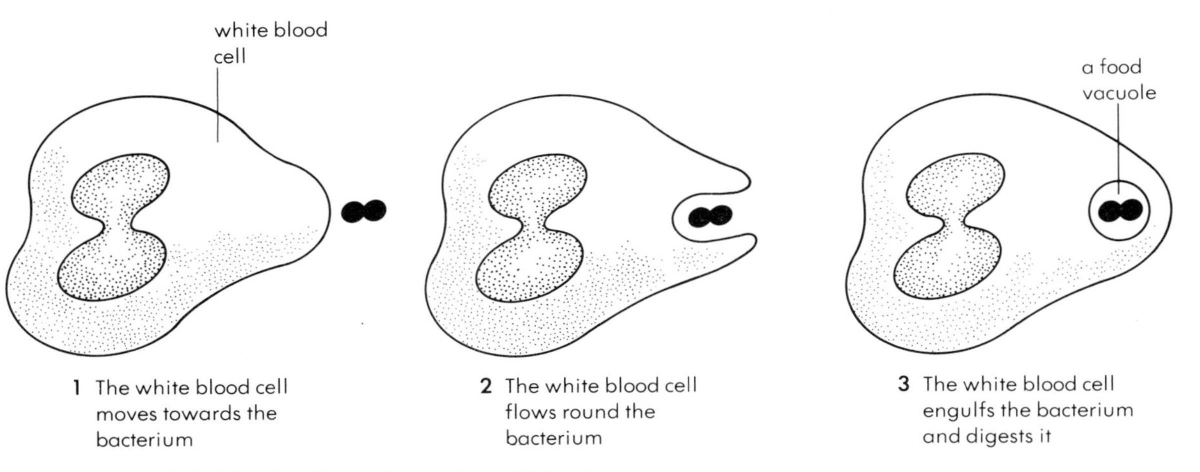

Figure 4 Some white blood cells can ingest (engulf) foreign particles such as bacteria. This is called **phagocytosis**

Questions

Recall and understanding:

1 The cell organelle in which energy is produced is the
A chloroplast
B mitochondrion
C nucleus
D lysosome

2 Mitosis is a form of
A growth
B cell movement
C cell division
D cancer

3 Chloroplasts are found in
A animal cells
B green plant cells
C plant roots
D bark

4 The release of energy from food is an example of
A photosynthesis
B anabolism
C decomposition
D catabolism

5 Metabolism does not take place in
A animal cells
B plant cells
C live cells
D dead cells

6 Which of the following uses energy?
A catabolism
B anabolism
C respiration
D diffusion

7 All enzymes are
A alive
B carbohydrates
C proteins
D fats

8 Match the pairs.

A organ	1 human egg
B cell	2 leukaemia
C organelle	3 heart
D cancer	4 nucleus

9 The actual size of a cell was 1/10 mm. When viewed through a microscope it appeared to be 1 cm. By what amount was the microscope magnifying the cell?
A 100 times
B 10 times
C 1000 times
D 10 000 times

10 Arrange in order simplest first
A organ
B tissue
C cell
D organism
E system

Interpretation and application of data

11 The results below show the progress of a cell dividing regularly to form a tumour.

Hours after first division	1	2	3	5	6	8
Number of cells	4	8	16	64	128	not counted

a) Plot the results on a graph.
b) How many cells were there 4 hours after the first division?
c) Predict how many cells there will be after 8 hours.
d) How many cells were there to start with? (zero hours)

12 Consider the chart

Number of deaths from different forms of cancer				
Cancer \ Year	1974	1979	1982	1984
Lung	33057	34759	34832	35739
Skin	1111	1184	1358	1432
Cervical	2068	2087	1932	1899
Breast	11319	12091	12405	13405

a) Which form of cancer shows the greatest increase?
b) State a possible reason for this.
c) Which form of cancer shows the greatest decrease?
d) State a possible reason for this.
e) What was the total number of deaths from these forms of cancer in 1984?
f) Why is it misleading to compare cervical and breast cancers with lung cancer?

13 The enzyme lipase catalyses the breakdown of fats. This reaction can be followed in a test tube by adding an indicator. The indicator is red to start with and turns yellow when breakdown is complete. The results of an investigation are shown below:

Test tube containing reactants	Temp-erature	Original colour	Colour after 12 hours
1	0 °C	Red	Red
2	10 °C	Red	Orange
3	40 °C	Red	Yellow
4	60 °C	Red	Orange
5	100 °C	Red	Red

a) From the results, at which temperature does lipase work best?
b) (i) Why was there no colour change in test tube 1?
(ii) Why was there no colour change in test tube 5?
c) Predict what would happen if both tubes 1 and 5 were warmed at 40 °C for 12 further hours.
d) Describe an experiment you could do to find the optimum pH for lipase.

14 A student wanted to examine blood through a microscope but she didn't know what concentration of salt solution she should place it in. She devised an experiment to find out. The results are shown below:

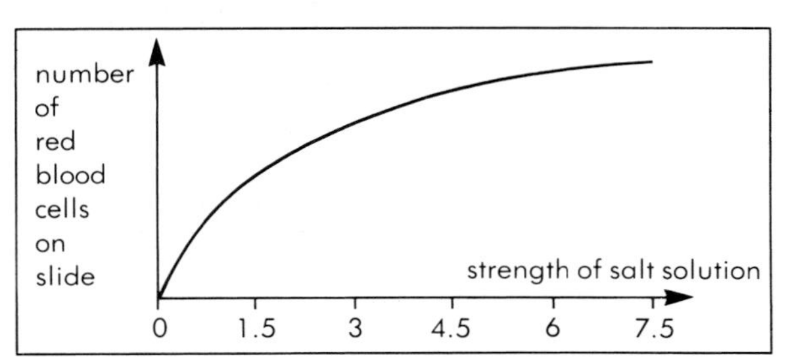

a) Which concentration of salt solution should the student use? Explain your reasoning.
b) Why did the number of red blood cells fall as the strength of the salt solution got progressively weaker?
c) Predict what the student would have seen on the slide had she used even stronger salt solutions.
d) What process is responsible for these changes?

15 The bar chart below shows the death rate due to lung cancer in the five social classes of the UK in 1970–72 for men.

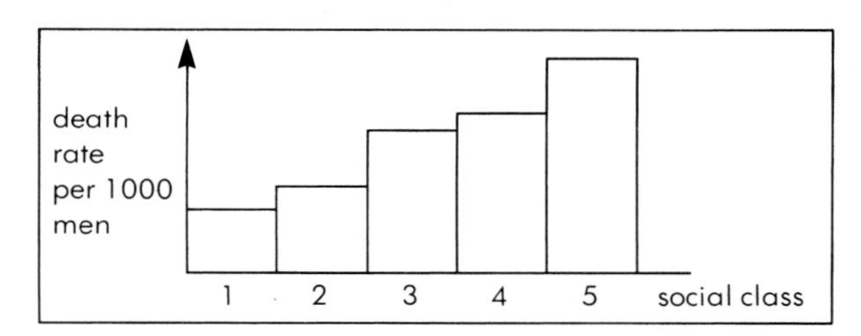

a) What is the relationship between social class and lung cancer?
b) State two possible reasons for this.
c) The figures below show the number of smokers per 100 people in each social class.

Social class	1	2	3	4	5
Number of smokers	21	34	46	46	48

i) Draw a bar chart using these figures.
ii) What do these figures tell you about the relationship between smoking and lung cancer?

CHAPTER 2: *NUTRITION*

You need a regular supply of food to stay alive and healthy. This chapter looks at the different types of food, and why you need a balanced diet. It explains what happens to food inside your body once you have eaten it.

The chapter also explores how food is produced, and the problems in countries like Ethiopia where there isn't always enough food.

New ways to feed the world

Is there a world food shortage?

Whilst many people in Africa are starving because they don't have enough to eat, others in places like Europe have too much. There is enough food in the world to feed the entire population. The problem is therefore not one of shortage, but one of uneven sharing out.

Why we need new sources of food

As the population of the world continues to increase, even the richer nations may find themselves short of food. Traditional methods of agriculture will not be able to cope with the demand. New ways of producing food will have to be found.

Bob Geldof formed Band Aid to help stop the people of Ethiopia dying of starvation. This is one of the concerts they gave to raise money.

Making protein

Much research has taken place over the last few decades into new methods of food production. In particular people have tried to invent new ways of making protein, because this is both difficult and expensive to obtain. Two of the most promising developments are outlined below.

Protein from micro-organisms

It looks like chicken. It tastes like chicken, and it even has the texture of chicken. It is in fact a fungus cleverly disguised as chicken.

Of all the new food sources, the most promising are the micro-organisms. These are easy to grow, reproduce very quickly and contain a lot of protein. One bacterium can give rise to a million bacteria or more in as little as 24 hours, and up to 80% of this new growth can be protein.

Perhaps the biggest clincher is that many of these micro-organisms can be fed on our wastes.

Food from fungus

The food produced from micro-organisms is sometimes referred to as **single cell protein** (SCP). The only one available at present for human consumption is produced by Rank Hovis McDougall. They call this product **mycoprotein** because it is made by a fungus. The fungus is fed on sugars left over from the company's other manufacturing processes. Table 1 compares mycoprotein to beef.

Table 1

Nutrient	Amount present in mycoprotein	Amount present in beef
Carbohydrate	10	0
Fat	14	30
Protein	47	68
Fibre	25	0
Other	4	2
All figures are percent of dry weight		

ICI is probably the worlds largest producer of SCP, over 70 000 tonnes per year. It calls this substance **pruteen**.

How is pruteen produced?

Pruteen is produced from bacteria grown in large fermenters using methanol and ammonia as the food. The bacterium has even been genetically altered to increase its yield.

At present pruteen is only used in animal feed as the high nucleic acid content is damaging to humans. However, scientists are hoping soon to solve this problem.

Protein from plants

Plants have of course been used as food as long as animals, but they have never seriously challenged meat as sources of protein. More often they are dismissed as sources of protein because most of them are deficient in some of the essential amino acids. However some are as nutritious as animals.

Table 2

Animal/plant	Protein (% dry weight)	Protein yield (kg/ha)
Beef	47	25
Pork	35	30
Chicken	50	40
Sunflower	32	380
Soya bean	43	510

Table 2 shows that **soya beans** not only contain as much protein as most meats, but also produce a much higher yield per hectare of land. Genetic engineering (see page 171) may even be able to increase this in the future.

In addition to protein, many of these plants also yield substantial amounts of edible oil.

Why can't we just make food instead?

The simple answer to this is that we *can* make food from its basic chemical 'building blocks', but it is still very expensive. It also tends to lack the texture, flavour and colour of 'real' food. Food scientists are working on this problem, however.

Will there ever be a tablet to replace food?

This is highly unlikely as just to get the energy we need every day, we must eat about ½ kg of food. This is an awful lot of pills.

2.1 Food

Why do you need food?

Food is necessary for life. You probably know that people are dying today in many parts of the world because they don't have enough of the right food.

All living things need a regular and balanced supply of food. The type of food, and the way they get it may differ but the way it is made use of is the same:

- Food provides body cells with the energy to keep living and to carry out their work e.g. muscle cells contracting to help an organism move.
- Food provides the raw materials for cells to grow bigger, multiply and repair themselves when damaged.
- Food provides materials that help metabolism take place in cells and in doing so keeps the body in a healthy state.

What is food?

Look at the label of any yoghurt and you will see something like figure 1. The substances listed are called nutrients and all foods are made up of different combinations of them.

Carbohydrates

Carbohydrates are compounds of the elements **carbon**, **hydrogen** and **oxygen**. The simplest forms of carbohydrates are called **sugars**. **Glucose** is an example. The more complex carbohydrates are built from these simple sugars. For example, plant starch is made from about 300 glucose molecules and animal starch (**glycogen**) is made from about 1500 glucose molecules.

In your body, carbohydrates are mainly used to supply energy but they all have first to be converted to glucose. Glycogen acts as a stored source of glucose. Plants contain a complex carbohydrate called **cellulose**. We sometimes call this **roughage**.

Proteins

Proteins are compounds of the elements carbon, hydrogen, oxygen and **nitrogen**. Some may also contain **sulphur**. These join together to form **amino acids**, which are the basic building bricks of all proteins. There are only twenty different amino acids, and a typical protein is made from a combination of hundreds of them. These amino acids are arranged in a particular order in different proteins, as shown in figure 2.

Plants make all their own amino acids from the raw materials. Our bodies can make some amino acids but we have to get others from the food we eat. Those we **must** obtain in our food are called '**essential amino acids**'. The rest are called 'non-essential', yet they are just as important to our cells. Most proteins from animals contain all the essential amino acids you need. Proteins obtained from plants are usually lacking in some.

There are many thousands of different proteins in your body. Many of these simply form structural components of cells and are therefore, needed for growth and repair. Others form the **enzymes** and **hormones** which are vital for the proper functioning of your body. Without these you will soon become ill.

Fats

Fats are made from simpler substances called **glycerol** and **fatty acids**. These both contain the elements carbon, hydrogen and oxygen. The difference between one fat and another is due to the kind of fatty acids present in the molecule. These fatty acids can be of two main types; **saturated** or **unsaturated**. Most animal fats contain mainly saturated fatty acids, whereas plant fats contain a lot of the unsaturated type. Some unsaturated fatty acids are often called '**essential fatty acids**' because your body cannot make them, yet you must have a regular supply.

The main use of fat in your body is as a source of stored energy. It is stored around some of your vital organs, where it acts as a protecting layer, and under your skin, where it doubles as an insulating layer.

Questions

1 Complete the table for carbohydrates, proteins and fats:

Nutrient	Elements it contains	Basic building bricks	Examples	Uses

2 What are the differences between:
a) essential and non-essential amino acids
b) saturated and unsaturated fatty acids
c) essential and non-essential fatty acids?

3 Use table 1 to calculate how much carbohydrate, protein and fat a cheese sandwich would contain. Assume 100 g is a normal size portion of each food.

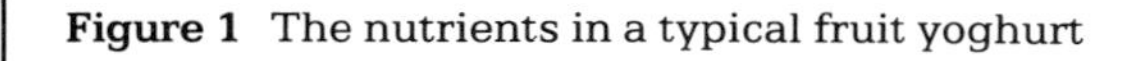

Figure 1 The nutrients in a typical fruit yoghurt

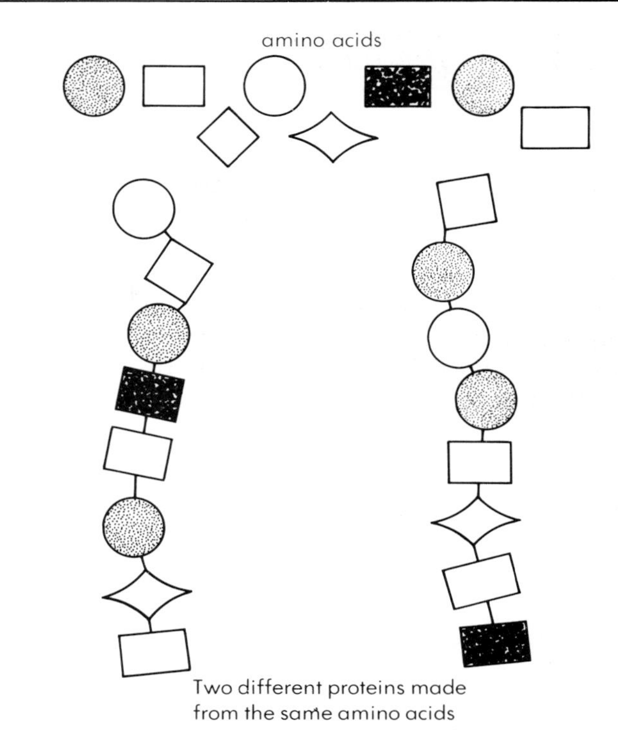

Figure 2 The building blocks for proteins are amino acids

Food	Carbohydrate/g	Protein/g	Lipid (fat)/g
Milk (whole)	4.6	3.2	3.9
Beef	0	16.6	27.4
Eggs	11.8	12.1	10.9
Potatoes (boiled)	18	2	0.2
Bread (white)	48.6	8.2	1.7
White fish	0	17.1	0.9
Butter	0	0.4	82
Cheese (cheddar)	0	26	33.5
Cabbage (cooked)	2.3	1	0
Peas (cooked)	18	7	0
Baked beans	15.1	4.8	0.5
Apples	11.9	0.3	0
Oranges	8.5	0.8	0
Bananas	19.2	1	0
Biscuits (chocolate)	66	7	24
Chocolate (milk)	59.4	8	30.3

Table 1 Biscuits are rich in carbohydrate and fat while fish is rich in protein. This table shows the average amounts of carbohydrate, proteins and fats in 100 g of some common foods

2.2 Minerals and vitamins

Why do you need minerals and vitamins?

Minerals are elements such as **calcium**, **iron** and **iodine**. These all have very specific functions in your body. For example, calcium forms part of bone, making it hard and is also needed to help one of the enzymes involved in blood clotting. If you body does not get a fairly regular supply of calcium your bones grow soft and your blood clots very slowly. These are called **deficiency symptoms**.

Most **vitamins** are complex chemicals. They all have names, but are usually just referred to by a letter. Like minerals, they are only needed in small amounts but deficiency leads to disease. Table 1 and figures 1 and 2 show some of the deficiency symptoms.

There are about thirteen vitamins and these can be split into two groups: those which your body can store and those which it cannot.

1 Vitamins your body can store. These are all found in the fat of fatty foods and are, therefore, called **fat soluble vitamins**. Examples are vitamins **A** and **D**.
2 Vitamins your body cannot store. These are all found in the water in foods and are, therefore, called **water soluble vitamins**. Examples are vitamins **B** and **C**. To prevent disease you must have a *daily* supply of these.

Is water an important nutrient?

Your body consists mainly of water so it is very important to life. **Water forms about 70 per cent of protoplasm** and is found in many other places, such as blood, tissue fluid, tears and digestive juices. It is used by your body in the following ways:

- to **transport food materials**, such as glucose around your body;
- to **transport waste materials**, like urea and carbon dioxide to the excretary organs;
- to **remove soluble waste substances**, such as urea;
- as a **solution in which metabolism** can take place;
- for **the diffusion** of oxygen;
- to **carry digestive enzymes**;
- to **cool down your body**, e.g. sweat
- to **protect parts of your body**, e.g. the central nervous system is protected by the cerebrospinal fluid;
- to **protect the developing baby** (amniotic fluid in the womb);
- to help as **a lubricant**, e.g. mucus and synovial fluid.

Water is so important to the normal functioning of a body that even a small amount of **dehydration** can cause death. The table below shows how daily water losses are replaced.

Daily losses (cm^3)		Daily gains (cm^3)	
evaporation from lungs	400	from food	850
in urine	1500	from drinks	1300
in faeces	100	from respiration	350
in sweat	500		

What are food additives?

Nowadays, many of our foods contain **additives**, especially if they come in cans. These have no nutritional value and are put in food to:

- change its flavour
- change its consistency
- colour it
- preserve it.

Food manufacturers are now supposed to list all additives contained in their food. Figure 5 shows some of the additives that can be put into a loaf of bread.

Questions

1 Which diseases result from a lack of:
  a) vitamin A,
  b) vitamin C,
  c) vitamin D?
2 Which minerals are required to help prevent the diseases anaemia and rickets?
3 For each of the following vitamins/minerals, name two good food sources:
  vitamin A, vitamin C, vitamin D, iron, calcium, iodine.
4 Explain why even a small amount of dehydration can cause death.
5 Using the information in figures 3 and 4 calculate the percentage of your daily requirements of calcium, iron, vitamin A and vitamin C that you would obtain from two boiled eggs (100 g each) and a glass of milk (200 g).

Vitamin or Mineral	Why your body needs it	Deficiency signs
Vitamin A	helps your eyesight	nightblindness – difficulty seeing clearly at night
Vitamin C	helps keep your skin and gums healthy	scurvy – bleeding, swollen gums, wounds heal slowly
Vitamin D	helps your body make use of calcium	rickets – rubbery and soft bones, dental caries.
Calcium	calcium combines with phosphorus to form the hard part of bone	rickets dental caries
Iron	forms part of the haemoglobin molecule	anaemia – tiredness breathlessness and irritability
Iodine	needed to form the hormone thyroxine	your thyroid gland swells, known as a goitre

Table 1 Vitamins and minerals are needed in very small amounts yet deficiency leads to disease

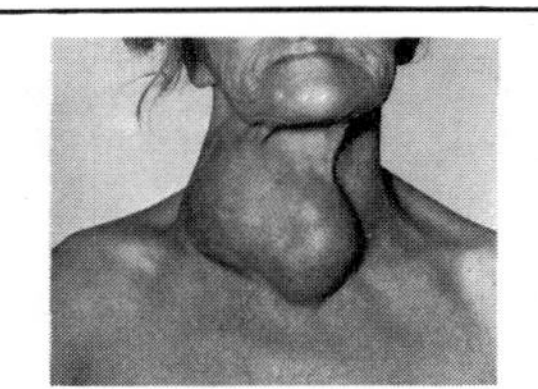

Figure 1 Lack of iodine can lead to goitre – a swelling of the thyroid gland

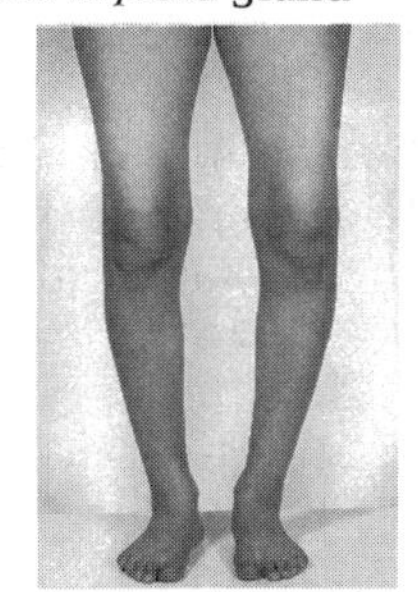

Figure 2 Lack of vitamin D causes rickets. The bones in the leg become too weak to stay straight

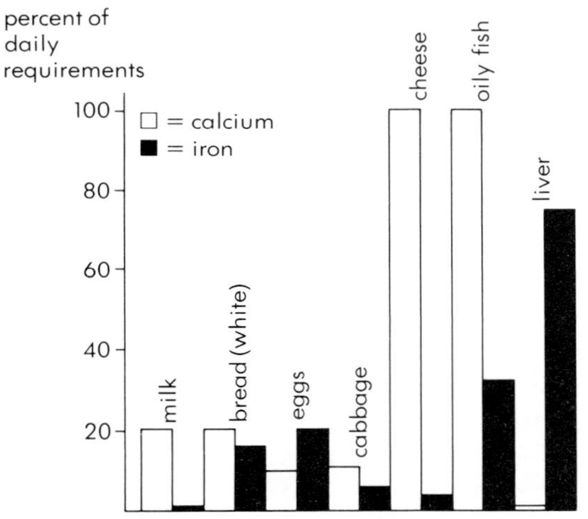

Figure 3 100 g of cheese provides all your daily calcium requirement but only a little of the iron. Compare this to the other foods

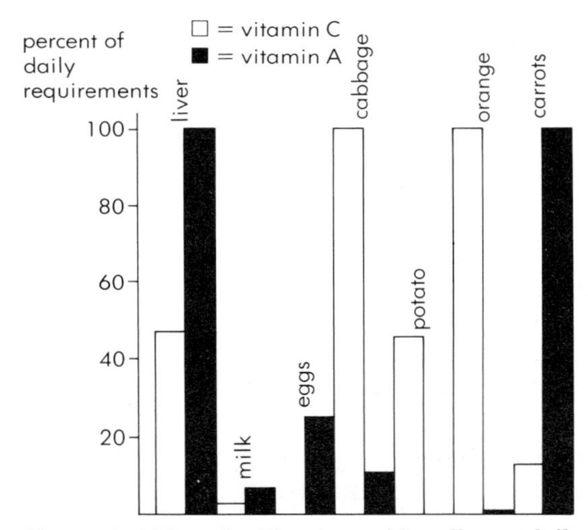

Figure 4 100 g of cabbage provides all your daily vitamin C requirement but only a little of the vitamin A you need. Compare this to other foods

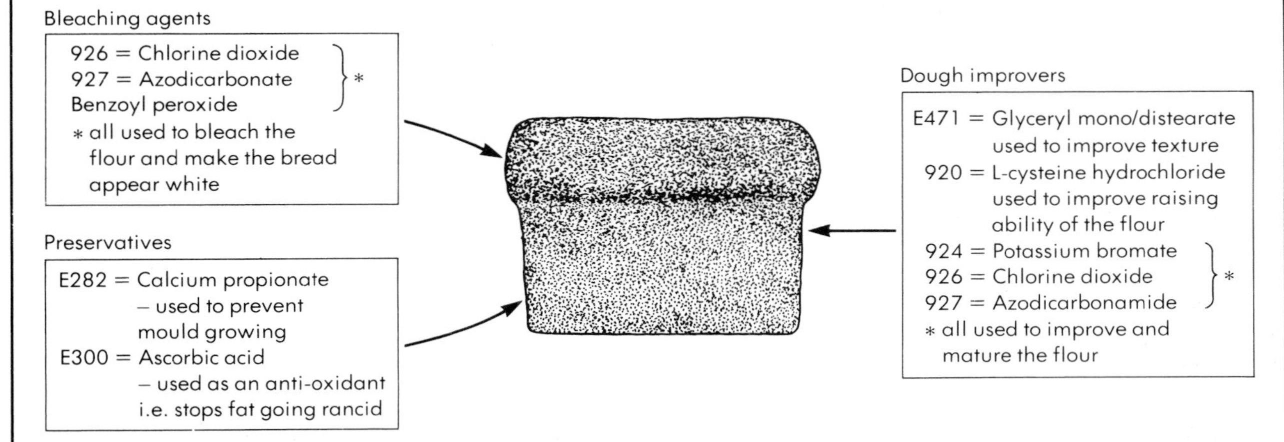

Figure 5 The permitted additives in a loaf of white bread

2.3 Food sources

Where does food come from?

All organisms except green plants (and some bacteria) get the food they need by eating other organisms. Green plants actually make their food from simple inorganic materials. They do this by a process called **photosynthesis**.

What is photosynthesis?

Photosynthesis is a complex process. The details are shown in figure 1. **Light energy** is absorbed by the **chlorophyll** (the green pigment) in the plant cells and is then used to combine water and carbon dioxide to form glucose. Oxygen is given off as a waste product e.g:

$$\text{water} + \text{carbon dioxide} \xrightarrow[\text{(absorbed by chlorophyll)}]{\text{light energy}} \text{glucose} + \text{oxygen}$$

Plants use this glucose in the following ways:

- As a source of energy.
- To convert into starch and store for future use.
- To make lipids (fats and oils) for the waxy cuticle of the plant.
- To make amino acids from which proteins are made.
- To make vitamins.
- To convert into cellulose and other carbohydrates.

Producers and consumers

Because they manufacture food, green plants are called **producers**. All other organisms are **consumers**. The animals which eat green plants are called **primary consumers**. Those which eat primary consumers are **secondary consumers** and so on. Thus organisms are linked together to form a **food chain**. In this way the original food made by the plant is passed on from one organism to another. For example

oak leaves
(producer)

↓ are eaten by

caterpillar
(primary consumer)

↓ is eaten by

thrush
(secondary consumer)

Often, the same organism will crop up in several food chains linking the chains together to form a **food web** (figure 2).

Decomposers

Decomposers are organisms that feed on dead organic matter such as dead plants and animals, fallen leaves or excreted waste. There are two main kinds:

- **Scavengers** which take in lumps of dead material and digest it inside their bodies.
- **Saprophytes** secrete digestive enzymes onto the dead material and then absorb the products of digestion. The result of saprophyte's action is known as **decay** or **decomposition**. All saprophytes are either fungi or bacteria.

Decomposers are essential organisms in nature because they stop dead organic matter building up. In addition, saprophytes release useful materials, such as minerals, for other organisms to use (see page 24).

Questions

1. What is photosynthesis? Write a word equation to summarise the process.
2. Explain why when you eat a beefsteak you are really eating food made by grass.
3. Consider the food web shown in figure 2:
   a) Which of the organisms is the producer?
   b) Draw one of the food chains involving this producer.
   c) Which of the organisms are primary consumers? Explain why.
   d) What order consumer is the hawk?
   e) Rearrange the following into a food web: thrush, hawk, oak tree, aphid.
   f) Which of the organisms in figure 2 are decomposers? What kind of decomposers are they?
4. Copy the following passage and fill in the missing words:
   ………… are essential organisms in nature because they stop the build up of ……… ……………. The action of the …………. also helps ……….. useful materials such as minerals. Examples of ……….. are………. and …….. .

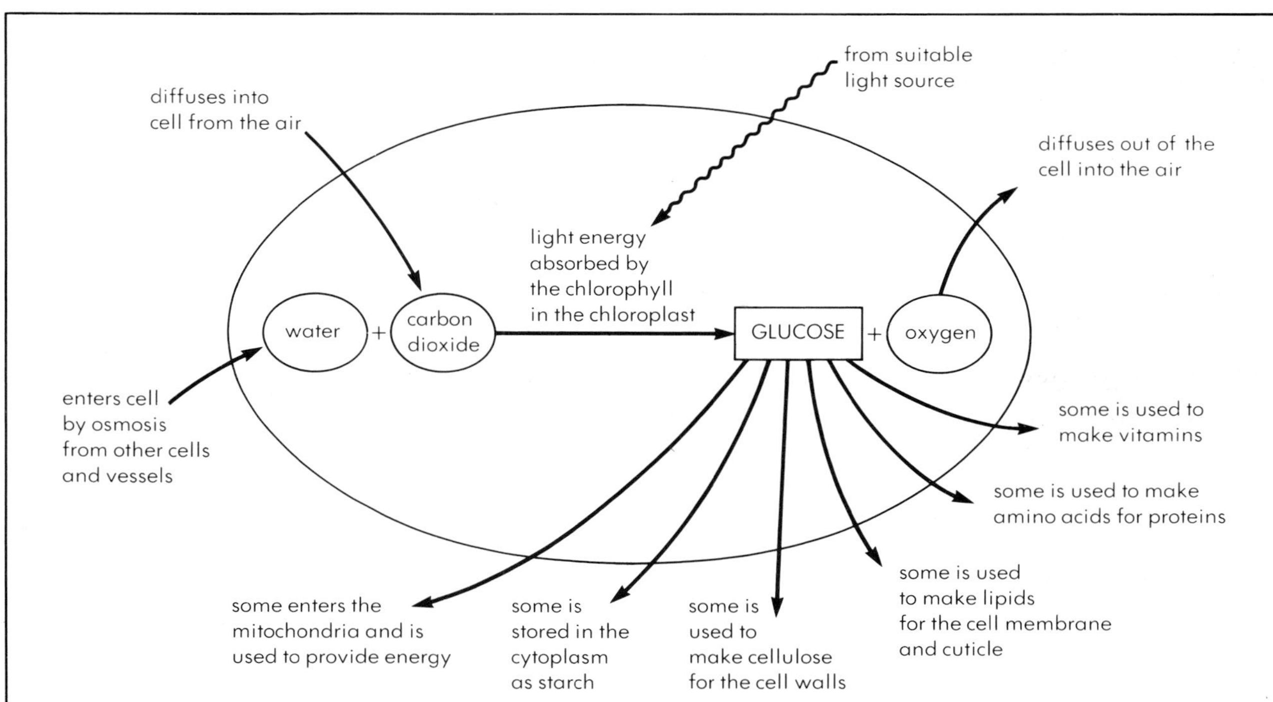

Figure 1 Photosynthesis takes place inside the chloroplasts of plant cells. It produces glucose which the plant uses in many different ways

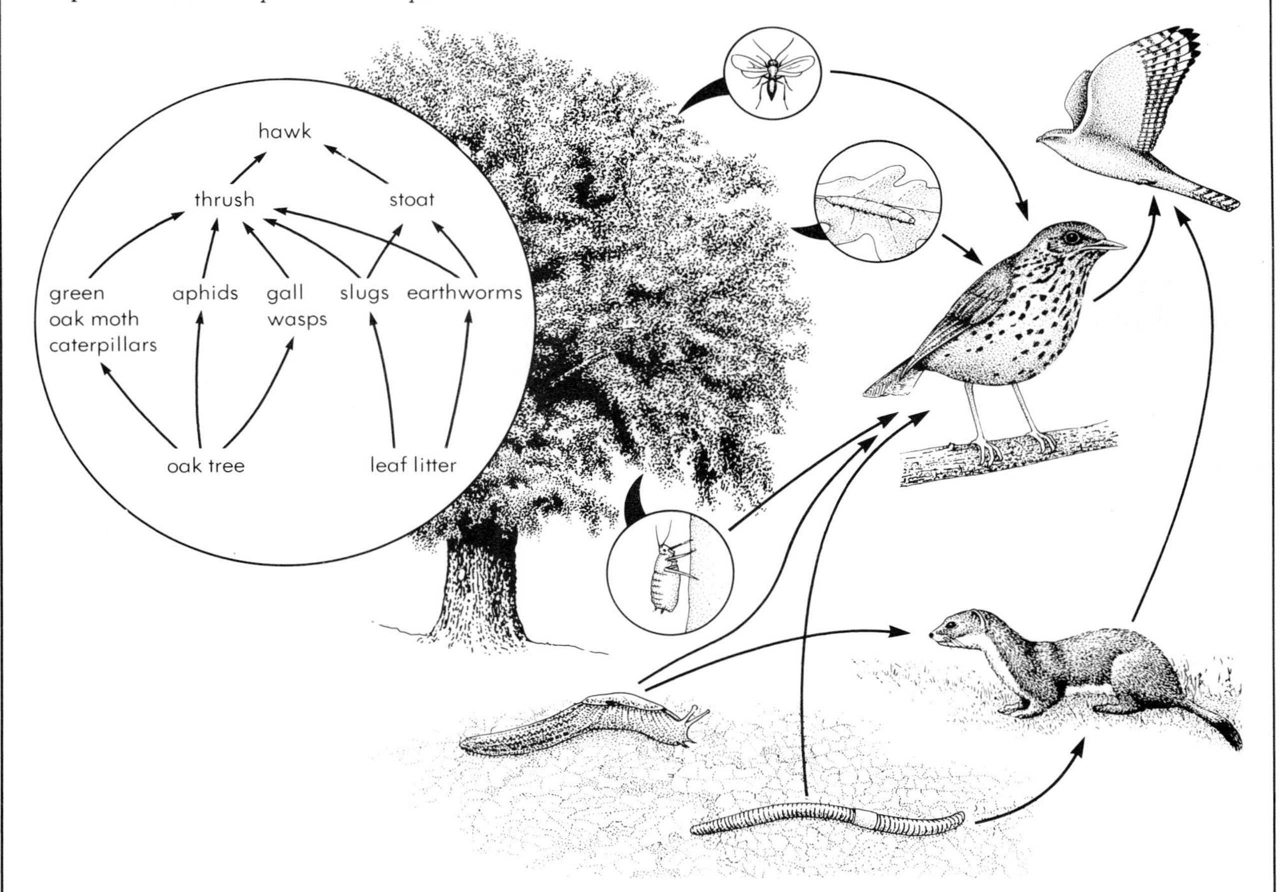

Figure 2 Part of a typical woodland food web. The arrows mean 'is eaten by'. The inset shows the different feeding levels

2.4 Recycling the nutrients

How do plants keep producing their food?

Plants must get all the substances they need to make food – carbon dioxide, water, nitrogen – from their environment. For example, they get carbon dioxide from the air; nitrogen (as **nitrates**) and water from the soil. So why doesn't this supply run out? The reason is that as some substances are removed, others are put back by the action of the saprophytes and other organisms. This is called recycling. All the basic nutrients that green plants need to make their food are recycled in this way.

How is water recycled?

Plants are very important to the recycling of water. An average sized oak tree can remove more than 500 litres of water per hour from the soil and lose 98 per cent of this into the atmosphere (a process called **transpiration**). Most of the remaining 2 per cent is used by photosynthesis.
Water also re-enters the environment by (figure 1):

- evaporation from animals bodies, e.g. lungs and skin.
- evaporation from animal excrement, e.g. urine and faeces.
- evaporation from rivers, lakes and the sea.

How is carbon dioxide recycled?

The recycling of carbon dioxide is very closely linked to the recycling of oxygen. Carbon dioxide is removed from the environment for photosynthesis whereas oxygen is replaced during it. Figure 2 shows this recycling. Carbon dioxide is replaced during:

- respiration,
- the burning of **fossil fuels** such as coal and oil,
- the decomposition of dead organic matter.

Oxygen is removed by the same three processes.

How is nitrogen recycled?

For green plants to turn some of the glucose they make into proteins, they require **nitrogen**. The most easily usable source is a compound of nitrogen and oxygen called a **nitrate**. This is found dissolved in soil water. As plants remove nitrate from the soil water, it is replaced by the actions of other organisms (figure 3).

Plants get nitrogen from the following sources.

- **Nitrifying bacteria**. When organisms die, decomposers break down their proteins and in doing so release ammonia into the soil. This ammonia is converted into nitrates by the actions of nitrifying bacteria which live in the soil. These bacteria get energy from this process and plants take in the nitrates created.
- **Nitrogen fixing bacteria**. These bacteria can incorporate atmospheric nitrogen into their proteins. Some live inside the roots of certain plants (peas, beans) where in return for the protection they get, they give up some of their proteins to the plant. This kind of relationship, where both organisms benefit is called **symbiosis**.

Some bacteria find it useful to turn nitrates back into nitrogen and oxygen which goes into the atmosphere. These bacteria live in the soil and are called **denitrifying bacteria**.

Questions

1 Copy these sentences and fill in the missing words:
a) During .... carbon dioxide is removed from the environment and .... is released into it.
b) During .... oxygen is removed from the environment and .... is released into it.
c) Carbon dioxide is produced when .... are burned.

2 List five ways by which water can enter the atmosphere.

3 All the words/statements in the left hand column can be matched with one of the words in the right hand column. Select which goes with which and explain the relationship.

Release of ammonia from proteins	Decay
Release of nitrogen from nitrates	Lightning
Combines nitrogen and oxygen	Nitrification
Conversion of ammonia into nitrates	Denitrification

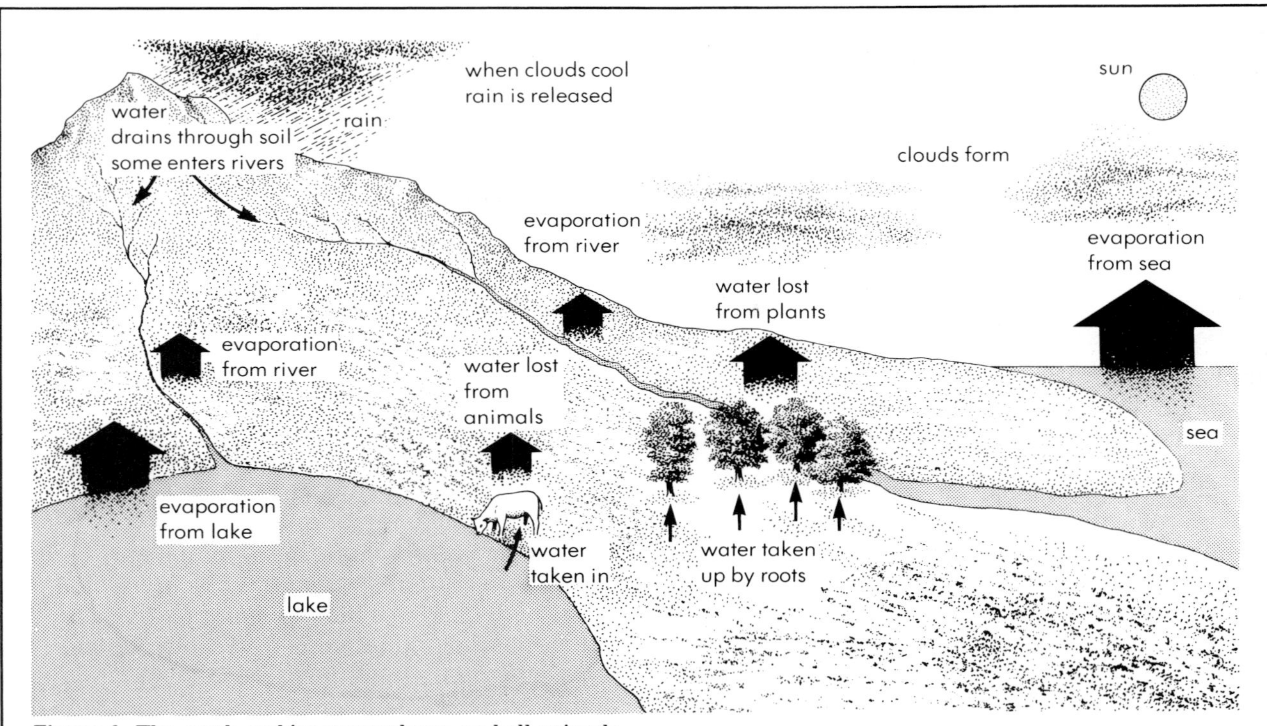

Figure 1 The earth and its atmosphere, and all animals and plants are involved in the recycling of water

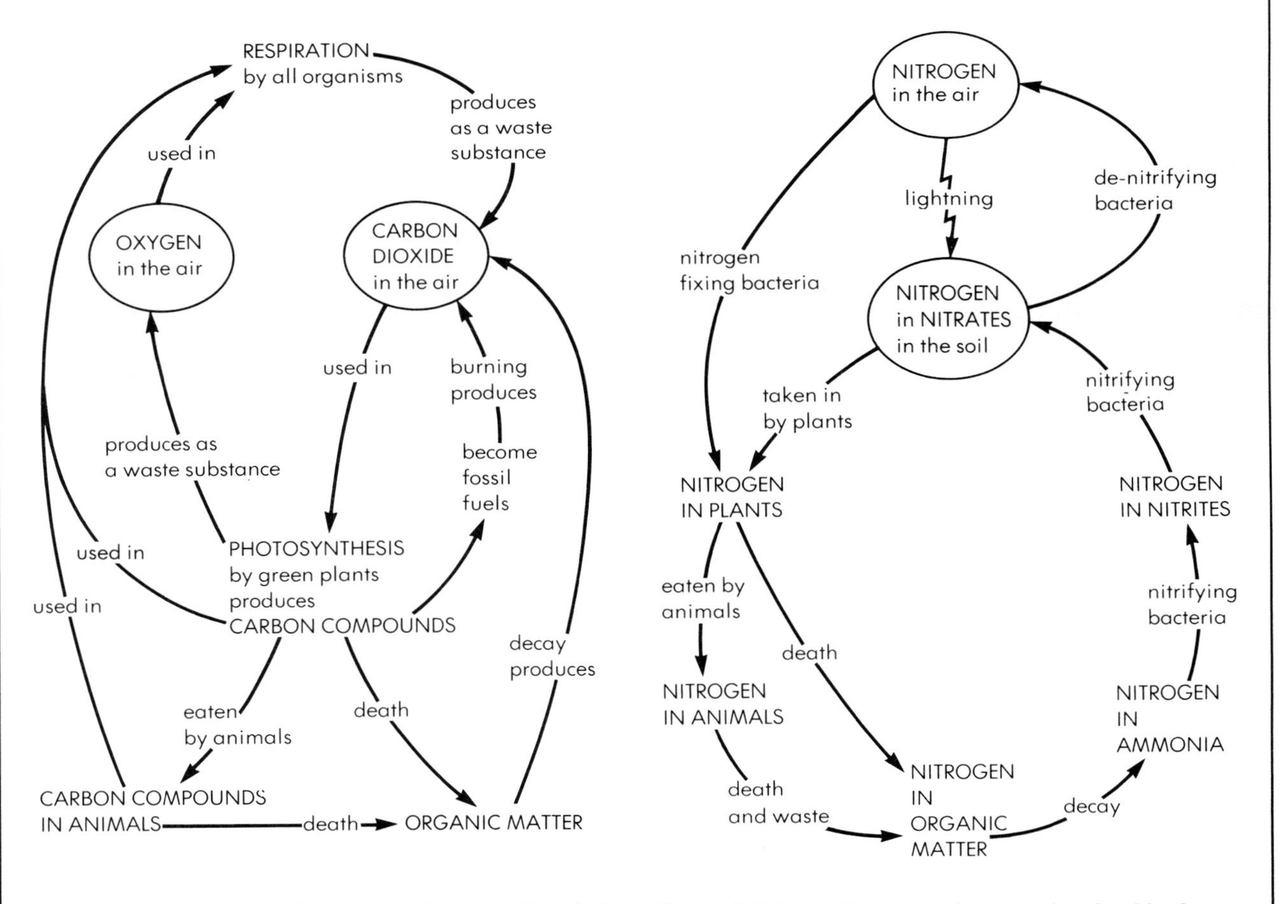

Figure 2 The recycling of oxygen and carbon dioxide is closely linked

Figure 3 Many micro-organisms are involved in the recycling of nitrogen

2.5 Growing food

Energy and food chains

Only a very small amount of the light energy striking a green plant is actually absorbed by the plant and used in photosynthesis to make food. Some of this food is used as material for growth and repair and is therefore said to increase the **biomass** (mass) of the producer. The rest is used in respiration to provide energy.

When a producer is eaten by a consumer, only part of the food is used for new growth, that is to increase the biomass of the consumer. Most of the food is used in respiration to provide energy for movement, excretion, transporting materials around the body, keeping the body warm and so on. This energy is 'lost' at this stage in the food chain. This happens again when one consumer is eaten by another consumer (figure 1).

This 'loss' of energy at each feeding stage puts a natural limit on the number of links a food chain can have. It is not worth having more than five because there is not enough energy left at this stage to pass on. Most food chains have only three links.

Humans and food chains

On modern large scale farms natural food chains are artificially altered so that certain animals are produced in large amounts to fit in with what people want to eat e.g. beef and lamb. This is not necessarily very sensible in terms of energy efficiency, as figure 2 shows. It takes far more energy to produce a pound of beefsteak than the steak itself will provide to the person who eats it.

In times of ever increasing population, agriculture will surely have to become more energy efficient in order to produce enough food for everyone.

How can more food be produced?

Economists have calculated that enough food is actually being produced to meet present demands. Unfortunately, most of this production is by countries in the developed world and they do not always share it with the developing world. There are very different problems in producing more food in the developing world because of the nature of the climate, economy, social structure and history. Most of the research which has gone into new food sources has been for the benefit of the richer nations and many of the solutions suggested are too expensive for poorer countries to try. Some of the ideas given below can be applied to the developing world, but most are specific to richer countries such as the UK.

- **Increase the energy input into agriculture**. At present in the UK, only about 7 per cent of the national daily energy usage is on food production. Much more than this is wasted. If more of this wasted energy was used to manufacture **fertilisers** and **pesticides**, food production would substantially increase (figure 3).
- **Recycle waste materials**. Animal and human excrement along with crop waste can be used in a **bio-gas plant** to produce fertiliser and **methane**. The methane can be used as a fuel for machinery. The fertiliser will help grow food.
- **Breed better plants**. Selective breeding programmes could result in plants which are capable of using more of the light energy striking them, or plants which can **fix** atmospheric nitrogen (page 24).
- **Make use of more of the land**. At present only about 10 per cent of the earth's land surface is used for agriculture. Of course some is useless for the sort of crops we grow at present, but new ones can be developed.
- **Make more use of the sea**, which in many respects is more productive than the land. Only a few per cent of food at present comes from the sea.

Questions

1. Look at figure 2. Explain, with reasons which of the two food chains is the most energy efficient.
2. Using too much fertiliser is a waste of money. Use the information in figure 3 to help you explain this.
3. Write a letter to the United Nations suggesting how food production can be increased to meet the demands of the ever increasing population.
4. Explain why there are rarely more than three links in a food chain.

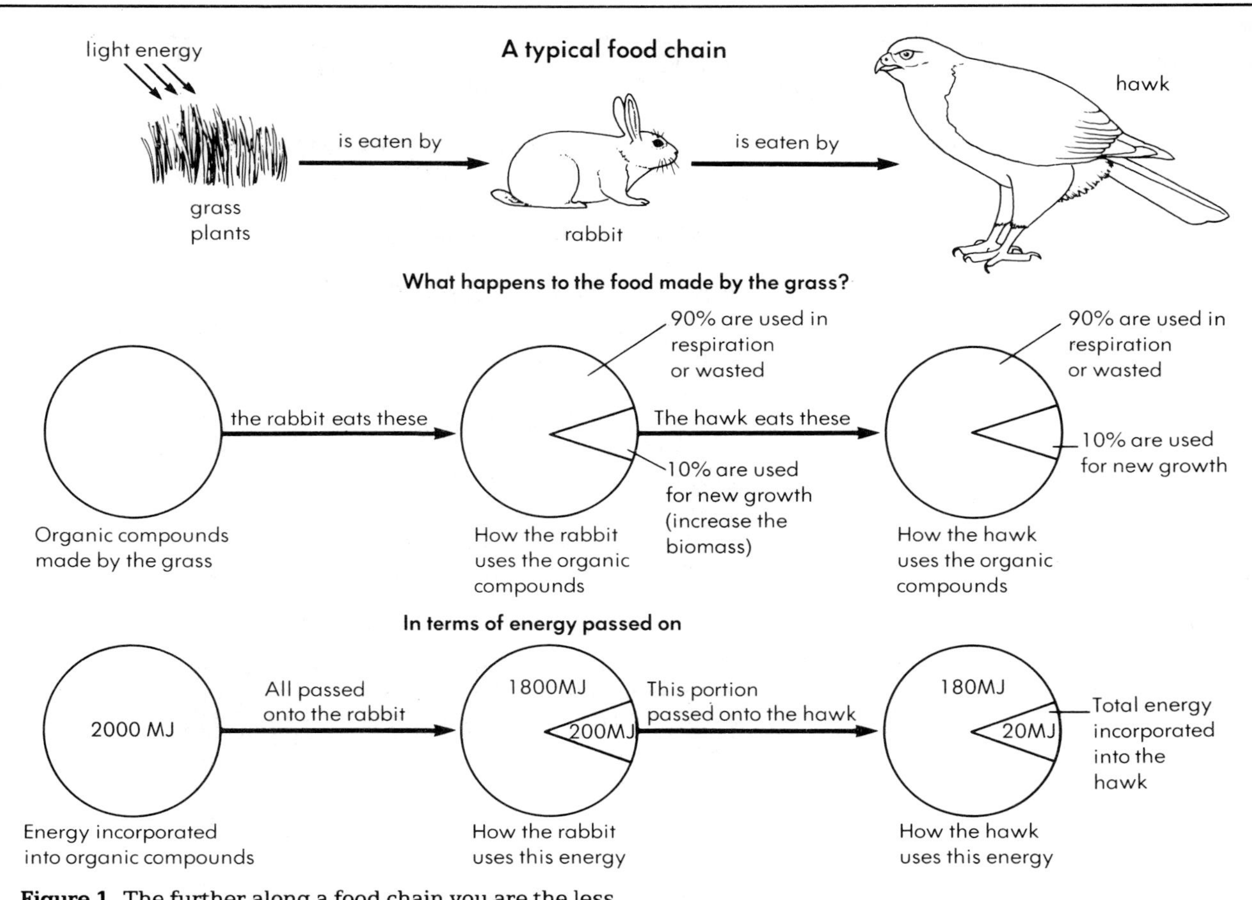

Figure 1 The further along a food chain you are the less energy is available to you. This is because each organism has to use up a lot of the energy it gets from the last organism

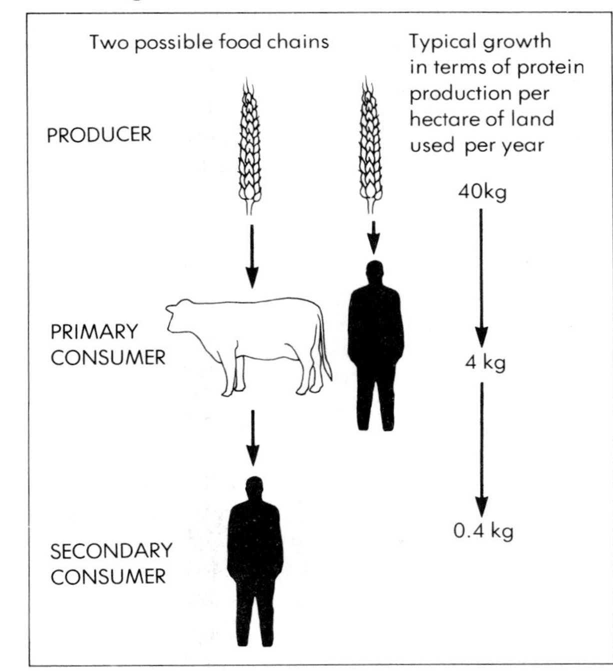

Figure 2 Cereals are a better food source than meat in terms of energy supplied to the consumer

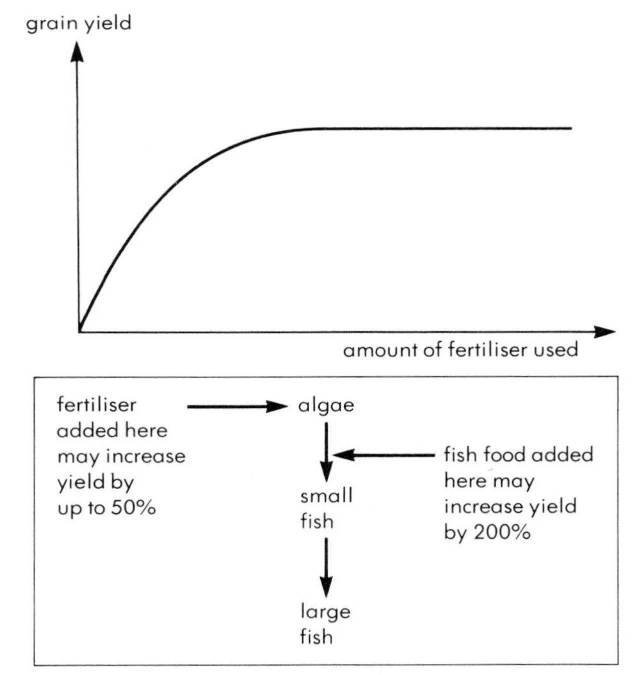

Figure 3 Fertilisers can be used to help grow food but a farmer must use the right amounts in the right places to get the best results

2.6 Agriculture

Is food production the same throughout the world?

Food production needs land. The amount of land available per person has largely determined the method used to produce food (table 1).

Some people use the land and available food simply as it is. They hunt and gather what is all around them and move to new areas when the food runs out, e.g. the nomads of the Kalahari Desert. This means using a lot of land – several hundred hectares per person.

Tribes like the Boro of the Amazon cultivate the land to some extent. They first clear a small area of trees and then use it to grow basic crops such as yams, sweet potatoes and cassava. They also raise wild pigs and chickens. After a couple of growing seasons the land is no longer productive and the people clear a new space.

The kind of food production we and other industrialised countries use has developed partly as a result of land shortage and partly because of the economic pressures to remove people from the land to do other jobs (table 2). Farmers in the UK make use of the same land time and time again. They do this by replacing the nutrients taken out each year with nutrients from fertilisers. Even the animals are raised for the intensive production of food. Most have been specially bred so that they produce more meat or milk, more quickly. High protein feed stuffs and vaccinations against disease all help to increase these yields.

Which method is best?

All ways of obtaining food have their pros and cons. Hunters and gatherers are constantly on the move preventing the development of a settled society. Cultivators and herders such as the Boro no sooner get settled than they also have to move.

Our own form of agriculture on the other hand, allows the development of large static social communities such as towns and cities. It also frees a lot of the people to do other jobs such as building houses, producing medicines and manufacturing other goods. In short it allows people to develop and use different skills – to divide their labour. This is how wealth is created.

In terms of energy produced compared to that put in (see page 26), the simpler forms of food production are quite efficient. A Kalahari nomad can obtain, in three to four days, enough food to feed himself and the rest of his family for a week. The food he has gathered contains about six times more energy than he had to use to get it. However, a modern European farmer given the same amount of land, will produce maybe 20 times as much food.

The problem in Europe and particularly the UK, is that there is a shortage of land. The farms are therefore much smaller yet still need to produce as much food. This is possible, but only by putting in more energy, e.g. by applying fertilisers (figure 2). With this energy taken into account, a European farmer uses twice as much energy to produce his crop as can be obtained from it. This unfortunately is one of the disadvantages of modern agriculture – the more productive it is, the more energy it requires.

The hazards of agriculture

Hunters and gatherers and to a lesser extent cutivators and herders obtain their food without upsetting the delicate balance of nature. More intensive methods of food production, however, threaten this delicate balance by altering the stability of the environment. Hedgerows have been removed, wild plants and animals that once lived on the land can no longer compete, and die. The use of fertilisers creates pollution. In short, the countryside of places like the UK has been changed beyond belief. (See chapter 14.)

One thing all forms of food production have in common is that they are all dependent upon the climate. An unexpected drought for example can seriously affect food production, whether simple or intensive. Some other factors which can affect productivity are shown in figure 1.

Questions

1. In the western world most of our food comes from intensive farming. What are the other methods which can be used to supply food, and who uses them?
2. Explain how the development of agriculture is linked to the development of towns and cities.
3. Why is it unlikely that many of the developing countries will ever prosper?
4. List as many factors as you can that affect agricultural productivity. Describe how these vary in different parts of the world.
5. The productivity of crop plants is far greater than that of crop animals (table 3). Explain why.

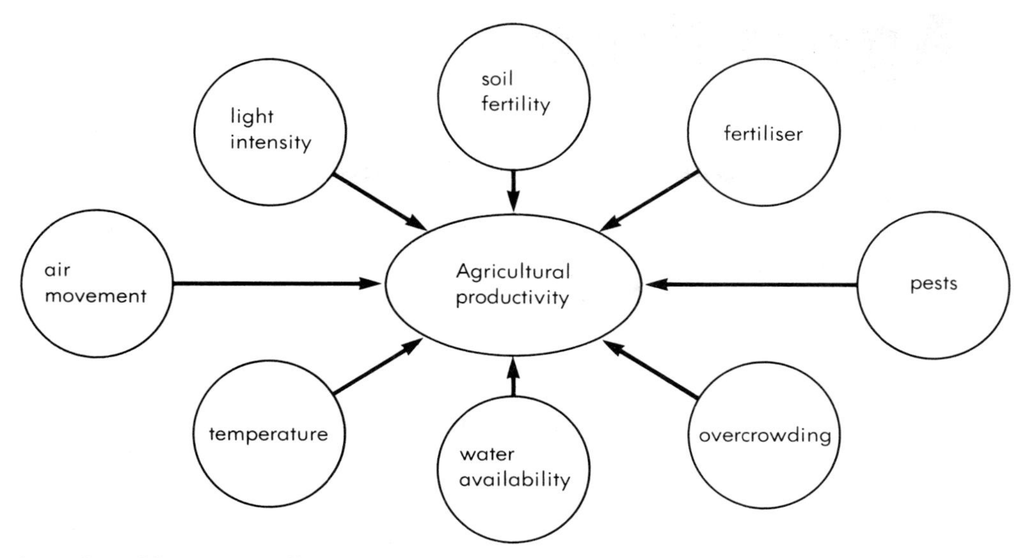

Figure 1 A number of factors can affect agricultural productivity

Area	Arable land (million hectares)	Arable land per person (hectares)
Africa	208.7	0.5
N. America	266.6	0.76
S. America	107.7	0.48
Asia	457.9	0.2
Europe	142.2	0.3
Oceania	46.5	2.13
USSR	232.4	0.9
World	1462.0	0.36

Table 1 The intensity of agriculture depends on the land available

Area	Percentage of active population engaged in agriculture
Africa	66.7
N. America	11.8
S. America	33.0
Asia	59.2
Europe	16.2
Oceania	21.8
USSR	18.1
World	46.2

Table 2 The more productive agriculture is the fewer people are involved in it

Type of farming	Energy produced (MJ) per hectare of land (approx)
Dairy	1300
Pigs	1000
Crops	9500
Beef	300
	Protein yield (kg) per hectare (approx)
Dairy	50
Pigs	30
Crops	150
Beef	25

Table 3 Some forms of farming are more productive than others

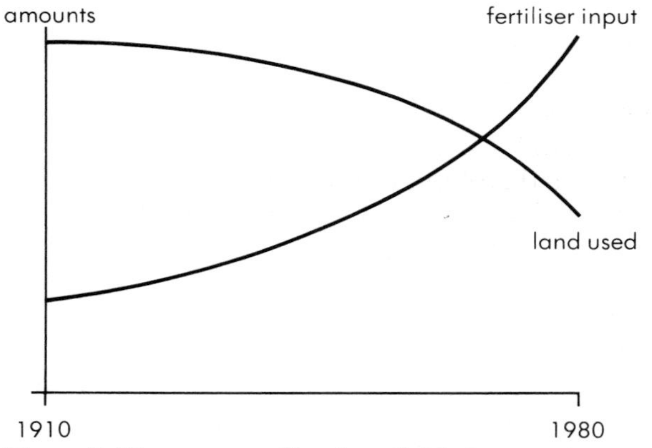

Figure 2 The amount of land available for growing crops is falling, but productivity levels have been maintained by using more fertiliser

2.7 A healthy diet

What should you eat?

You, like most people, probably have favourite foods. It would be nice if you could eat these all the time. But would these give you the full range of nutrients your body needs? Most probably not. A healthy person needs to have a **healthy, varied diet**. This should include, every day:

- **Some carbohydrates and fats** to meet your body's energy needs. Your body's main source of energy is glucose (a carbohydrate). This is the only source the brain cells can use. Muscles on the other hand, can also use fatty acids, so it is best to include some of these in your diet (figures 1 and 2).
- **Proteins** for growth and repair and to produce your body's vital enzymes. To make sure you get all the essential amino acids (see page 18), about 60 per cent of these proteins should come from animal products (table 1). These contain more first class proteins than most plant material. They also contain a lot more protein. Vegetarians can get enough protein including all the essential amino acids they need, but only by eating a wide variety of plant material and in very large amounts.

  Any proteins taken in which are surplus to requirements are used to provide energy, mainly in the form of heat. Proteins already incorporated into your body are only used in extreme emergencies, such as after a long period without food.
- **A selection of minerals and vitamins** to allow your body to function properly and therefore remain healthy. Mineral deficiency is rare in the UK as even tap water contains them. Hard water contains calcium and magnesium, and soft water contains sodium, copper, iron and zinc. Some water authorities also add fluoride to the water to help combat tooth decay.

  Vitamin deficiency is more common, but only in a mild form. The general symptoms of this are tiredness and poor skin condition. The law requires that some foods must have vitamins added to them. Margarine for example has vitamin A and D added.

  When working out how much food is needed to supply your daily vitamins, you must take into account vitamin losses during cooking and storing (table 2 and figure 4).
- **Water** to replace that which is lost each day. Most people can go without carbohydrates, proteins and so on for weeks, but without water they would die within a few days. It is better to take in too much water rather than too little, as the excess will be excreted.
- **Some fibre** to add bulk to the food in the intestine making it easier to move the food along and expel the waste (figure 3). Lack of fibre can result in constipation.

To get the best from your body, you need to supply it with the full range of nutrients. The amounts of these will depend upon your exact body requirements. For example, a growing child needs more protein than a middle-aged person and a pregnant woman needs more calcium than other women. The factors that seem important in determining these amounts are your size, sex, age, state of health and activeness, as well as the environmental temperature.

Do we all need to eat the same foods?

A healthy diet can be achieved by many combinations of different foods. The choice of these foods is a very individual thing and is influenced by many factors; availability, cost, appearance and so on. We are very lucky in this country in that most foods are available to us should we want them. This means we can also base our choice on what food gives us pleasure. People in many other countries do not get this choice.

A good way of planning your diet is to use the following three food groups.

1 Meat and animal products
2 Fruit and vegetables
3 Cereals and cereal products

By selecting something from each group for every meal, in the right amounts of course, you are well on the way to getting a full range of nutrients, that is, achieving a healthy, varied diet.

Questions

1 What should a healthy diet contain?
2 Explain each of the following:
a) You need more food in cold weather.
b) Some animals can survive without actually drinking any water.
3 It is almost impossible to achieve a healthy, varied diet in some countries. Suggest why.

Essential amino acids	mg in 100g of the food				Requirements in mg per day		
	Eggs	Milk	Beef	Wheat flour	Infant	Child	Adult
A	54	47	53	42	80	28	12
B	86	95	82	71	128	42	16
C	70	78	87	20	97	44	12
D	93	102	75	79	132	22	16
E	47	44	43	28	63	28	8
F	66	64	55	42	89	25	14

Table 1 For good health we require daily amounts of the essential amino acids. Some foods contain more of these than others

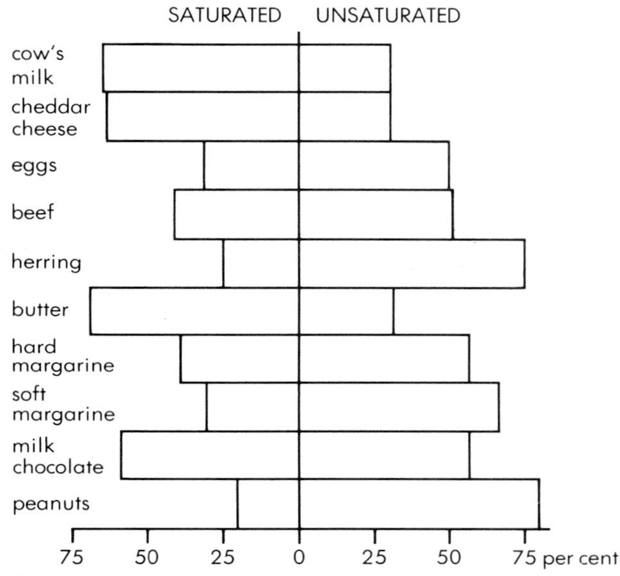

Figure 1 Butter and margarine contain about the same amount of fat but in buttter more of this is of the saturated type. Only about ⅓ of your fat intake should be saturated fat

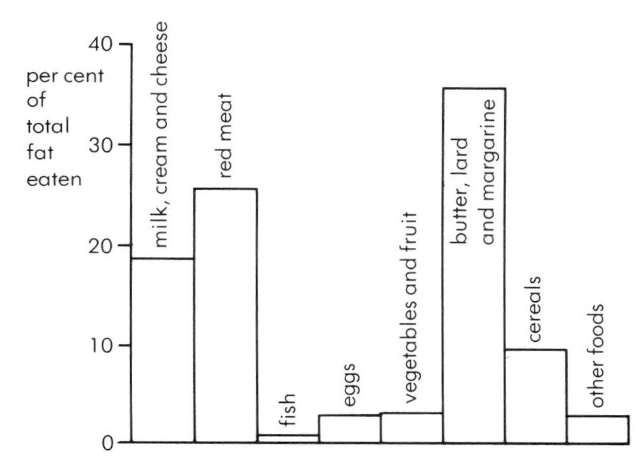

Figure 2 Red meat provides over 25% of the fat we eat. This is nearly as much as butter and margarine supply

Nutrient	Heat	Air	Water (cooking or soaking)
Proteins	✓		
Minerals			✓
Vitamin A	✓	✓	
Vitamin C	✓	✓	✓

Table 2 How food is stored and prepared can affect its nutrient content

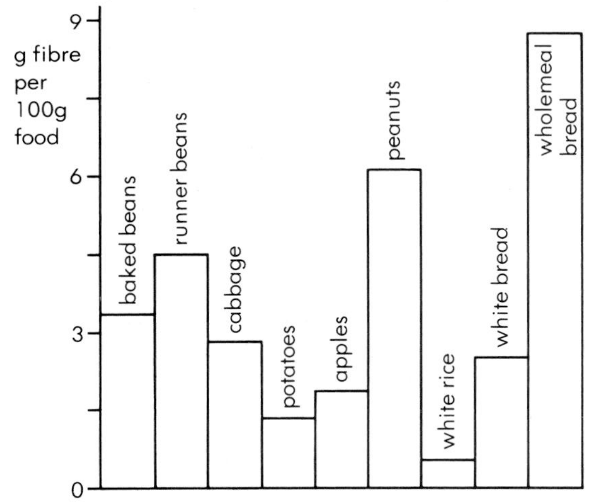

Figure 3 Fibre is very important in our diet. Foods like white rice have most of the fibre taken out during processing

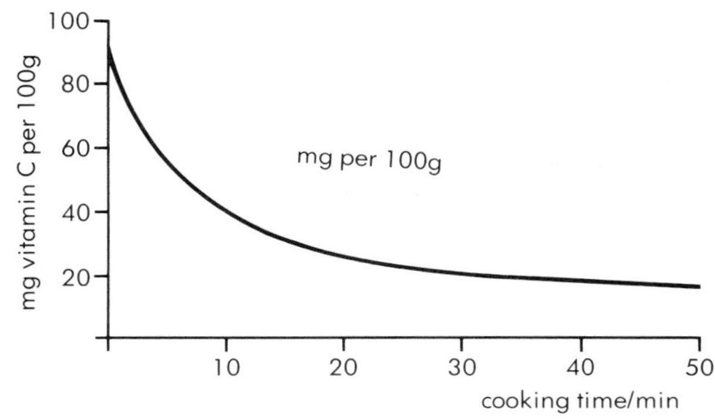

Figure 4 Cooking reduces the vitamin C content of food. This graph shows the loss of vitamin C during the cooking of sprouts. The table shows which factors reduce the nutrients in foods

2.8 Too much or too little

What will happen if you eat too much food?

If you eat more food than your body requires, the excess food is turned into fat and stored. You will become quite overweight, a condition often described as **obesity**. Table 1 shows the approximate amounts of nutrients needed by some average Europeans.

Fat should make up between 10 per cent and 25 per cent of your body weight. This fat is stored in special cells under your skin and around some of your vital organs. Many scientists believe that the number of fat cells your body contains is determined while you are a baby. Overfed babies develop more fat cells and these people put weight on easier later on in life.

You can easily check if your body contains too much fat by doing the pinch test. Pinch the skin at the back of your upper arm halfway between your shoulder and elbow. If the fold of skin is more than about three centimetres thick, then you are probably too fat!

Obesity is far more common in the developed countries such as the UK, the USA and Australia. The main reasons are probably the amount of refined sugar we all eat and our great liking for red meat and dairy produce.

Obese people are generally less healthy than slim people (figure 1). The extra body weight puts a strain on some vital organs such as the heart, lungs and kidneys. Medical statistics show that the countries which contain the most obese people also have the highest incidence of heart disease. Is this purely a coincidence?

What will happen if you eat too little?

Eating too little food will result in poor health and deficiency diseases. Eating too few energy foods will eventually result in **starvation**. To start with your body will use its fat reserves to produce energy. When these are exhausted, body proteins start to be used and this can quickly lead to death.

Many people starve themselves on purpose, in an attempt to lose weight. This is the idea behind slimming diets, i.e. reduce your consumption of the energy foods slightly so that your fat reserves are used. If this is to work without causing serious illness, it is important that the rest of the diet remains the same so that you still get enough of the other nutrients.

It is possible to slim too much. Some people are so keen to slim that they make themselves detest food and once they reach the desired weight they cannot eat at all. They then become dangerously underweight. These people are suffering from **anorexia nervosa** and need urgent medical care. This is now quite common in the UK, particularly among young people (figure 2).

When should you slim?

Slimming has become very big business in the UK partly due to fashion, and partly due to the desire of most people to improve their health. Whatever the reason, it is always best to first consult a doctor before starting a slimming diet. You never know, you may be better off staying just as you are.

An alternative to cutting down on food is to increase the amount of exercise you do. This works by using more of the food you take in for energy. But remember, to do any good the exercise periods must be regular.

So-called **slimming foods** contain less nutritious food and more packing material, such as fibre. Once inside your stomach, this swells and makes you feel full.

Should you alter your diet?

You may not need to change your diet. However, recent surveys do indicate that the average person in the UK does need to. These clearly show that most of us are eating:

- too much fat, especially fats containing a lot of saturated fatty acids,
- too much refined sugar,
- too much salt,
- too little fibre.

Questions

1. The average person in the UK eats far too much fat. Suggest five ways by which you can reduce your fat intake.
2. Fats containing a lot of saturated fatty acids have been linked with coronary heart disease. Name three such foods that you regularly eat.
3. Eating a lot of refined sugar can result in obesity and tooth decay. Suggest five ways of reducing your sugar intake.
4. Most of us would benefit from eating twice as much fibre. Name four foods which we could include in our diet to help us do this.

Age and sex (years)	Energy kJ	Protein/g approx	Fat/g * maximum	Vitamin A /μg	Vitamin C /mg	Calcium /mg	Iron /mg
Baby	3250	19	25	450	20	600	6
MALES							
5–6	7250	43	59	300	20	600	10
12–14	11 000	66	86	725	25	700	12
18–34 } fairly	12 000	72	94	750	30	500	10
35–64 } active	11 500	69	90	750	30	500	10
65+	10 000	60	78	750	30	500	10
FEMALES							
5–6	7000	42	59	300	20	600	10
12–14	9000	53	71	725	25	700	12
18–54 fairly active	9000	54	71	750	30	500	12
55+	8000	47	63	750	30	500	10
pregnant	10 000	60	78	750	60	1200	13
breast feeding	11 500	69	90	1200	60	1200	15
Diet of a typical UK adult	approx 13 000	90	126+	1500	123	1100	16

* These figures are based on 30% of energy requirements coming from fat.

Table 1 How much of the main nutrients do you need every day? The Table shows what the DHSS recommends

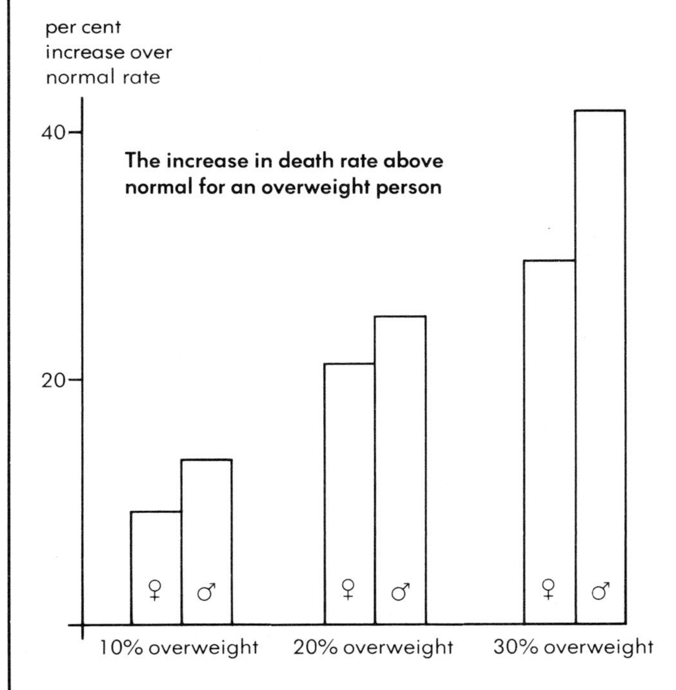

Figure 1 The fatter you are the greater the risk to your health

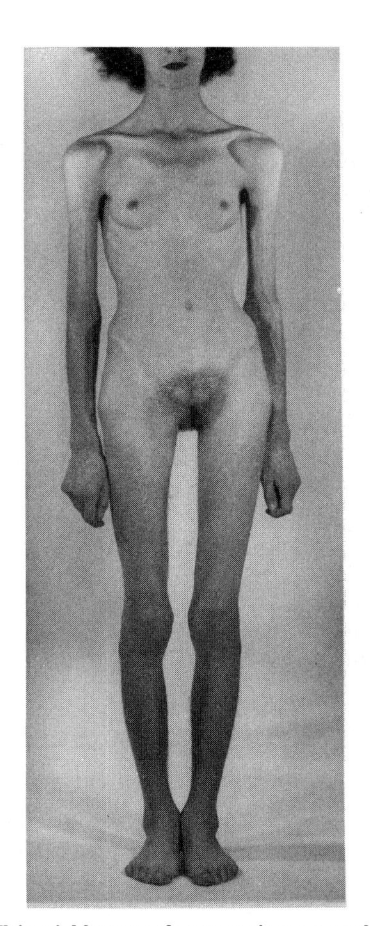

Figure 2 This girl has not been eating enough food because she has anorexia nervosa

2.9 The world's food problem

Is there enough food in the world?

In 1984 the pop singer Bob Geldof was so distressed by a BBC report about famine in Ethiopia that he started a movement, Band Aid, which over the last few years has focused the world's attention to two things:

- There are millions of people in the world who are suffering from **malnutrition** – that is, do not have enough to eat or do not eat the correct foods (figure 1 and table 1). The last World Health Organisation estimate put the figure at about one eighth of the world's population. This includes 200 million children. Most of these people are in the developing countries.
- The world is divided into two halves: the 'haves' and the 'have nots'. There is enough food to go round , but it is not shared out evenly. Some of the richer countries, rather than give food away to those in need, store it in warehouses, forming the notorious food mountains (figure 2).

Why are so many people suffering from malnutrition?

There is no simple answer to this question as there are so many factors which can combine to create the problem. Some of these are:

- **Overpopulation**. The population of many countries has been rising faster than food production, (figure 3).
- **Health**. Sleeping sickness and malaria are **endemic** (always present) in many of the countries. These diseases do not usually kill people but leave them so weak that they are unable to work. People who cannot work cannot provide adequate food for themselves or their families.

  The chances of someone recovering from a disease are not helped by the serious lack of medical facilities. In Ethiopia, for example, there is one doctor to every 100 000 people, compared to 1 to every 1000 in the UK.
- **Poor soil**. To grow good crops, the soil has to contain nutrients. If the same soil is used year after year, the nutrients are all gradually removed, leaving the soil unable to support life.
- **Lack of water**. Crops need adequate water to grow. Long periods without it (droughts), such as happened in Ethiopia in 1972–74 and 1980–84, can devastate food production. Even a 10 per cent reduction in food for some countries can result in millions of people having to go without.
- **Pests**. A single locust swarm can contain millions of insects. A swarm of 1000 million can eat in one day an amount of food which would feed 100 people for a year.
- **Poverty**. Poor people often eat the cheapest foods, e.g. rice, yams, sweet potatoes, and very little else. Unfortunately these are usually lower in protein and are missing essential minerals and vitamins. These deficiencies can result in diseases.

  Poor countries do not have the money to invest in modern agricultural machinery or to buy the more expensive disease-resistant seeds.
- **Lack of knowledge**. If you don't know what foods are best to provide, you cannot be expected to provide them.

What can be done to help?

The resources and wealth of the world can be used in many different ways to provide immediate help to places in desperate need, and to create long term changes. Some ways Band Aid helped were by:

- **Providing food** bought from other countries
- **Providing pesticides** and pest-resistant seeds
- **The building of irrigation systems** to provide water
- **Providing fertilisers** to restore soil nutrients
- **Digging of wells** to supply water
- **The building of hospitals** and the staffing of them
- **Providing drugs** to fight disease
- **Providing appropriate agricultural machinery**
- **Educating the people** as to their basic needs and how to achieve them

Questions

1 What is malnutrition?

2 Based on the figures in figure 1

a) Which countries are most likely to contain a lot of fat people?

b) Which countries are most likely to contain a lot of starving people?

3 Explain why food production in general is going up yet the amount of food available per person is going down.

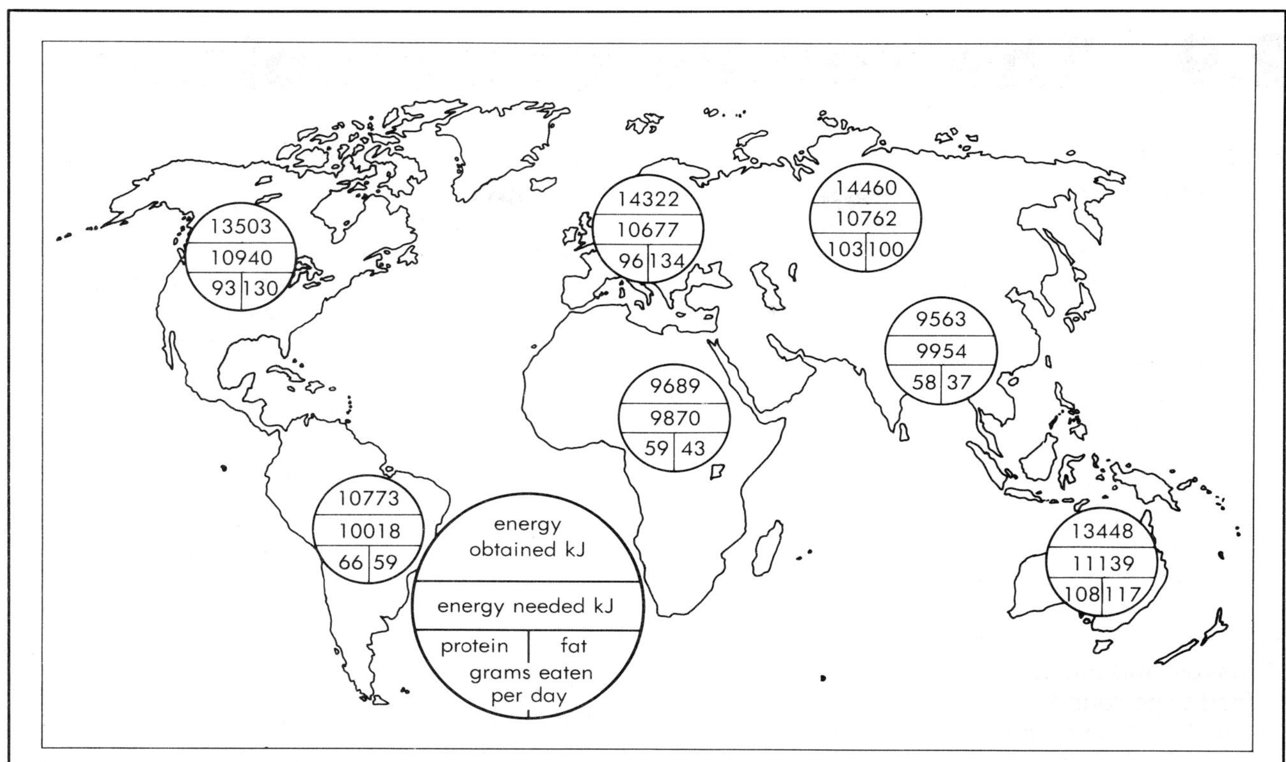

Figure 1 The amount of food available across the world varies considerably. For example people in Asia and Africa do not get enough energy foods

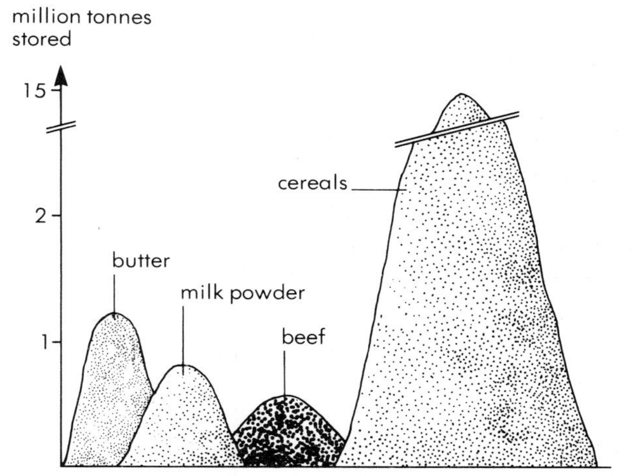

Figure 2 Some countries have too much food. The EEC food mountains, November 1987

British child		African child (Ghana)
Breakfast cornflakes + milk toast + butter orange juice		beans nuts
Lunch stewed steak carrots potatoes rice pudding milk		fish peppers cassava leaves
Evening meal beans on toast biscuit apple milk		cocoyam plantain peppers
approx energy	6350 kJ	2100 kJ
protein content	52 g	12 g

Table 1 The diet of a child in Africa has less variety and is lower in both energy and protein than the diet of a child in the UK

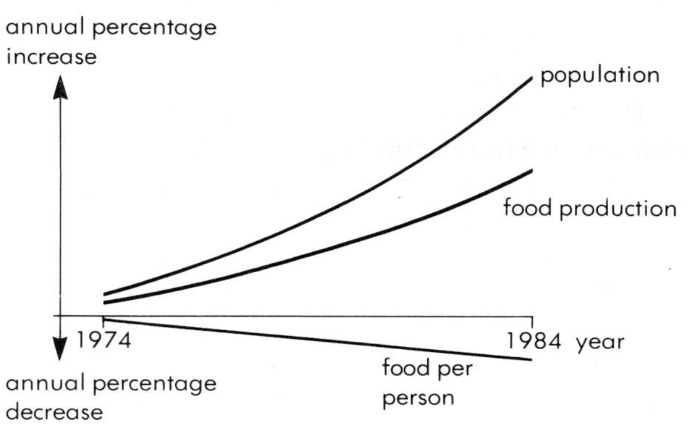

Figure 3 In many developing countries the food available per person is decreasing as the population rises

2.10 The digestive system

What is digestion?

You now know that all cells need a regular supply of nutrients so they can carry out their daily activities. They get these nutrients from the blood and tissue fluid. The blood gets them from the food we eat. It is the role of the **digestive system** to make these nutrients available to the blood in the right form. It breaks down the large lumps of food you eat into pieces that are small enough to pass through the gut wall and dissolve in the blood. Foods are broken down in the following ways:

- all **carbohydrates** are broken down into **simple sugars** like glucose.
- **proteins** are broken down into **amino acids**.
- **fats** are broken down into **fatty acids** and **glycerol**.
- **minerals, vitamins** and water are released from their packaging.

This breaking down process is called **digestion**.

The structure of the digestive system

The digestive system consists of (figure 1):

- The **alimentary canal.** This is the muscular tube in which digestion takes place. It is about eight metres long, running from your mouth to your anus and has several distinct parts, each one fulfilling an important role in digestion.
- The **salivary glands**, **liver** and **pancreas**. These produce **enzymes** and other chemicals which pass into the alimentary canal and help with digestion.

How does digestion take place?

Digestion is brought about in two ways:

- **Mechanically**, by your teeth, tongue and muscles in the wall of the alimentary canal, and
- **Chemically**, by the enzymes and other chemicals made by the salivary glands, liver and pancreas.

Both these processes go on at the same time. The mechanical breakdown of food actually helps the enzymes to do their work by (a) mixing the food with the enzymes and (b) increasing the surface area of the food for the enzymes to work on. It also helps move the food through the gut.

What do the muscles do?

The wall of your alimentary canal contains two sets of muscle fibres (figure 2). One set is arranged in rings (**circular muscle**), and the other set is arranged lengthwise (**longitudinal muscle**). These muscles work in opposition to one another. When the circular muscle contracts, the diameter of the gut is reduced and the food in it is squashed. When the longitudinal muscle contracts, this increases the diameter of the gut so the food can move into this area. By careful co-ordination of these muscles the food can be kept moving through the alimentary canal and at the same time be broken down into smaller pieces. The co-ordinated actions of these muscles is called **peristalsis**.

How long does it take to digest a meal?

It takes on average 24 hours to complete the digestion of a typical meal. For the first five hours the food is in your stomach. This is followed by five hours in the small intestine, seven hours in the colon and seven hours in the rectum.

Questions

1 Copy and complete the following:
During the process of digestion all .... are broken down into simple sugars, proteins are broken down into .... and .... are broken down into .... acids and .... Only in these forms are they .... enough to pass through the wall of the .... canal and enter the ....

2 a) Draw and label the digestive system.
b) Under each label, describe in one sentence the function of that part.
c) On your diagram, colour the alimentary canal yellow and the accessory glands green.

3 Digestion is brought about by both mechanical and chemical means. Explain fully how mechanical digestion takes place and how it helps to speed up chemical digestion.

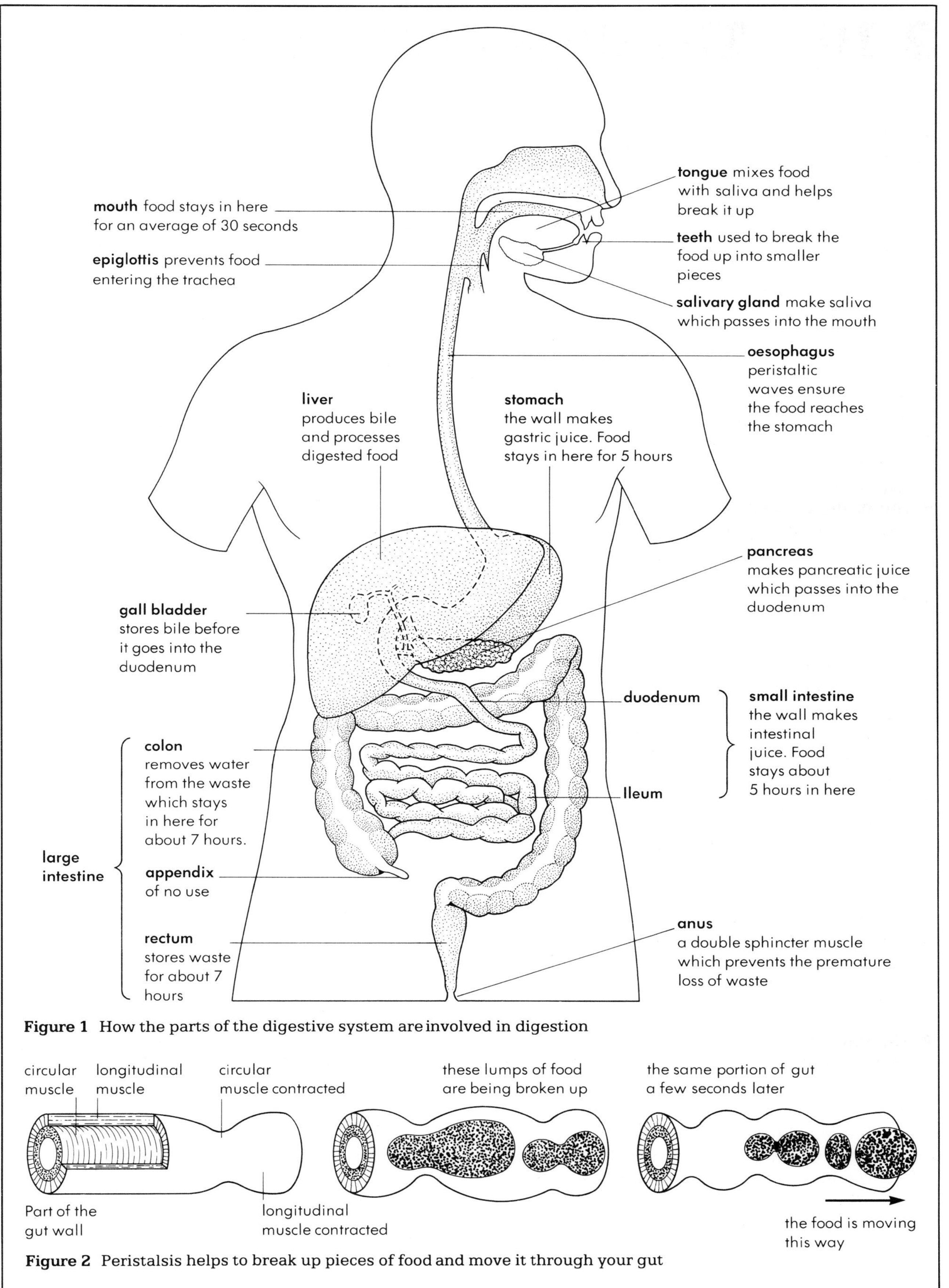

Figure 1 How the parts of the digestive system are involved in digestion

Figure 2 Peristalsis helps to break up pieces of food and move it through your gut

2.11 Digestion

What happens to food in your mouth?

Most people chew their food before swallowing it – unless it is liquid food or you are particularly greedy! This chewing breaks the food down into smaller pieces and mixes it with the digestive juice (saliva) produced by the salivary glands. Saliva contains two main substances:

- the enzyme **amylase** which starts the digestion of starch
- mucus which wets the food and makes it easier to swallow (figure 1).

What happens to the food in your stomach?

Once food is in your stomach, it is sealed by two strong circular muscles, one at either end. These are the **sphincter muscles**. For the next hour or so **peristaltic waves** pass along the wall of the stomach. These continue the mechanical breakdown of the food and mix it with the **gastric juice** made and secreted by cells in the stomach wall.

Gastric juice contains the enzyme **pepsin** which starts the breakdown of proteins. It also contains **hydrochloric acid** which provides the acidic conditions that pepsin needs and kills any germs present in the food. The mixture that eventually leaves your stomach is a creamy white pulp called **chyme**. A little of this leaves every time the lower sphincter muscle relaxes (figure 2).

What happens in your small intestine?

Your small intestine is about six metres long and has two regions. The first 25 cm is called the **duodenum** and the rest is the **ileum** (see page 37, figure 1). Muscles in its wall continue the mechanical breakdown of the food, whereas chemical digestion is brought about by three new juices:

- **Bile** which is produced in the liver and stored in the **gall bladder** before entering the duodenum via the bile duct. Bile neutralises the acid in the chyme and breaks up any lipids (fats and oils) present into small droplets. This process is called **emulsification**.
- **Pancreatic juice** produced by the **pancreas** and secreted into the duodenum. This contains the enzymes **trypsin**, **amylase** and **lipase**. Trypsin continues the breaking down of proteins which was started by the pepsin in the stomach. It is helped by the alkaline conditions created by the bile. The amylase continues the breakdown of starch and the lipase breaks down the lipids. This can proceed much more quickly because of the earlier emulsification of the lipids by bile.
- **Intestinal juice** produced by special cells in the intestine wall. Intestinal juice contains **protease** enzymes to complete the digestion of proteins and **carbohydrase** enzymes to complete the digestion of carbohydrates.

What happens in your large intestine?

The first part of the large intestine is the **colon**. By the time the food reaches this, all the useful substances, except most of the water and a few minerals, have been removed. As this watery waste travels through your colon, a lot of the water and most of the remaining minerals pass into your blood. By the time it reaches the rectum the remaining **faecal matter** is quite dry and can be moulded into **faeces**. These faeces can be stored for a day or so, but must eventually be passed out through the **anus** (a double sphincter muscle).

If the movement through your colon is too fast, a lot of the water remains in the waste resulting in **diarrhoea**. If the movement is too slow, far too much water is removed leaving hard, dry faeces which are difficult to get rid of. This is **constipation**.

Questions

1. Imagine you are eating a chip. Describe what happens when you swallow a piece of it.
2. Describe what happens to this chip as it passes through your alimentary canal.
3. Copy and complete the table:

Digestive juice	Where made	Place it works in	Enzymes it contains
Saliva Gastric juice Pancreatic juice Intestinal juice			

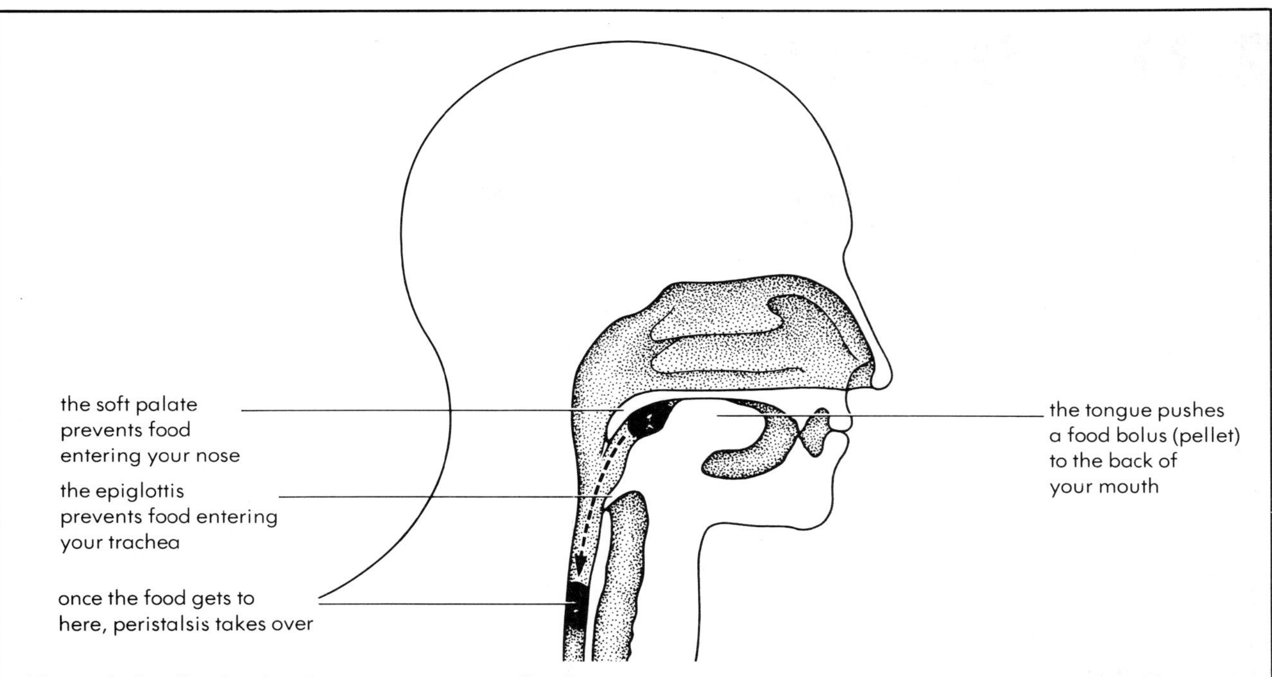

Figure 1 Swallowing involves your tongue, soft palate and epiglottis

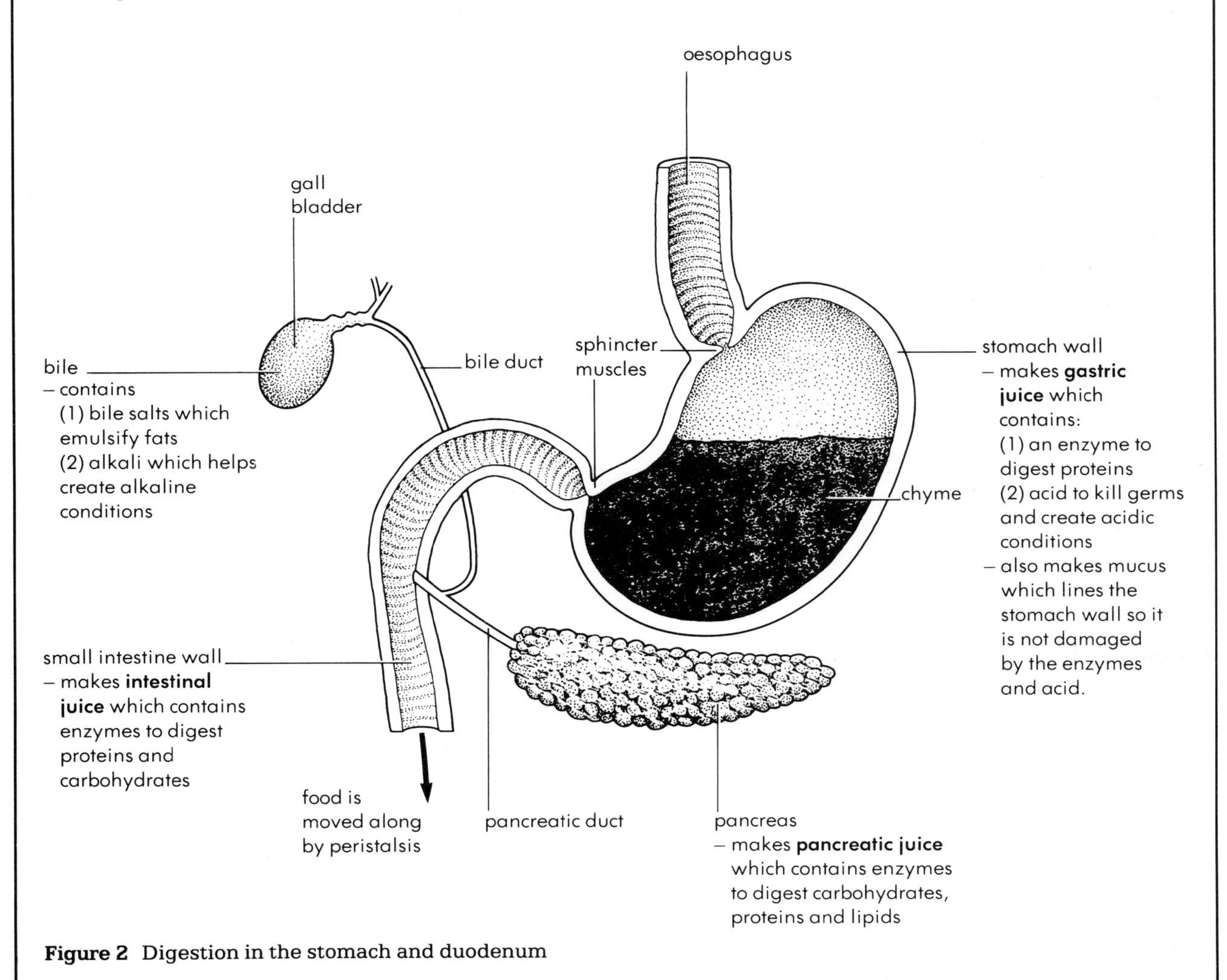

Figure 2 Digestion in the stomach and duodenum

2.12 Absorption and assimilation

How do nutrients get into the blood?
The process of digestion breaks food down into smaller and smaller pieces. When these pieces are small enough, they can pass through the intestine wall and dissolve in the blood. This process is called **absorption**.

Where does this absorption take place?
Most absorption takes place in the intestines. Sugars, amino acids, fatty acids, glycerol and most of the vitamins and minerals are absorbed from the small intestine into the blood. Water and the remaining minerals and vitamins are absorbed from the large intestine.

Most absorption takes place by diffusion and will therefore be much quicker and more efficient if there is a large surface area over which it can take place. The internal surface of the small intestine is made larger by millions of tiny finger-like projections, each called a **villus** (figure 1). Each villus is well supplied with blood for the nutrients to enter and also contains a lymph vessel (called a **lacteal**) into which fatty acids and glycerol can pass (figure 2).

The large intestine has many folds, so it has a large surface area - hence its name!

Where do the nutrients go now?
All the nutrients which enter the blood from the intestine are taken in the **hepatic portal vein** straight to the **liver** (figure 3). The liver deals with them in the following ways:

- The blood glucose level (see page 122) is adjusted and any remaining glucose is converted into **glycogen** and stored in the liver cells. Some can also be stored in the muscle cells. When both these stores are full, surplus glucose is converted into fat by the liver cells.
- Some amino acids are removed by the liver cells and used to make **plasma proteins** (proteins present in plasma, the liquid part of the blood). Some are left in the blood for the body cells to use and the rest are **deaminated** by the liver cells. Deamination is the splitting of an amino acid into the part which contains the nitrogen, that is the **amino group**, and a part which can be converted into glucose. The amino group is then combined with carbon dioxide to form **urea**. This urea is taken to the kidneys where it becomes part of **urine**. Deamination is necessary because our bodies have no storage area for amino acids.
- Surplus iron, and vitamins A and D, are removed and stored.

Fatty acids and glycerol by-pass the liver and eventually rejoin the blood at the **thoracic duct** (figure 4). Some are used to make cell membranes and the rest are reconverted to fat and stored in **adipose tissue**.

How do the cells use the nutrients?
Glucose is used to provide energy for the cell to do work. Amino acids are used for the production of new protoplasm, the repair of damaged parts and the formation of enzymes, plasma proteins and hormones. Some of the fats are used to build structures like cell membranes, and the rest are used to provide energy.

Questions

1. Most absorption is by diffusion. Explain how the lining of the small intestine helps to make sure that the process is as quick and efficient as possible.
2. Copy and complete the table:

Nutrient	Form in which it is absorbed	Part of the gut in which it is absorbed
Carbohydrates		
Proteins		
Fats		
Water		

3. Absorbed carbohydrates and amino acids are taken to the liver. Which blood vessel are they carried in and how does the liver make use of them?
4. What happens to the absorbed fatty acids and glycerol?
5. For each of the following nutrients, select the statement which best describes how it is used by the cells:
   glucose, amino acids, fatty acids and glycerol.
   a) to make new protoplasm, enzymes, hormones and plasma proteins,
   b) to provide energy,
   c) to make cell membranes and provide energy.

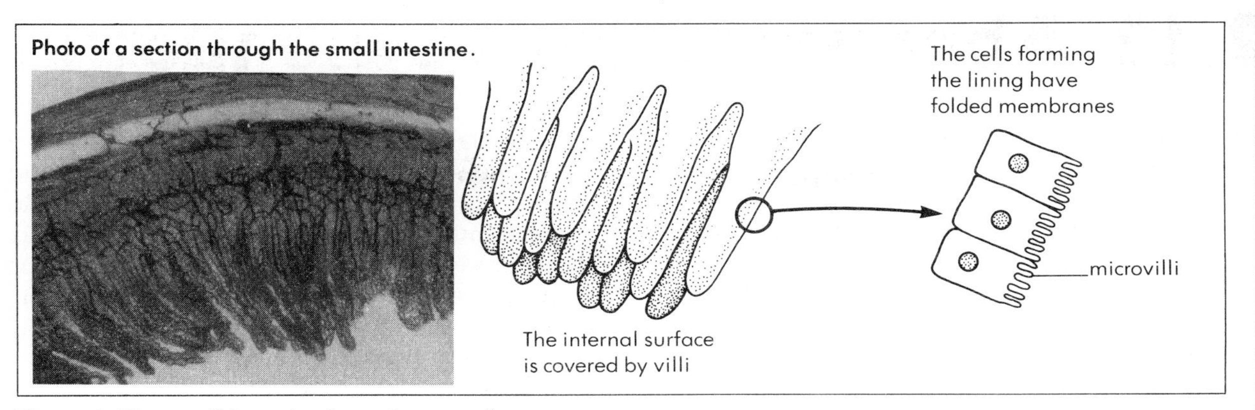

Figure 1 The small intestine has a large surface area because it is covered by millions of villi

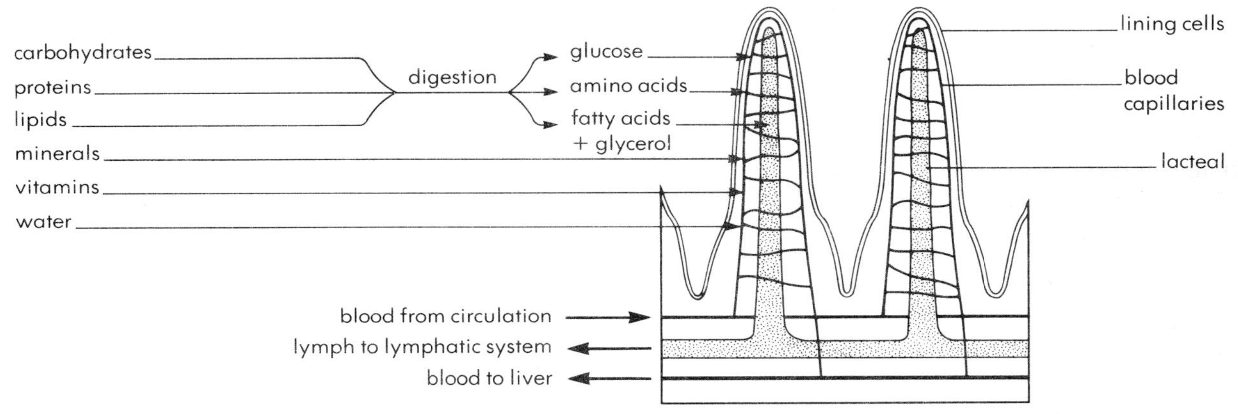

Figure 2 Most of the absorption of nutrients takes place in the small intestine

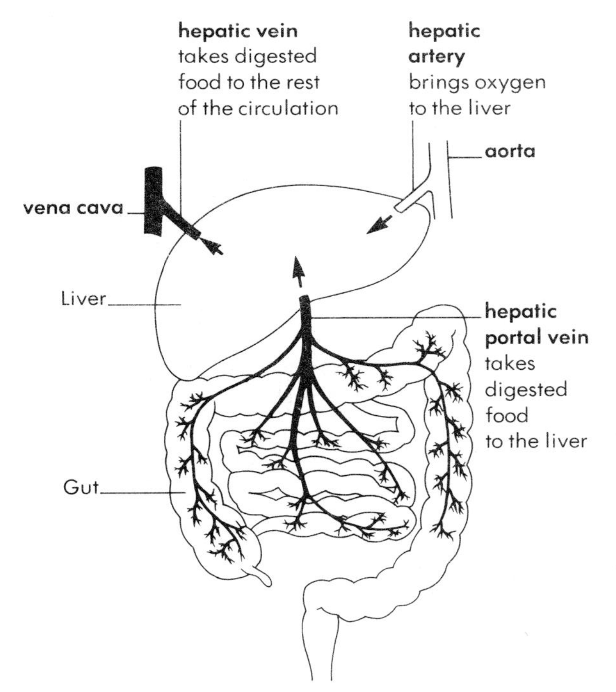

Figure 3 All the nutrients that pass into the blood from the small intestine are first taken to the liver

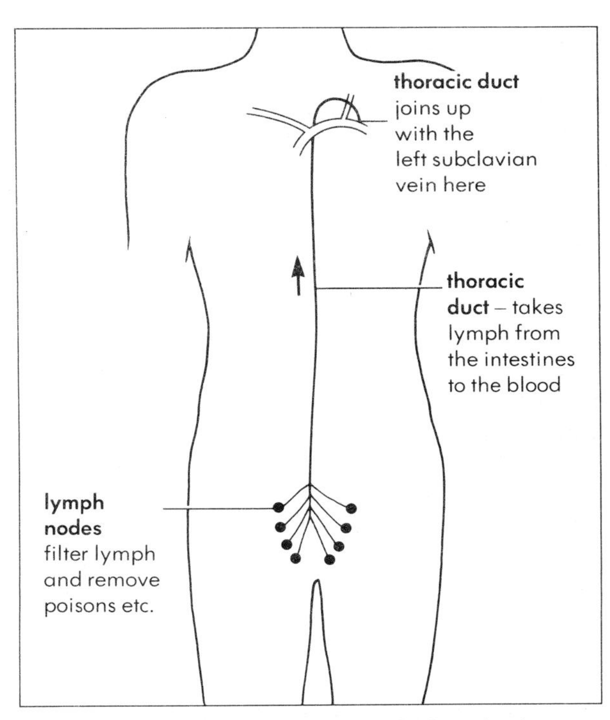

Figure 4 Absorbed fat passes into the lymphatic system. This eventually joins up with the blood system

2.13 The structure of the teeth

Why do you have teeth?

Your teeth are very important and are worth looking after. They are necessary for the first stage in breaking down food. For example, when you eat an apple you use your teeth to bite pieces out of it and then to crush these pieces before swallowing them. This makes the digestion which follows easier.

Well-cared-for teeth help you speak better and look much nicer than discoloured, rotten teeth!

What is a tooth made from?

A tooth has two distinct parts. The part which sticks out from the gum is called the **crown** and the part which is set in the jawbone is called the **root**. A section through a tooth shows several different layers (figure 1):

- The **enamel** covering the crown of the tooth. This is made mainly of calcium and phosphorus. It is very hard but once damaged, it will not be replaced.
- The **cement** covering the root of the tooth. Tough fibres are embedded in this and the other end of these are attached to the jawbone.
- The **dentine** which forms the bulk of the tooth. This is very similar to bone in structure, but considerably harder.
- The **pulp** at the very centre of the tooth. This is the soft part of a tooth and is made up of living cells, blood vessels and a nerve. Extensions of the cells and tissue fluid enter channels in the dentine.

Are all teeth the same?

Humans have four types of teeth (figure 2), each adapted to a different job and quite different in shape.

- **Incisors** are chisel-shaped, with sharp edges which can be used for cutting and biting food.
- **Canines** are pointed teeth, the top of the point being called a cusp. They are used to tear food.
- **Premolars** have two cusps and are used to tear and grind food.
- **Molars** can have up to five cusps. These are used to chew, crush and grind food.

Have a look at your own teeth in a mirror, or those of a friend, and see if you can see the different types.

When do we get our teeth?

You have two sets of teeth during your lifetime. A baby starts to get its first set, called the **milk**, or **deciduous teeth**, when it is about six months old. We say the baby is **teething**. It may take up to three years for the complete set of twenty teeth to appear (figure 3).

The milk teeth start to be replaced by the **permanent teeth** from about six years old. The milk teeth simply come loose and fall out and new teeth grow. By 13 or 14 years, all the milk teeth have gone and have been replaced by 28 of the permanent teeth: eight incisors, four canines, eight premolars and eight molars.

The four remaining molars may, if your jaw is long enough, appear in your late teens or early twenties. Because they are late to appear, they are sometimes called **wisdom teeth**.

Why take care of milk teeth?

If you do not have a full, well-positioned set of milk teeth, you will never have a good set of permanent teeth. This is because the roots of the milk teeth guide the permanent teeth into position. Parents therefore should make sure that young children take as much care of their milk teeth as they would their permanent teeth.

Questions

1 Describe the structure of a typical tooth.
2 Draw a map of the teeth in your own mouth. Count the number of incisors, canines, premolars and molars you have.
3 Explain how you would use each type of tooth when eating an apple.
4 What are the milk teeth? State two differences between a set of permanent and a set of milk teeth.
5 What should parents do to help make sure their children develop a well positioned set of permanent teeth?

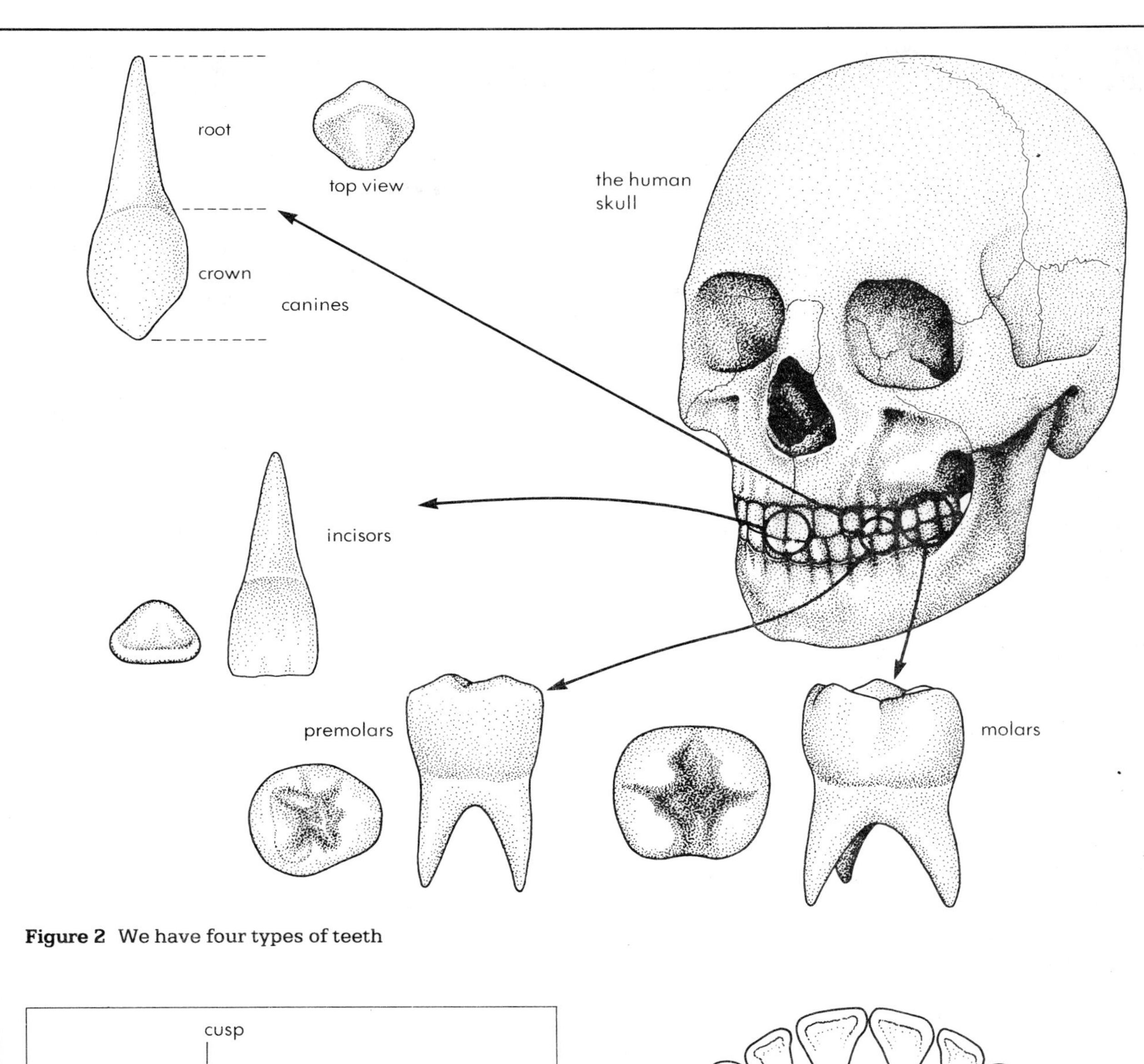

Figure 2 We have four types of teeth

Figure 1 A typical tooth has three main layers: the enamel, dentine and pulp

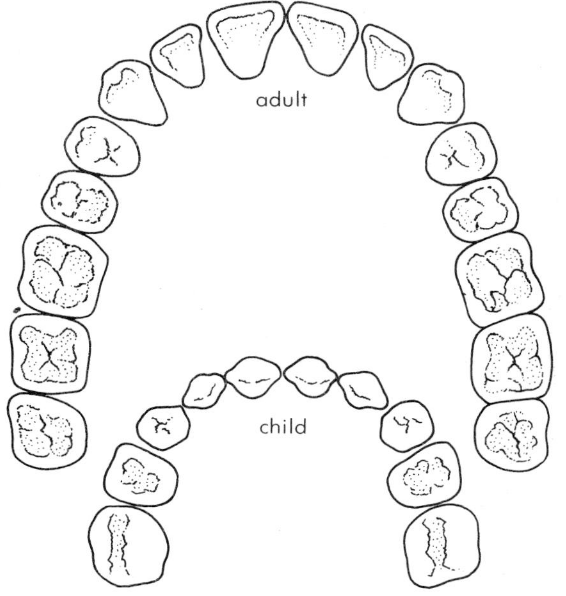

Figure 3 A 4 year old child has 20 milk teeth. These are replaced in the adult by 32 permanent teeth

2.14 Care of teeth

What happens if you do not look after your teeth?

In the UK over half of all 40 year olds have *no* teeth. The main reason for this is the food we eat. Much of it contains far too much sugar which the natural **bacteria** in your mouth feed on. In doing so they produce **acid**. This acid dissolves the enamel and irritates the gums, causing **tooth rot** and **gum disease**. Eventually, if not treated, the teeth fall out.

Tooth rot

Tooth rot is the main cause of tooth loss in children and teenagers. If your teeth are not cleaned properly, the acid produced by the bacteria dissolves a hole in the enamel and eats through the dentine. When the pulp is reached, the nerve is exposed and this causes the pain of **toothache**. Infection can quickly spread through the pulp, forming an **abscess**, which can be very painful.

Gum disease

Gum disease is more common in older people. The acid irritates the gums, making them sore. They may swell up and bleed easily. At this stage the disease is called **gingivitis**. If not treated, the acid eats down between the tooth root and the jaw bone, destroying the cement and fibres holding the tooth in place. So the tooth falls out.

Can we see the bacteria which produce the acid?

If bacteria have been allowed to build up on your teeth they form a thin film called **plaque**. You can see this very clearly if you first chew a **disclosing tablet** (figure 2). This contains harmless vegetable dye, which colours the plaque. An excessive amount of plaque is called **calculus**. This shows up as a yellow deposit.

How can you prevent tooth rot and gum disease?

The only way to do this is to prevent acid being produced. This can be achieved in two ways:

- By avoiding sugar in the food you eat.
- By preventing the build up of plaque.

The easier of the two options is to prevent plaque forming. You can do this by cleaning your teeth properly (figure 3) at least once a day and preferably after every meal. You should also visit the dentist for regular check-ups. In the UK you can have one free check-up every six months.

Can dentists cure tooth decay and gum disease?

If you look after your teeth you should not need much dental surgery. However, if you do get toothache, a dentist can stop it by removing the bad part of the tooth and replacing it with a new substance. This is known as **filling** a tooth (figure 4). Figure 1 shows the number of fillings the average 15 year old has.

Gum disease can often be cured by removing the calculus which has built up on the teeth. This process is known as **scaling**.

Dentists can even straighten teeth. Many teenagers have crooked teeth because they did not look after their milk teeth.

You should visit the dentist for regular check-ups to make sure all is well.

Why should I use a fluoride toothpaste?

Fluoride helps your body make stronger enamel so it is very important that children get enough of it. Some Local Authorities add fluoride to their water supplies.

Questions

1 Table 1 shows the percentage of children with decayed teeth.
a) What causes this decay?
b) Suggest reasons for the improvement between 1973 and 1983.

2 Draw a map of the teeth in your mouth. Mark the fillings with a cross. Compare this with your friend's mouth.

3 What causes gum disease? Explain why your teeth may eventually fall out if gingivitis is not treated.

4 Prevention is better than cure. Write down four immediate changes you can make to your lifestyle which will help you keep your teeth throughout your life.

5 Can you explain the data in figure 1?

Age (yrs)	Per cent children with decay		Average number of teeth with decay	
	1973	1983	1973	1983
5	71	48	3.4	1.7
10	93	80	4.9	3.4
15	97	93	8.4	5.7

Table 1 Tooth decay in children has fallen in recent years

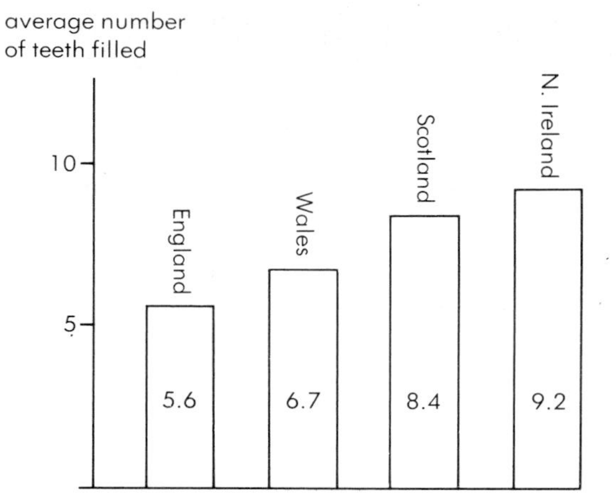

Figure 1 There are regional differences in the number of fillings that 15 year olds have

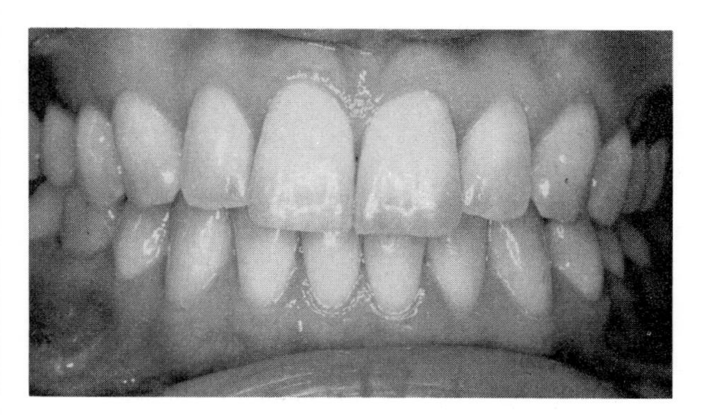

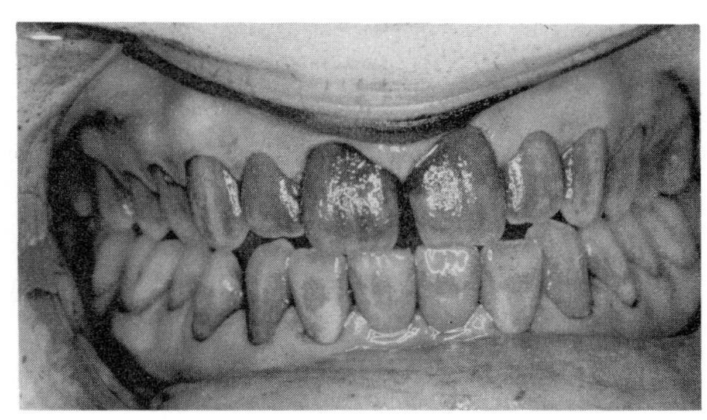

Figure 2 Disclosing tablets show up plaque on the teeth

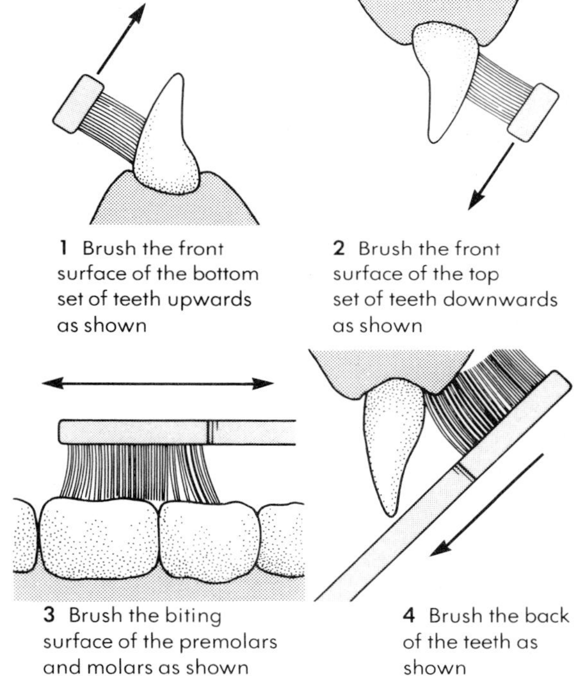

Figure 3 It is important to brush your teeth correctly

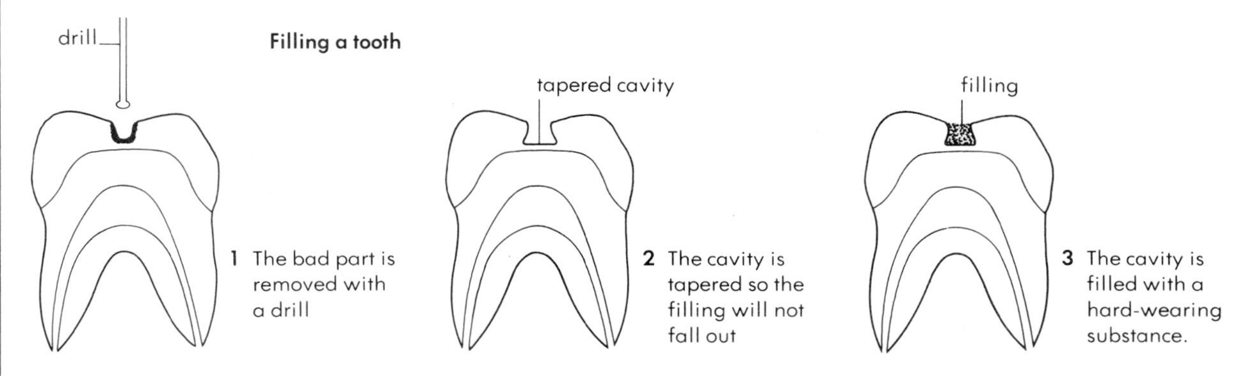

Figure 4 It is easy to fill a tooth when the decay has not gone further than the enamel

Questions

Recall and understanding

1 Which nutrient is missing from the following list: carbohydrates, proteins, lipids, minerals, vitamins:
A glucose C water
B fat D roughage

2 The villi of the small intestine:
A help push food along
B increase the surface area for absorption
C aid digestion
D prevent the food moving too fast

3 Green plants make food by the process of
A respiration C biosynthesis
B procreation D photosynthesis

4 Food must be digested before it can be:
A absorbed C decomposed
B excreted D egested

5 Gastric juice is produced in the:
A mouth C stomach
B liver D small intestine

6 Match the minerals or vitamins with their deficiency diseases:
A vitamin A 1 scurvy
B iron 2 anaemia
C calcium 3 night blindness
D vitamin C 4 rickets

7 Surplus amino acids are deaminated in the:
A stomach C gall bladder
B liver D spleen

8 Which of the following food chains is correct:
A grass → sheep → human
B human → sheep → grass
C human → grass → sheep
D sheep → grass → human

9 When one organism eats another only about 10 per cent of the energy it gets is used for growth. Most of the rest is:
A left behind in the waste
B used in respiration
C lost as heat
D used for eating

10 The acid which causes tooth decay is produced when:
A sugary foods are eaten
B fatty foods are eaten
C teeth are not cleaned for 24 hours
D plaque bacteria feed on sugar

Interpretation and application of data

11 The graph below shows the percentage of adults with no natural teeth:

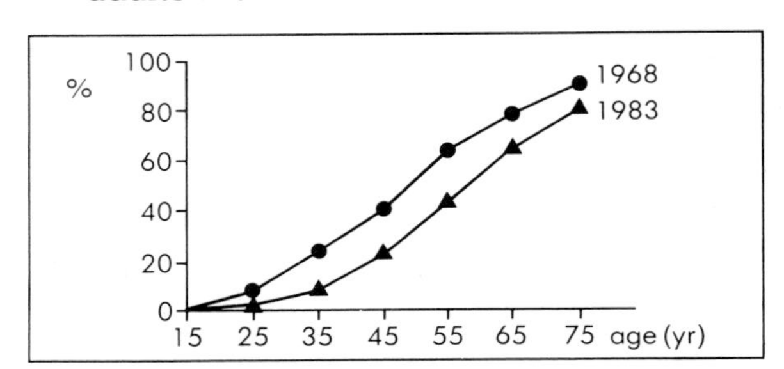

a) Use the graph to fill in the table below:

Age (yrs)	No teeth 1968 (%)	No teeth 1983 (%)	Difference
25			
35			
55			
65			

b) Suggest three possible reasons for the differences between 1968 and 1983.

c) The data below is from a recent study on fluoridation of water.

Concentration of fluoride in water (ppm)	0	0.5	1.0	1.5	2.0
Average number of decayed teeth per child	8	4	3	2.5	2.5

i) What do you conclude from the data about the effect of fluoride on tooth decay?
ii) There is an optimum flouride level. Explain.

12 The data below relates to potatoes:

Storage time (months)	Vitamin C (mg/100g)
freshly dug	30
2	20
4	15
6	10
8	8

a) Plot this data on a graph.
b) Explain what the graph shows.
c) Predict how much vitamin C potatoes stored for one year will contain.
d) Vitamin C is also lost when potatoes are cooked:

Cooking method	percent vitamin C lost
Boiled after peeling	20–50
Boiled in skins	20–40
Baked in skins	10–40
Roasted	15–50
Chipped and fried	15–30

i) Which method of cooking wastes the least vitamin C?
ii) Boiling can lose 20-50 per cent of the vitamin C. Why is there such a wide range?

13 The table below shows how efficient in terms of energy production different types of farming are.

Type	Typical energy production in kJ per hectare
cows for milk	1300
pigs	1000
crops	9500
sheep	340

a) Which type of farming is the most energy efficient?
b) During the winter, the pigs are fed on crops but their energy production does not alter. Explain why.
c) The table below shows typical protein yields per hectare.

Organism	Protein yield per hectare (kg)
beef cattle	25
fish	30
poultry	40
pigs	30
wheat	150
soya	510

i) Which animal is the most productive in terms of protein?
ii) Many farmers feed crops to the animals. Why is this wasteful?
iii) If you were a farmer, which organism would you grow?

14 The graphs below illustrate the principle of modern farming.

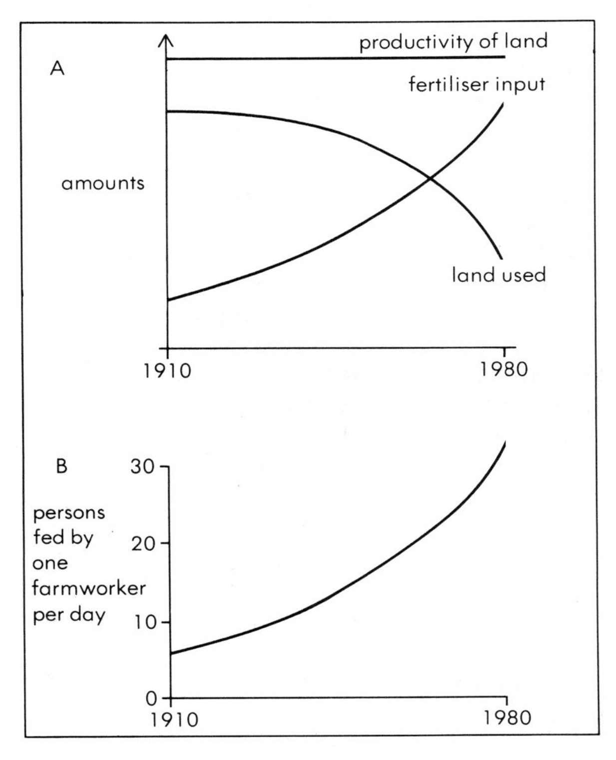

a) What does graph A show?
b) Why has the number of people fed by one day's work by a farm worker increased?
c) Compare this to an African Bushman.

	Energy produced from days work	Energy put in
Bushman	72 000 kJ	12 000 kJ
Modern farmer	552 000 kJ	1 104 000 kJ

i) Calculate the input/output energy ratios.
ii) Which type of agriculture is the most energy efficient?
iii) What accounts for the enormous amount of energy a modern farmer puts in?
iv) How can modern farming become more energy efficient?

CHAPTER 3: *TRANSPORT*

If you cut yourself it soon becomes quite obvious that you have blood in your body, but do you know what blood is made of or why you need it?

This chapter describes what blood is and why you cannot live without it. It also describes how it travels round your body and why your heart is so important to this.

Some of the people injured in this accident will probably have lost blood and need a transfusion.

How to help save a life

Become a blood donor

The blood used in transfusions has to come from other people. The National Blood Transfusion Service organises the collection and storage of blood. **Blood donors** are ordinary people who give blood regularly – usually every 6 months. Any healthy adult can be a blood donor.

Blood taken from donors is stored in a blood bank until it is needed. The blood can be kept for up to 3 weeks if it is stored at 4°C. It can be kept for several years if it is stored in liquid nitrogen (that is, at a temperature of minus 196°C). However, this long-term storage is only used for rare blood types.

Blood groups

The first successful blood transfusion from one person to another was carried out in 1818. That it worked was more due to luck than judgement, as we now know that there are many blood groups, some of which will not mix.

Your blood can be grouped by the kind of **antigens** (page 143) you have on the surface of your red blood cells. The most common groups are those of the **ABO system** and the **Rhesus system**. All donated blood is routinely tested for these groups.

The ABO groups

In the ABO system, there are two antigens, A and B. These create four groups: A, B, AB and O.

Blood containing antigen A will also contain an antibody (anti-B) to destroy antigen B. Blood group B will contain anti-A (table 1). In blood transfusions, the donor's blood must not contain antigens that match the antibodies in the recipient's blood. This prevents certain transfusions as shown in table 2.

The Rhesus system

The red cells of about 85% of people in the UK have antigen D. These people are said to be Rhesus positive. The remaining 15% are Rhesus negative.

If a Rhesus negative person is given Rhesus positive blood in a transfusion, he or she will produce antibodies (anti-D) to destroy it.

Cross matching

For transfusion, blood of the same ABO and Rhesus groups as the recipient is chosen. The bloods are then **cross matched**.

Cross matching involves incubating the donor red blood cells with the recipient's plasma. If a clot forms, the bloods are not **compatible** and the blood must not be given to that person.

In emergencies when there is not even enough time to test the patient's blood group, O/Rhesus negative blood is used. Can you think why?

Is blood now tested for AIDS?

All blood collected in this country is routinely tested for AIDS, Hepatitis B, Syphilis and Cytomegalovirus. Any signs of these and the blood is rejected.

Table 1 The ABO blood groups

Blood group	Antigen(s) present	Antibodies present
A	A	anti-B
B	B	anti-A
AB	A and B	none
O	none	anti-A and anti-B

*Table 2 The ticks show possible transfusions. Dangerous combinations are marked with a cross. Group O is the **universal donor** because it can be given to all the other groups. Group AB is the **universal recipient** because it can receive all the other types.*

Donor group	Recipient group			
		B	AB	O
A	✓	✗	✓	✗
B	✗	✓	✓	✗
AB	✗	✗	✓	✗
O	✓	✓	✓	✓

Using blood components

Blood is sometimes separated into its components before being used.

- **Red blood cells** These are given to patients suffering from anaemia.
- **Platelets** These are given to patients suffering from diseases which destroy their platelets.
- **White blood cells** These are given to patients who have a bacterial infection which is not responding to antibiotics.
- **Plasma** Whole plasma is given to serious burn victims. Parts of the plasma are used for other disorders. For example, factor VIII is used to treat haemophilia.

3.1 The composition of blood

Your own transport system

If you live near a big town you will know that it needs a good transport system to keep all the traffic flowing smoothly. In the same way your body depends on a good transport system to distribute substances to its cells. The transport system of the human body is the **blood circulatory system**. This is made up of a fluid tissue called blood which circulates around the body in a series of tubes called **blood vessels**. The heart acts as a central pump which keeps the blood moving round.

What is blood?

You may think blood is just a red liquid, but in fact about 45 per cent of blood is made up of solid particles. It is a tissue, really, in which the cells float in a liquid (figure 1).

The parts of the blood

Blood is made up of:

- **Plasma**. This is the liquid part of the blood. About 90 per cent of it is water. The remaining 10 per cent is a mixture of **plasma proteins, mineral ions, hormones** and various food and waste substances.
- **Red blood cells**. These are made continuously in the **red bone marrow** at the rate of about $2\frac{1}{2}$ million every second. They live for about four months and are then destroyed by the liver or spleen. Red blood cells are the only cells in your body that do not contain a nucleus. This disintegrates very early on in the cell's formation and is replaced by a red pigment called **haemoglobin**. It is this haemoglobin that gives your blood its red colour.
- **White blood cells**. White blood cells are larger than the red cells and there are fewer of them. There are two main kinds:
  1. **Phagocytes**. These are made in the red bone marrow and only live for a few days. They are irregular in shape with an unusual shaped nucleus and cytoplasm containing granules. Phagocytes are capable of engulfing foreign substances such as bacteria (the process is called **phagocytosis** — see page 13).
  2. **Lymphocytes**. These are made in the **lymphatic system** and may live for several years. They are spherical in shape, with a very large nucleus and a small amount of non-granular cytoplasm. Lymphocytes make special chemicals called **antibodies**.
- **Platelets**. Blood platelets are tiny fragments of cells made in the red bone marrow.

How much blood does the average person have?

The average adult man has about five and a half litres of blood in his body at one time. Women often have slightly less, depending on their size. The amount you contain can be roughly calculated from the formula:

Blood volume = (1/12 x body weight in kg) litres

Is everybody's blood the same?

There are in fact four kinds of blood. These are **type** A, **type** B, **type** AB and **type** O. These are called **blood groups**. It is important to know which blood group you are, because not all of them can be mixed together.

Blood transfusions are covered in more detail on pages 48 and 49.

Questions

1. In every 100 cm^3 of blood how much is plasma? Calculate how many red and white blood cells are present.
2. How much blood does the average adult man contain? Calculate your own blood volume.
3. Describe the structures of all the blood components.
4. Copy and complete the table:

Type of blood cell	Where made	Lifespan	Other information
			do not contain a nucleus
			are capable of phagocytosis
			make antibodies

5. What are the four blood groups and why is it so important to know which of these you are?

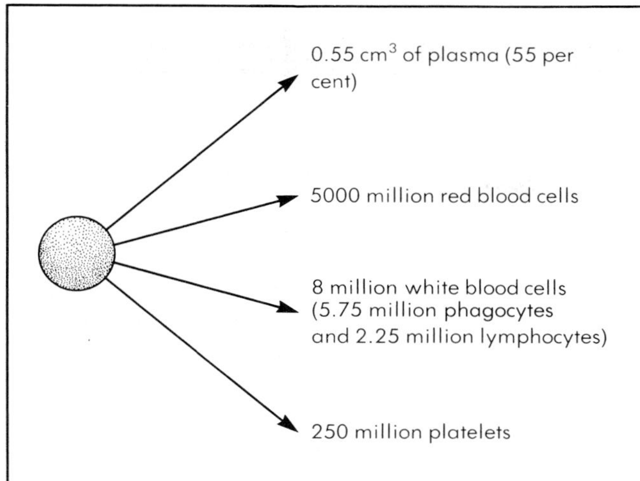

Figure 1 A drop of blood this size (1 cm^3) contains all these things

Blood group	Can give blood to	Can receive blood from
A	A and AB	A and O
B	B and AB	B and O
AB	AB	A, B, AB and O
O	A, B, AB and O	O

Table 1 Not all blood can be mixed. This table shows which groups can and which groups must not be mixed during transfusions

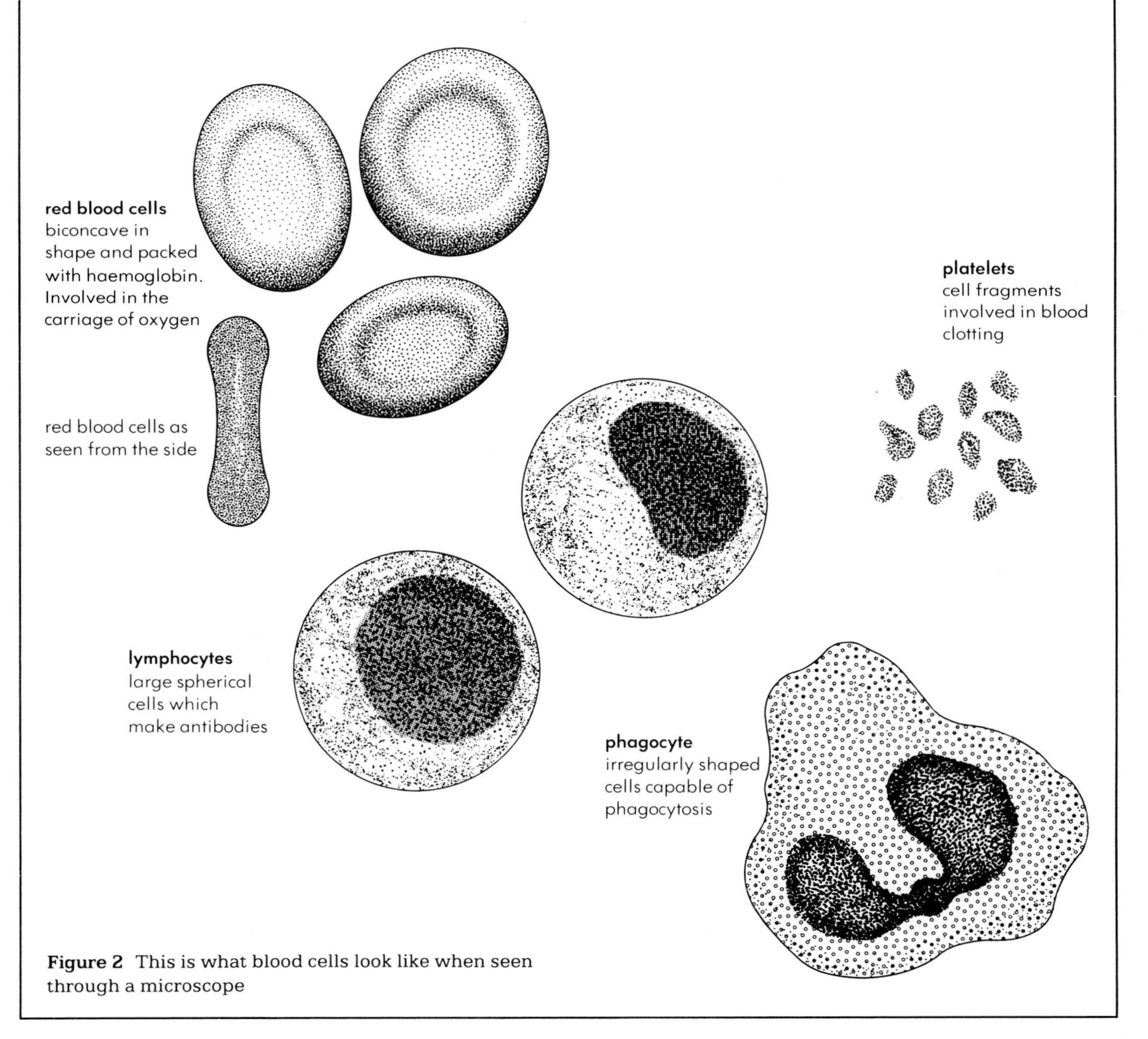

Figure 2 This is what blood cells look like when seen through a microscope

3.2 The circulation of the blood

The blood vessels

There are three main kinds of blood vessel; **arteries**, **capillaries** and **veins** (figure 1). Together these form a network of tubes in your body through which your blood can flow (figure 3):

- **Arteries**. Arteries take blood away from the heart towards the tissues. They have thick muscular, elastic walls because they have to withstand the very high pressure caused by the heart pumping blood into them. Near to the tissues, the arteries divide up into smaller vessels called **arterioles**.
- **Capillaries.** Capillaries are found at the ends of arterioles. They are extremely tiny vessels for carrying the blood through the tissues. Every tissue, therefore, contains a large capillary network so that *every* cell is less than 0.1 mm from a capillary. As the blood passes through these capillaries, some of the plasma passes out and enters the spaces between the tissue cells. Here it is known as tissue fluid. The cells remove food and oxygen from this **tissue fluid** and excrete their waste into it.
- **Veins**. The capillaries reunite to form slightly larger vessels called **venules**. These join up to form the veins. Veins take the blood back to the heart and have much thinner walls than arteries, because they do not have to withstand quite so much pressure. Most veins are sandwiched between muscles and when these contract, the veins are squashed, forcing the blood in them to move (figure 2). To make sure it moves in the correct direction, the veins contain one way valves (**pocket valves**).

What happens to the tissue fluid?

As new tissue fluid is formed some of the old fluid returns into the capillaries. The rest enters another type of vessel called a **lymph vessel** where it becomes known as **lymph** (figure 4).

The lymphatic system

All tissues contain lymph vessels. These are part of the **lymphatic system** which is connected to the blood circulatory system near the heart, (the thoracic duct). The lymph eventually rejoins the blood here, but before this it is filtered several times to remove foreign substances. The filters are called **lymph nodes** or **lymph glands**.

Any foreign substances filtered out are eventually eaten by phagocytes.

Lymph glands also supply your blood with lymphocytes. These pass into the lymph before it rejoins with the blood.

The lymph is kept moving in the lymph vessels in much the same way as the blood is kept moving in the veins (figure 2).

Where are your lymph glands?

The main lymph glands in your body are in your neck, armpits, abdomen and groin. Your tonsils and adenoids are lymph glands.

If a lymph node (gland) gets blocked much of the tissue fluid surrounding the cells has nowhere to drain and so it stays in the tissues, causing them to swell. This swelling is called an **oedema**. The disease elephantiasis is caused by worms blocking the lymph nodes in the groin. In this case tissue fluid builds up in the leg making it swell.

Tonsillitis is an infection of the tonsils by bacteria or viruses. This makes your tonsils swell, causing pain and discomfort.

Questions

1 What are the three kinds of blood vessel? Draw these.

2 State one functional and two structural differences between arteries and veins.

3 What are the names of the main artery and the main vein?

4 For each organ in column A, select the artery from column B which supplies it with blood:

Column A	Column B
Liver	Pulmonary
Kidney	Hepatic
Lung	Renal

5 Draw up a table showing the differences between blood, tissue fluid and lymph.

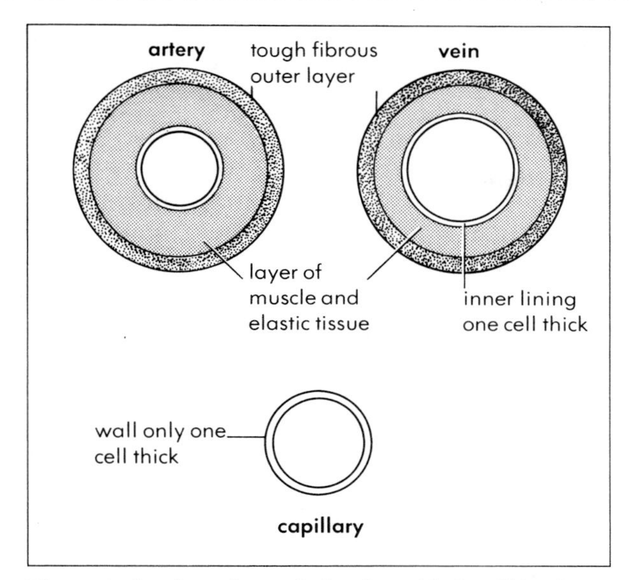

Figure 1 Sections through the three kinds of blood vessel. Look at the differences in thicknesses of the layers

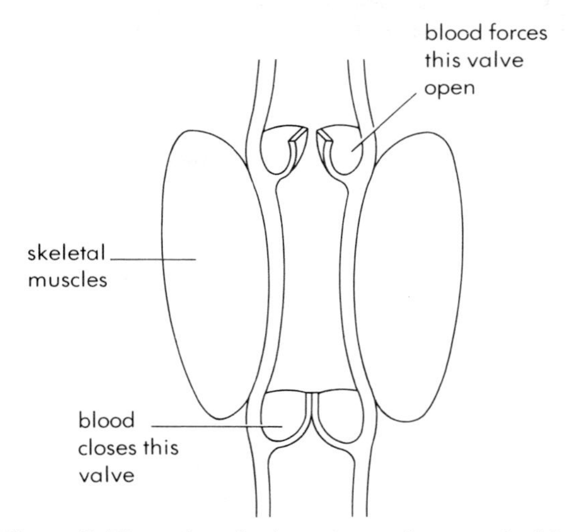

Figure 2 The valves in the veins make sure the blood flows in the right direction

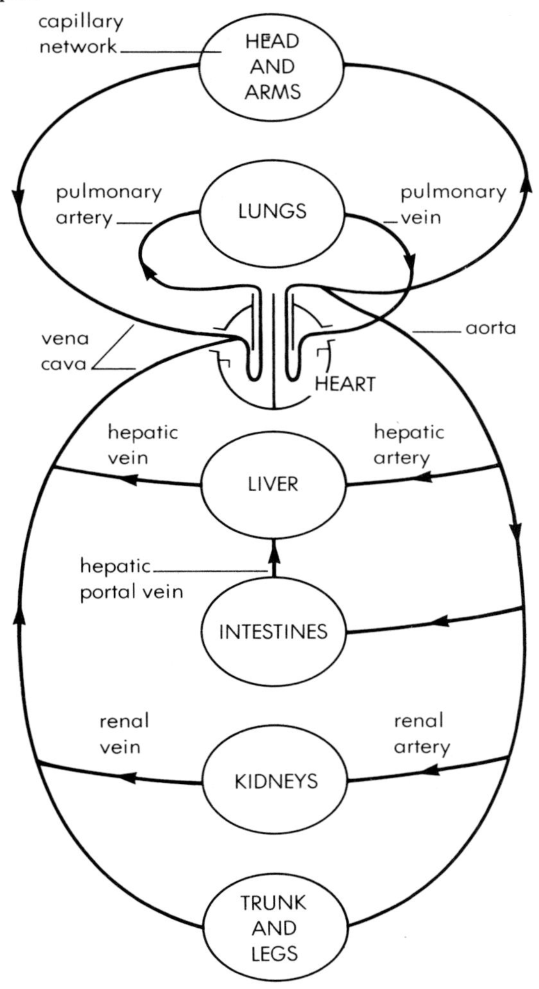

Figure 3 Some of the main blood vessels in the circulatory system. The arrows indicate the direction of blood flow

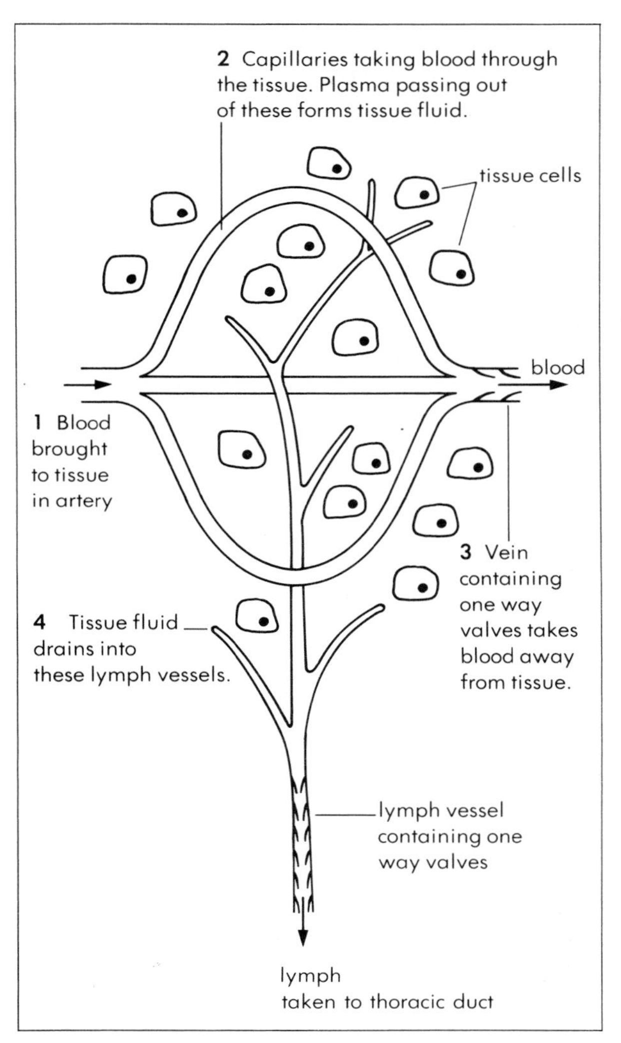

Figure 4 How tissue fluid and lymph are formed

3.3 The heart

Your heart as a pump

Your heart is the most important organ in your circulatory system. Its job is to pump your blood through the blood vessels and around your body. It is situated in the centre of your thorax, partly surrounded by the lungs and protected from injury by the ribcage.

The blood is also moved through the veins by contractions of the **skeletal** muscles.

What does your heart look like?

The heart is a hollow ball about the same size as an adult's clenched fist. Its inside is divided into two separate halves. Each half has an upper chamber called an **atrium** (plural atria) and a lower chamber called a **ventricle**. These two chambers are connected by a **cuspid** (flap) **valve**. There are also valves in the beginning of the blood vessels leaving the ventricles. These are called **semi-lunar valves** and, together with the cuspid valves, they make sure that the blood only flows through the heart in one direction, (see figure 1), like a one-way traffic system.

The walls of the heart are made from **cardiac muscle**. This is a very special muscle. It has to be special to be able to contract on average 70 times a minute for 70 years or more without tiring! It gets the supplies of food and oxygen it requires (to produce energy) from tiny blood vessels which run through it. These are the **coronary vessels** (page 57). Many heart problems are due to blockages in these vessels.

How does your heart work?

The action of your heart is illustrated in figure 2. Notice how the cuspid and semi-lunar valves ensure that the blood flows in one direction only. Notice also that the left ventricle is much more muscular than the right. This is because it has further to pump the blood.

The full cycle as shown here is called the cardiac cycle. During the cycle, each chamber goes through two phases:

1. **systole** = a period of contraction; and
2. **diastole** = a period of relaxation (resting).

The timing of these phases is controlled by the pacemaker which is a small patch of tissue on the wall of the right atrium. You may know someone who has had an operation to fit an artificial pacemaker. People who have this sort of surgery can live for their usual life span.

What is the pulse?

The pulse is not the same as the heartbeat. When the heart forces blood into the arteries (during ventricular systole) they expand slightly so that the blood will fit in. This expansion is followed by a contraction of the walls (elastic recoil). When you press an artery against something solid like a bone, you can feel this slight expansion followed by the contraction. It happens once every heartbeat and is known as the pulse. The easiest place to feel your pulse is on your wrist.

What is blood pressure?

This is the pressure of the blood inside your arteries. It is the result of both the pumping action of the heart and the narrowness of the arteries. Because of this, anything that makes the heart beat faster (excitement, anger, exercise, stress), or reduces the diameter of the arteries (fat deposits) raises the blood pressure. Permanent high blood pressure can be very damaging to many of your body organs.

The instrument used to measure blood pressure is a **sphygmomanometer**. Two readings are usually taken, the maximum and minimum pressures. These are taken during ventricular systole and ventricular diastole respectively. A normal reading would be around 120 mm/80 mm mercury.

Questions

1 a) Draw and fully label a section through the heart.
b) Draw arrows on your diagram to show the route that the blood takes through the heart.

2 What makes sure the blood travels in the correct direction through the heart?

3 Write the following in the correct sequence to show the route taken by a red blood cell from the kidney to the aorta;
right ventricle, left ventricle, right atrium, lungs, left atrium, pulmonary artery, pulmonary vein, renal vein, vena cava.

4 A person's blood pressure was 100 mm/60 mm mercury. Explain what this means.

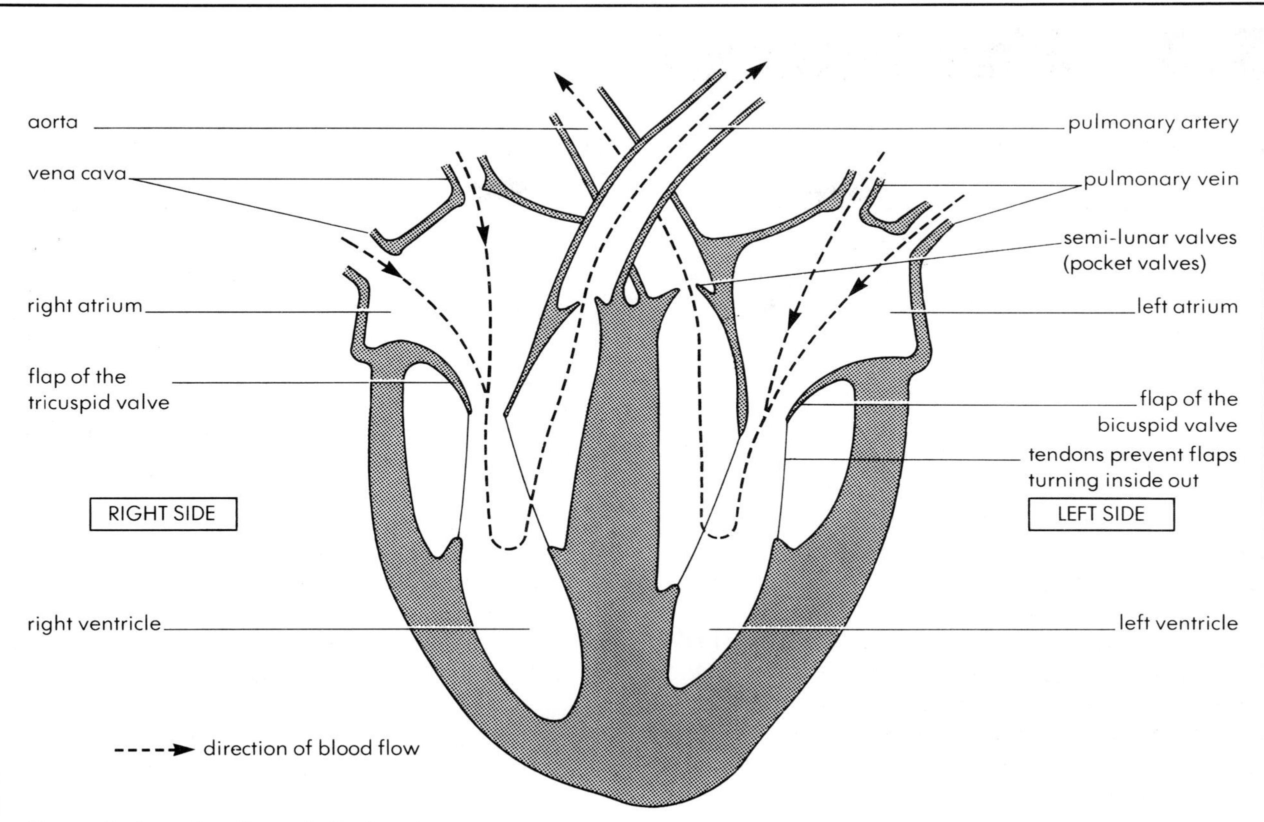

Figure 1 A section through the heart

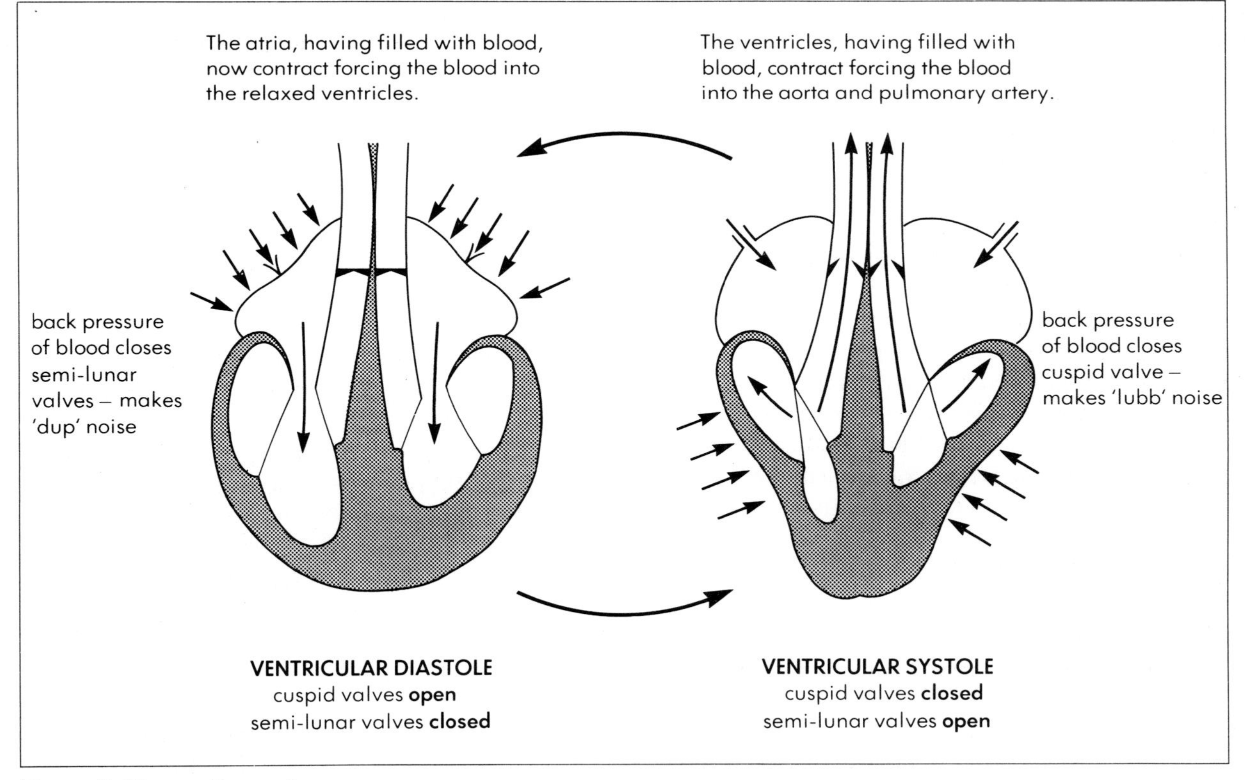

Figure 2 The cardiac cycle

3.4 Heart disease

What is coronary heart disease?

Coronary heart disease (CHD) is the most common cause of premature death in the Western world (figure 1). In 1984 in the UK about one person every three minutes died of a heart attack; this about 180 000 people in that year alone.

CHD is really a disorder of the coronary arteries that supply blood to the heart muscle (figure 2). If these are damaged or blocked, the oxygen supply to the heart muscle is reduced, the heart stops contracting, and the person has a heart attack. This can cause instant death, if a large part of the heart is affected, or if it is less serious the person may recover.

What can block the coronary vessels?

Most CHD usually starts with the build up of a fatty substance called **cholesterol** on the inside of the coronary vessels. This reduces the diameter of the vessels and therefore, the flow of blood to the heart muscle. The pain which accompanies this is called **angina** and is often the first sign that something is wrong.

The cholesterol may eventually completely block the coronary vessel or more usually cause **blood clots** to form, and stick in the narrowed areas. This cuts off the blood supply altogether and the heart stops (figure 3).

What sort of people get CHD?

Anybody can suffer from CHD, but doctors have now identified those who are a high risk. These are people who:

- **Smoke**. The link between smoking and CHD has been firmly established for many years now. Fifty per cent of smoking related deaths are from CHD (see page 70).
- **Have high blood cholesterol levels**. The amount of cholesterol in a person's blood seems to be determined partly by genes (see page 172) and partly by diet. Saturated fatty acids in a diet will usually increase the blood cholesterol level.
- **Have high blood pressure**. The higher the blood pressure, the more work the heart has to do and therefore the more oxygen it needs. Blood pressure can be raised by smoking, stress and obesity. It can be lowered by regular exercise.
- **Have a family history of CHD**. The part played by the genes is not fully understood, but CHD does seem to stay within a family.

How can you prevent yourself getting CHD?

You probably cannot *prevent* yourself from getting CHD, but you can *reduce* your risk by:

- **not starting to smoke**
- **eating less fat** and making sure that what you do eat contains mostly unsaturated fatty acids
- **exercising regularly**

In general the best thing to do is to live as healthy a life as possible.

Questions

1 A heart attack can be caused by:
a) an atheroma,
b) hypertension,
c) old age or disease.
Explain how.

2 Which of the following groups of people are considered to be high risk categories with respect to heart disease?
smokers
workaholics
drug addicts
promiscuous people
obese people
people who are under stress
people who belong to families with a history of coronary heart disease

3 Write a letter to a friend who you know eats a lot of fat and explain how the Framingham study shows that he or she could be increasing his or her chances of dying from coronary heart disease (see figure 1).

4 Write a list of the changes you can make to your own lifestyle to reduce your risk of suffering a heart attack.

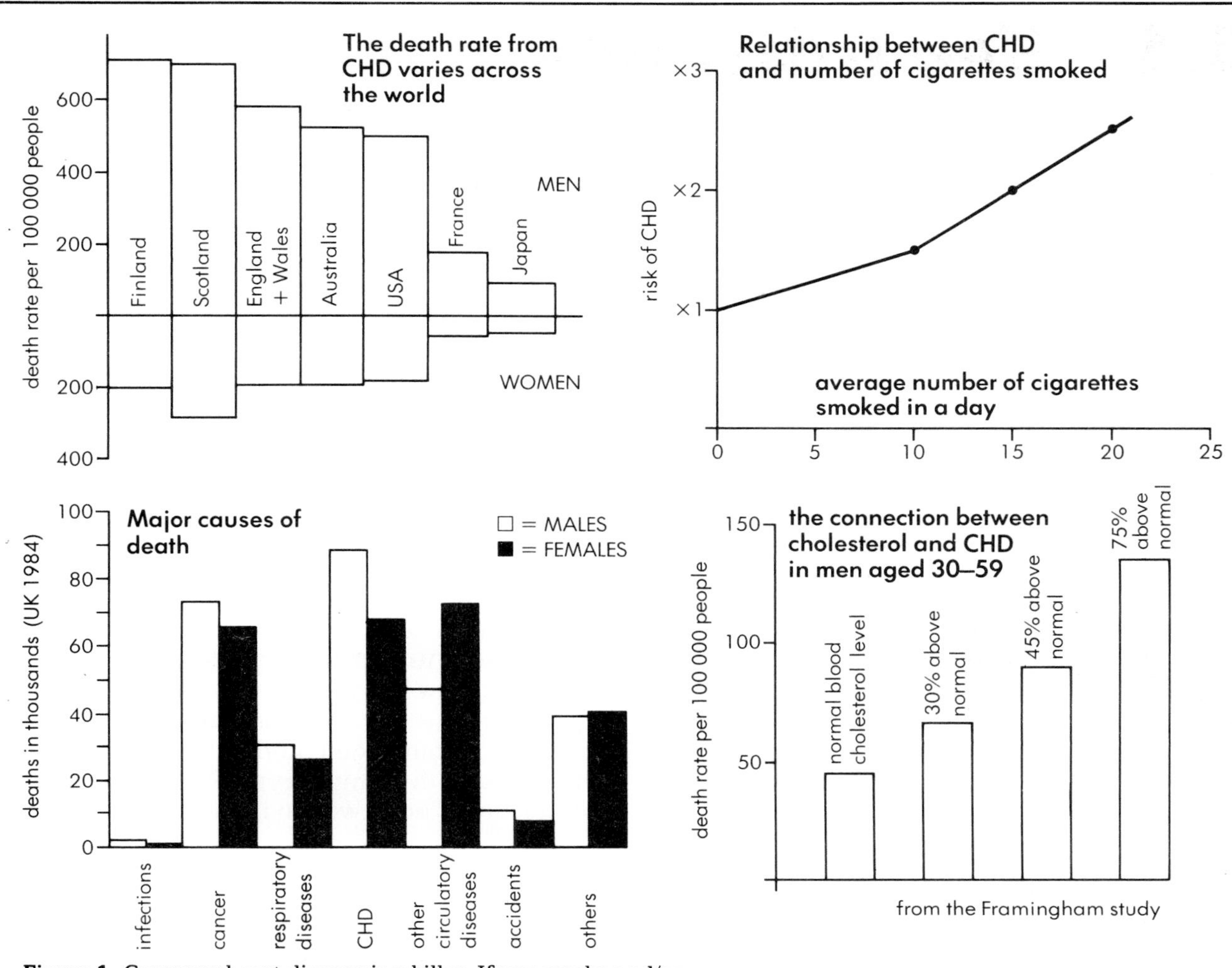

Figure 1 Coronary heart disease is a killer. If you smoke and/or eat a lot of animal fat, you are more likely to suffer from it.

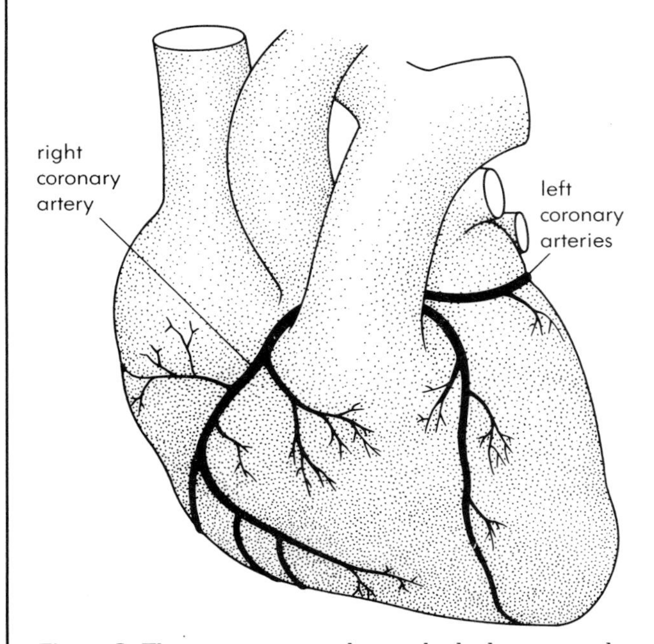

Figure 2 The coronary vessels supply the heart muscle with blood containing food and oxygen

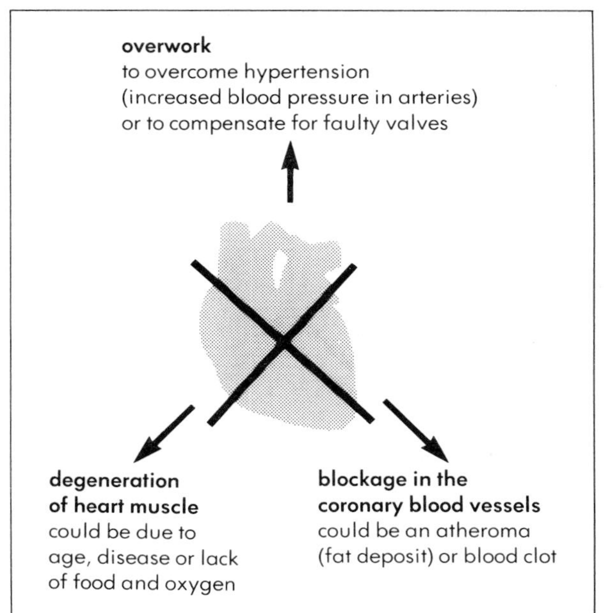

Figure 3 The main causes of a heart attack

3.5 The functions of the blood

Why do you need blood?

Blood is important for two reasons:

- **it transports substances around your body** (figure 1)
- **it defends your body against germs** (figure 3)

If your blood was to stop circulating, you would die within minutes.

Transport

The plasma, the fluid part of your blood, transports:

- **food substances**, such as glucose, amino acids, fats, mineral salts and vitamins from your intestines to the cells of your body. Here they are used immediately or stored for future use.
- **waste substances**, such as carbon dioxide, water and urea from your cells to your excretory organs. If these were allowed to build up, they would soon kill you.
- **heat**, made in your muscles and liver, to all areas of your body. This helps keep your body temperature constant.
- **hormones** from your endocrine glands to the correct organs. These hormones control and influence many important processes which take place in yoúr body.

The red blood cells transport **oxygen** from your lungs to your other body cells. As the red blood cells pass through your lungs oxygen combines with the **haemoglobin** in them. This oxygen is released as the red blood cells pass through active tissues:

$$\text{Oxygen} + \text{Haemoglobin} \underset{\text{in active tissues}}{\overset{\text{in lungs}}{\rightleftharpoons}} \text{Oxyhaemoglobin}$$

Defence

Many of the things which go wrong with your body are the result of foreign substances (germs) entering it. The skin and **mucus membranes** act as barriers to prevent these germs entering, but if these are damaged, your blood takes over. The first sign of this damage is **bleeding**.

Why do we bleed?

Bleeding is useful because it carries germs away from the wound, but it must not go on for too long. To prevent serious loss of blood and the entry of more germs, the wound is naturally plugged by the blood clotting. The production of a blood clot is complicated and is outlined in figure 2.

Even while a clot is forming your body also begins to tidy up the damage and begin repairs. The damaged tissues release chemicals which cause the nearest blood capillaries to **dilate** (increase in diameter) and become leaky. This results in an increase of blood flow to the area and to the formation of extra tissue fluid. This extra blood also supplies extra **phagocytes**. These pass into the tissues around the wound and engulf the germs. **Pus**, the yellow substance that collects at wounds, is a mixture of these phagocytes, tissue fluid and the germs.

What do the lymphocytes do?

Some of the germs must first be acted on by special chemicals called **antibodies** before the phagocytes can engulf them. These antibodies are made by the lymphocytes. Their action and production is dealt with in more detail in chapter 13.

Germs which penetrate deeper than the immediate wound area are filtered out of the tissue fluid as it passes through the lymph nodes and are then eaten by more phagocytes.

Questions

1 Use the following headings to draw up a table summarising the transport functions of your blood:

substance transported	part of the blood involved	transported from	transported to

2 Describe how your body defends itself against invading germs.

3 Explain why wounds always turn red, appear warm and swell slightly.

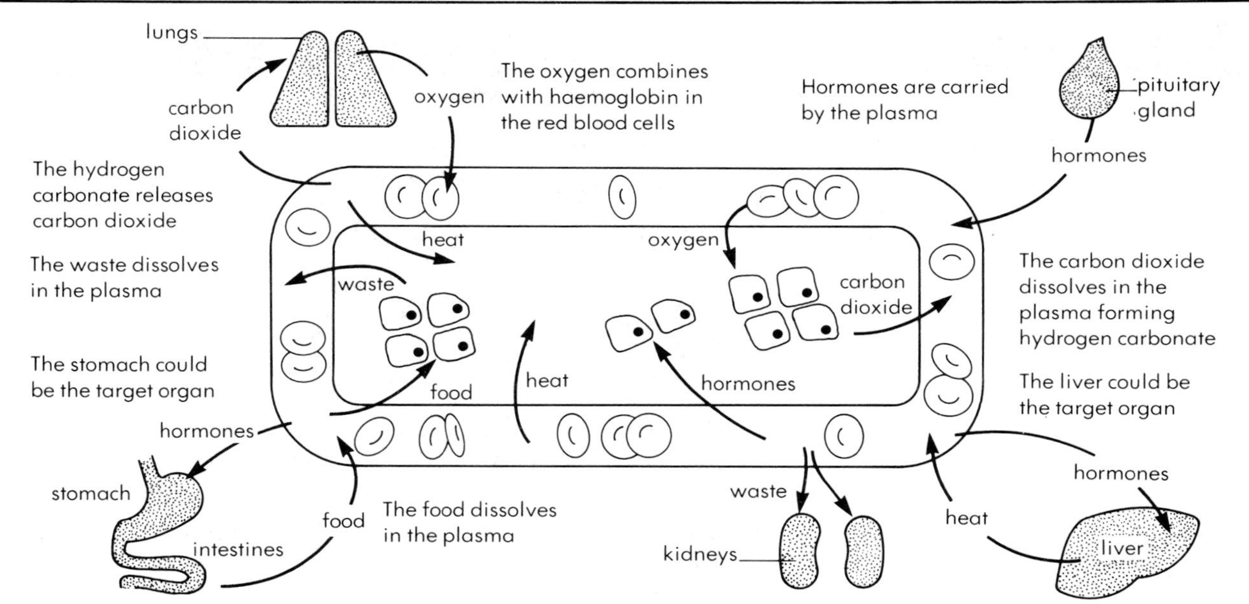

Figure 1 One of the main functions of blood is to transport substances around the body

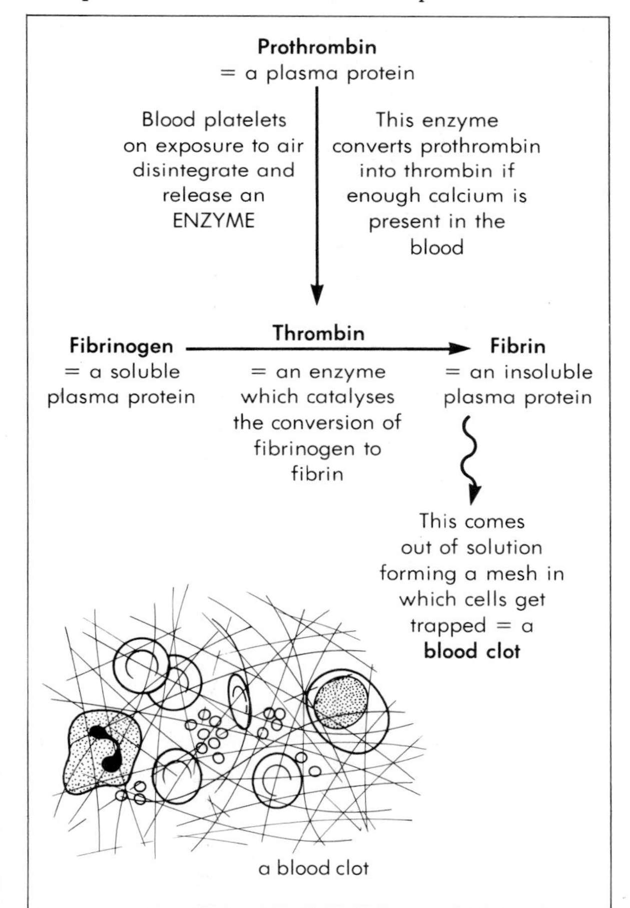

Figure 2 Blood clots are formed by a series of complex reactions involving enzymes. Fibrinogen (which is soluble) is changed to fibrin (which is insoluble) by enzymes produced when blood is exposed to air. Fibrin forms a mesh of fibres across the wound, trapping cells to form a clot

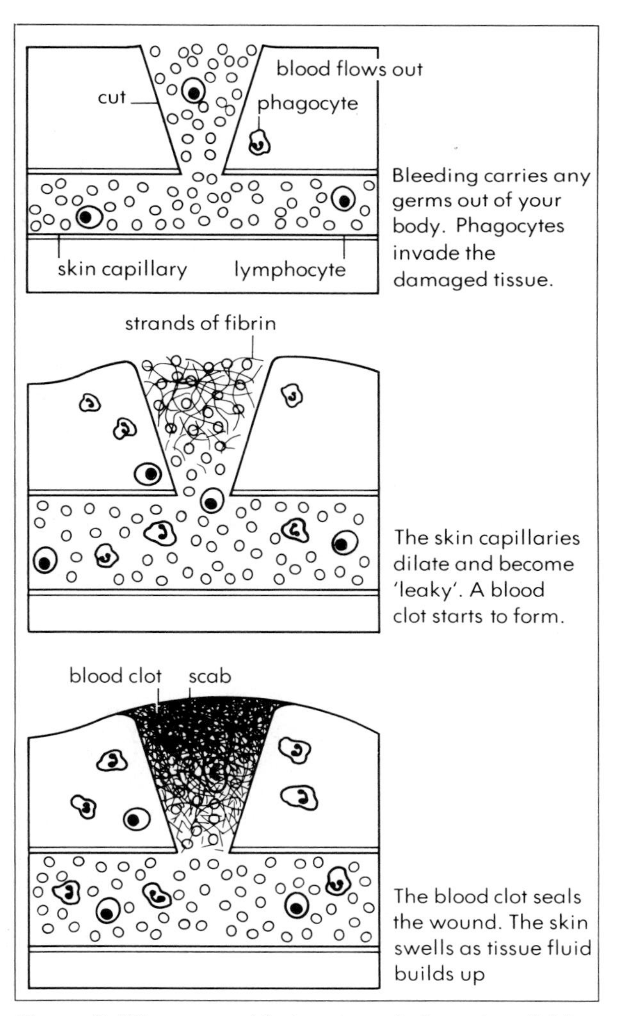

Figure 3 When your skin is cut, a whole series of things happens to prevent both germs entering and you losing too much blood

Questions

Recall and understanding

1 About what percentage of blood is liquid?
A 75 C 45
B 55 D 90

2 Phagocytes are a kind of:
A red blood cell C germ
B protein D white blood cell

3 Lymphocytes are made in the:
A red bone marrow C lymph nodes
B liver D yellow bone marrow

4 A sphygmomanometer is used to measure:
A pulse rate
B breathing rate
C the number of red blood cells
D blood pressure

5 Pus is a mixture of:
A white blood cells and germs
B plasma and phagocytes
C white blood cells and fibrin
D white blood cells, tissue fluid and germs

6 Which of the following is transported as hydrogen carbonate?
A oxygen C food
B carbon dioxide D heat

7 In which organ is oxyhaemoglobin formed?
A the heart C the muscles
B the lungs D the liver

8 The area around a cut feels warm because:
A warm air gets in
B there is an increased blood flow
C damaged cells release heat
D of the activities of the phagocytes

9 The pulmonary artery contains:
A oxygenated blood
B blue blood
C dirty blood
D deoxygenated blood

10 Ventricular systole is always followed by:
A atrial diastole
B ventricular diastole
C a rest
D fibrillation

Interpretation and application of data

11 The graph below shows the variation in blood pressure in the circulatory system.

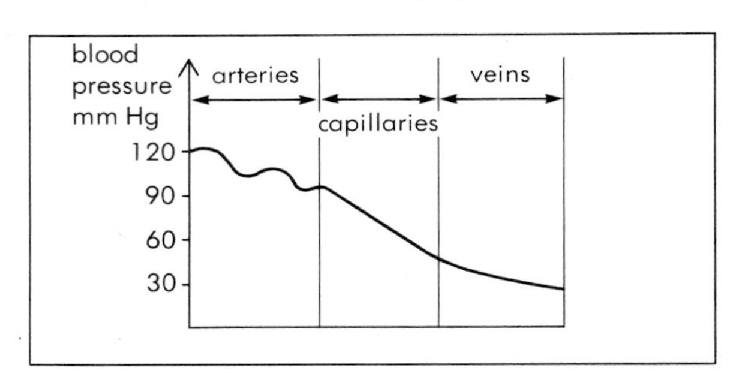

a) What range of blood pressures are found in the arteries?
b) Explain why the blood pressure in the arteries fluctuates.
c) What is the pressure of the blood as it enters the heart?
d) In which vessels does the blood pressure fall the most?

12 The pulse rates of a class of school children were measured. The results are shown below:

Pulse rate	Pulse rate
80	71
72	70
70	73
73	80
70	70
79	72
80	68
68	71
71	70
72	71

a) Explain what the pulse is.
b) State two places in the body where you can measure the pulse.
c) Use the results to copy and complete this table:

Pulse rate	Number of children

d) Use these results to plot a bar graph.
e) What was the average pulse rate?
f) What was the modal pulse rate?
g) Name two circumstances when the pulse rate might increase.

13 The data below is for 1984:

	England and Wales	Scotland	Britain
Deaths from all causes	566 881	62 345	629 226
Deaths from heart diseases	278 849	31 489	310 338
Deaths from heart disease as % of all deaths			

a) Work out the approximate percentages for the last row.
b) There are 525 600 minutes in a year. Approximately how often does a death from heart disease occur in Britain?

	England and Wales	Scotland	Britain
Deaths from CHD	157 506	18 107	175 613

c) i) What per cent of deaths from heart disease are from CHD?
ii) What is the overall proportion of all deaths due to CHD?

14 The number of red blood cells in 1 cm^3 of blood was counted for several people living at different altitudes.

Red blood cell count	Height above sea level
5000 million	sea level
6000 million	1000m
6500 million	2000m
7500 million	3000m
8000 million	5000m

a) Plot these results on a line graph.
b) What red blood cell count would you expect at 4000 m above sea level?
c) What is the role of red blood cells in your body?
d) Why do you think people living at high altitudes have more red blood cells?
e) Would you expect athletes who train at sea level to do well at high altitude? Explain.
f) Athletes often go early to high altitude locations in order to acclimatise. What do you think is happening in their bodies during this acclimatisation?
g) Which mineral would you advise the athletes to eat more of?

15 The table below shows the percentage of blood types found in the UK population.

Blood type	per cent
A	40
B	10
AB	5
O	45

a) Use this information to decide:
i) The percentage of the population who can receive type O blood.
ii) The percentage of the population who must not receive type AB blood during a transfusion.
iii) The percentage of the population who can receive type A blood.
b) The graph below shows the incidence of anaemia in children of different social classes.

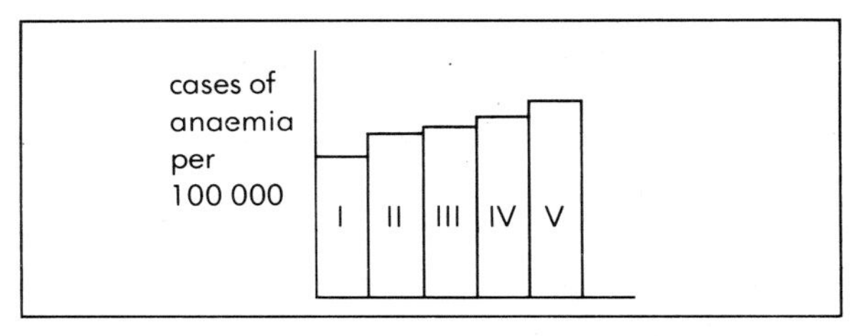

i) In which social class are there more anaemic children?
ii) Suggest a reason for this.
iii) What are the main causes of anaemia?
iv) Suggest a treatment for each of these causes.

CHAPTER 4: *RESPIRATION*

Everything you do requires energy. To produce this energy you need a constant supply of oxygen. In this chapter you will learn about how you get this oxygen and what happens if you cannot get enough. It also looks at how some people damage their breathing systems by smoking.

Breathing in strange places

This section looks at the problems faced by people who live and work in places where there is little or no natural oxygen available.

This climber on Mount Everest has his own supply of oxygen.

Breathing at altitude

At sea level, the air pressure is about 100 kiloPascals (kPa). As you go higher it falls below this.

The pressure exerted by the air depends on how **dense** it is. Therefore a fall in pressure also means a fall in density.

Table 1

Altitude (m)	**Air pressure** (kPa)
sea level	100
1000	90
3000	70
6000	47.5
9000	32.5

Whether you are at sea level or at (say) 6000 m, the *proportion* of oxygen in the air is the same – about 21%. But because the air is more dense at sea level, there is *more* oxygen than at 6000 m. In fact there is more than twice as much.

This reduction in the amount of oxygen available presents problems for people who live, work and travel at high altitude.

Climbing mountains

The summit of Mount Everest is about 9000 m above sea level. The air there contains less than one third of the oxygen of air at sea level. Mountaineers attempting to climb at this height need to take a supply of oxygen with them. It is possible to do it without, but very risky.

How do people who live at altitude cope?

People who live at high altitude breathe more deeply and more quickly. They also have an increased oxygen carrying capacity because they have more blood and a higher red blood cell (RBC) count.

Table 2

Altitude (m)	**Blood volume** (litres)	**RBC/cm^3 blood**
sea level	5.5	5000 million
1000	5.65	6000 million
3000	6.0	7500 million
5000	6.25	8000 million

Athletes and altitude

In 1968 the Olympic Games were held in Mexico City which is 2000 m above sea level. This was the first time they had been held at altitude. Not surprisingly many of the medal winners were athletes who lived and trained at altitude.

The only athletes from lowland countries who managed to compete successfully, particularly in long distance events, were those who had spent some time living and training at altitude. These athletes had **acclimatised**, that is their bodies had made the adjustments for living in reduced oxygen.

A long distance athlete's performance depends to a large extent on his or her oxygen carrying capacity. Living and training at altitude can increase this.

Nowadays many athletes train for all major events at altitude.

Breathing under water

Breathing under water is complicated by the water pressure. At a depth of only 10 m, the water pressure is double what the air pressure was at the surface. The air in your lungs only occupies half its original volume here.

Table 3

Depth (m)	External pressure (kPa)	Volume of air in lungs (%)
sea level	100	100 (full)
10	200	50
20	300	33
30	400	25
40	500	20

Before diving rapidly to a depth of even 10 m, divers must completely fill their lungs to prevent them collapsing. Before returning to the surface, they must empty their lungs to prevent them rupturing.

The pressure differences can also cause the diver to lose consciousness on returning to the surface. If the ascent is too rapid, oxygen actually passes *out* of the blood into the lungs, i.e. the wrong way.

These divers will have to return to the surface slowly although they are using breathing apparatus.

Scuba diving

Scuba divers can go to greater depths and stay under water longer because they take a supply of oxygen with them. For deep diving (100 m or more), this oxygen is not supplied in a normal air mixture, because at this depth the nitrogen in the air is poisonous. The mixture the divers breathe is 98% helium and 2% oxygen. This is supplied to the lungs at the same pressure as the environment.

What are the bends?

If a scuba diver returns to the surface too quickly, nitrogen or helium bubbles form in his or her body tissues. This condition is known as **the bends**. Apart from being very painful, this can cause structural damage.

All deep divers must go through a slow **decompression** to prevent helium bubbles forming.

4.1 Energy

What is energy?

Energy is often described as the power to do work. Your body cells are working all the time and, therefore, must have a continual supply of energy. In fact, without this energy, they would soon die.

Energy comes in several forms; as **chemical** energy, **light** energy, **sound** energy, **heat** energy and **mechanical** energy. It is possible to change one form into another. For example, when coal is burnt, the chemical energy contained within it is turned into heat energy and light energy.

In what form does your body get energy?

Our energy comes from the food we eat. This food contains chemical energy which our cells are able to release, and use to perform work. The process which releases energy from food is called **respiration** (figure 1).

Respiration is really a series of chemical reactions which take place in the cytoplasm and mitochondria of cells. The main food substance used is glucose, but other carbohydrates, lipids and even proteins can also be used. For all the chemical reactions to take place, oxygen needs to be present. For this reason it is often called **aerobic respiration**.

Much of the energy released is lost as heat, but the rest is caught by another compound called **ATP** (adenosine tri-phosphate). In this form it is more readily available to the cell and can be transported more easily to those parts where it is most needed.

Does all food give the same amount of energy?

Different types of food release different amounts of energy when they are used in respiration (figure 2). Energy is measured in **joules**. One thousand joules is known as one **kilojoule**. Carbohydrates and proteins supply 17 kilojoules per gram whereas lipids supply 38 kilojoules per gram. These values were obtained using a calorimeter, like the one shown in figure 3.

How much energy do you need?

The amount of energy you need will depend upon your metabolic rate, which is the rate at which your cells consume energy. This will vary with your age, size, sex, activeness and state of health (table 1). The minimum amount of energy the average adult man needs just to keep his cell's life-sustaining reactions going is about 7000 kilojoules per day. At this energy consuming rate, his body is said to be operating on **basal metabolism**. For the cells to do more work, more energy must be provided.

What is anaerobic respiration?

Anaerobic respiration is respiration without oxygen. In animals, this can be summarised as follows:

glucose = energy + lactic acid

The glucose is not completely broken down to release all its energy. In fact, only about one tenth of it is released so it is not very efficient for humans. However, sometimes our muscle cells are forced into respiring anaerobically because they need more energy than is being produced by aerobic respiration. Unfortunately, this always leads to **cramp**.

The symptoms of cramp are hard, rigid, painful muscles which refuse to contract. This is due to the build up of **lactic acid**. The muscles will only return to normal when all this lactic acid has been removed, but this requires oxygen. The oxygen required to get rid of this lactic acid is called the **oxygen debt**.

Questions

1 Energy comes in many forms. State these forms and for each name one source.

2 The energy we use is supplied by our food.
a) What is the process which releases this energy from the food called?
b) Write a word equation to summarise this process.
c) Where does this process take place?

3 Rewrite in order of energy needs; pregnant woman, male labourer, female office worker, teenage boy, teenage girl, male pensioner.

4 Explain the difference between aerobic and anaerobic respiration. Why does anaerobic respiration produce less energy?

5 An athlete often builds up an oxygen debt? Explain why.

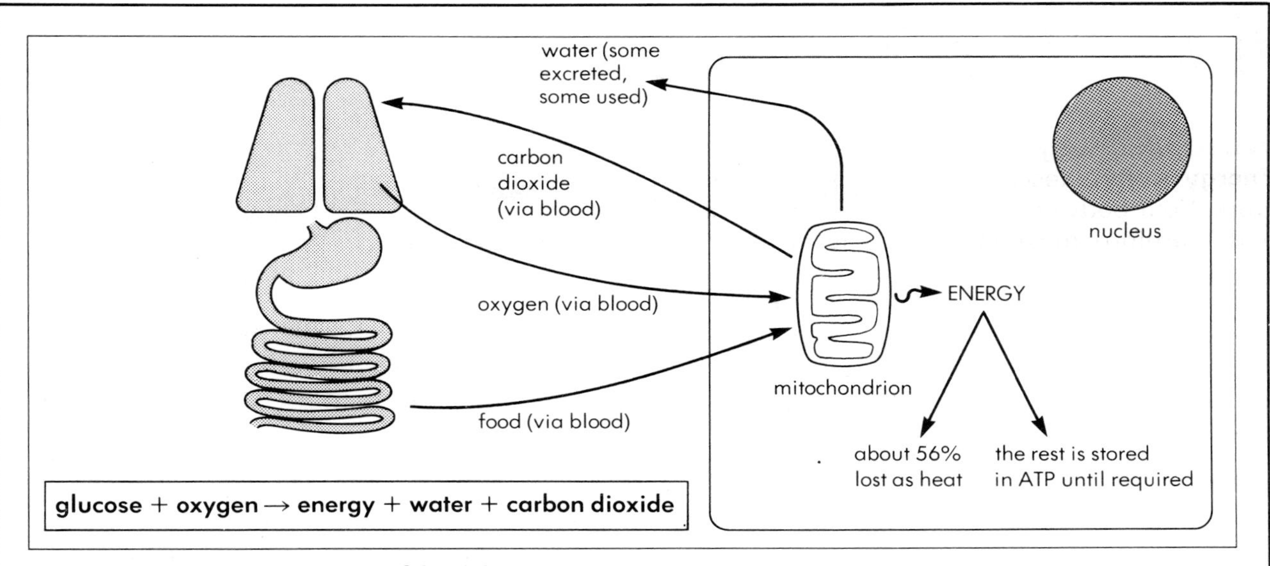

Figure 1 Energy is only produced if the right substances are present

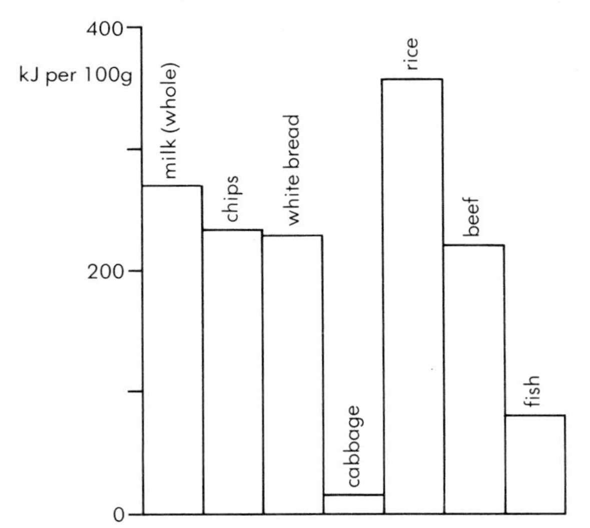

Figure 2 Foods do not contain the same amounts of energy

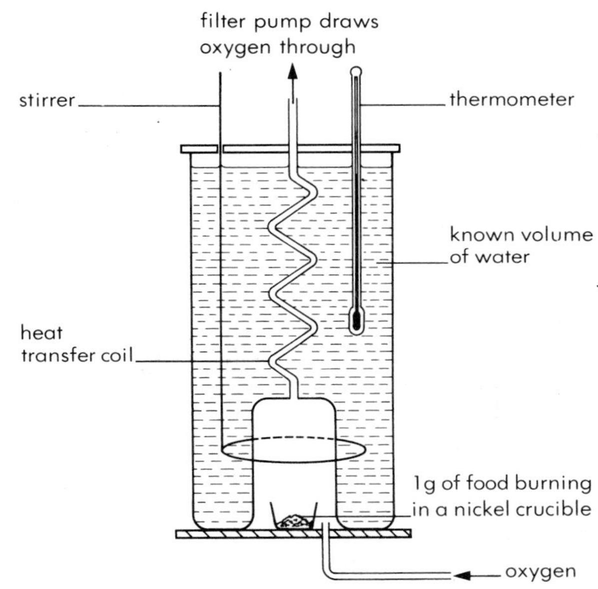

Figure 3 A calorimeter is used to measure the energy value of a food

Daily energy requirements of the average European adult			
Age/years	**Sex**	**Occupation or status**	**Energy required/kJ**
Under 1	boy or girl	child	3250
5–7	boy or girl	child	7500
12–15	boy	child	11700
12–15	girl	child	9000
18–34	man	office worker	12000
18–54	woman	office worker	9000
18–34	man	labourer	14000
18–54	woman	keep fit teacher	10500
35–64	man	officer worker	11500
65+	man	pensioner	10000
55+	woman	pensioner	8000
pregnant	woman	housewife	10000

Table 1 People of different ages and occupations require different amounts of energy

4.2 The air you breathe

Getting oxygen

To keep burning strongly, a fire needs a constant supply of oxygen. In the same way, to get the maximum amount of energy from food, our cells also need a continuous supply of oxygen. The process of obtaining this oxygen is called **external respiration** and the body system responsible is the **respiratory (breathing) system**. This makes oxygen available to the blood which picks it up and then transports it to the cells. The respiratory system also gets rid of much of the carbon dioxide made in the cells.

The structure of the respiratory system

The respiratory system consists of (figures 1–3):

- A series of air passages connecting the respiratory surface with the air. These passages are formed by the nose and nasal cavity, the throat and mouth, the **trachea**, the two **bronchi** (singular bronchus) and many **bronchioles**.
- The surface over which oxygen and carbon dioxide are exchanged with the blood. This surface is formed by millions of tiny air sacs called **alveoli** (figure 4). Each is surrounded by a network of capillaries and together with the bronchioles these form the core of the **lungs**.
- The ventilating structures which support the lungs and move air in and out of the alveoli and air passages. The structures involved are the **diaphragm, ribs, intercostal muscles** and **pleural membranes**.

What happens in the lungs?

This table shows the content of the air we breathe in (**inhaled** air) and the air we breathe out (**exhaled** air). Can you see three major differences?

Gas	Inhaled air	Exhaled air
Nitrogen	79%	79%
Oxygen	20%	16%
Carbon Dioxide	0.04%	4%
Water Vapour	a little	a lot

The differences are:

- There is about 4 per cent less oxygen in exhaled air. This missing oxygen has diffused into the blood to be transported away to the body cells.
- There is about 100 times more carbon dioxide in exhaled air. The extra carbon dioxide has diffused out of the blood into the air in the alveoli and is then breathed out (figure 5). This exchange of gases across the respiratory surface is often called **gaseous exchange**.
- Exhaled air is warmer and saturated with water vapour. The water vapour and heat come from the mucous membranes lining the air passages. Alveoli are very delicate structures and for them to work properly their linings must be kept moist, free of dust and not exposed to extremes of temperatures. Air we breathe in is usually fairly dry, very dusty and often very cold. Before it gets to the alveoli it must therefore be moistened, filtered and warmed. This is done as it passes through the trachea and other air passages (figure 3).

Questions

1 Describe in words the structure of the respiratory system.

2 Copy the name of the structure from the left hand side of the list below. Select from the right hand side, the correct description of its role in the respiratory system and write this next to it.

Alveoli	Moist air passage
Trachea	Ventilating structure
Diaphragm	Respiratory surface
Intercostal muscles	
Bronchi	
Nasal cavity	
Pleural membranes	

3 a) Make a list of differences between inhaled and exhaled air.
b) Explain these differences.
c) Where do these changes take place?

4 Explain why it is better to breathe through your nose rather than your mouth, especially in a dusty, dry environment.

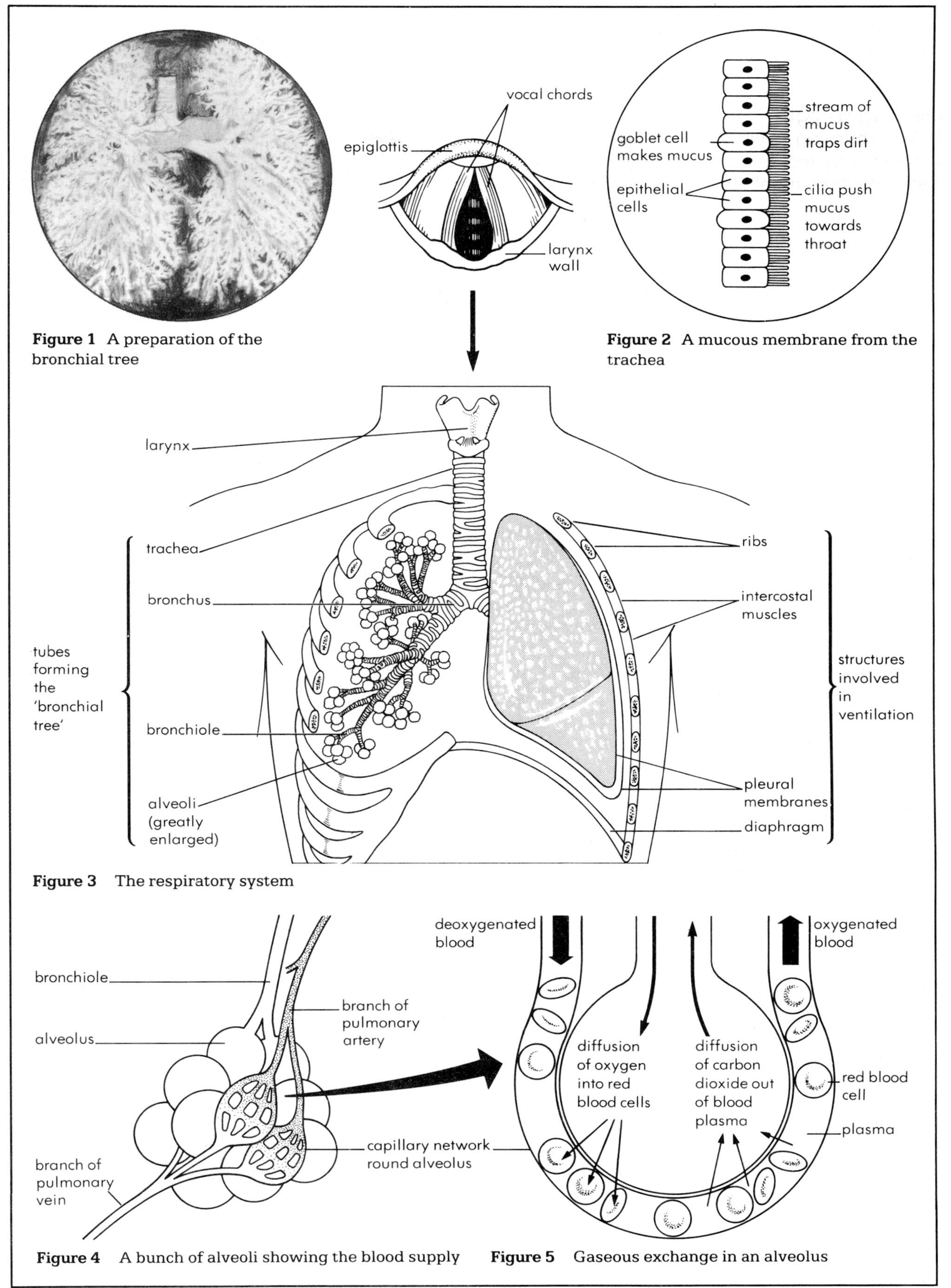

Figure 1 A preparation of the bronchial tree

Figure 2 A mucous membrane from the trachea

Figure 3 The respiratory system

Figure 4 A bunch of alveoli showing the blood supply

Figure 5 Gaseous exchange in an alveolus

4.3 Breathing

How do your lungs work?

If you place your hand on your chest after exercising, you can feel it moving up and down. These movements are brought about by your own diaphragm and intercostal muscles. They are to move air in and out of your lungs.

The diaphragm is a sheet of tendon surrounded by muscles. It separates your thorax from your abdomen. At rest, the up and down movements of the diaphragm are often all that is needed to move sufficient air in and out of your lungs. It is only when you start to exercise and more oxygen is needed that the intercostal muscles are used to move the ribcage. This is why you can only feel and see the ribcage moving during and immediately after exercising.

You have two sets of intercostal muscles between the ribs, an outer set (external) and an inner set (internal). When the external set contract, your ribcage pivots on the backbone upwards and outwards. When the internal set contract your ribcage pivots downwards and inwards.

Figure 1 shows how the intercostal muscles and diaphragm work together during breathing. Air moves in and out of the lungs because of the pressure changes involved.

What is the role of the pleural membranes?

Each lung is surrounded by two smooth membranes called the **pleural membranes**. In between these is a fluid-filled space called the **pleural cavity**. When you breathe one of the membranes slides over the other. The **pleural fluid** acts as a lubricant, reducing friction. If for any reason air gets into the pleural cavity, breathing becomes impossible.

How much air do your lungs hold?

The size of your lungs depends upon your age, sex and physique. If the average man was to fill his lungs, they would contain about five and a half litres of air.

If this same man then tried to empty his lungs, he would only be able to breathe out about four litres of this air. The maximum amount of air which a person can breathe out from full lungs is known as the **vital capacity**. The one and a half litres of air which has remained in the lungs can never be expelled and is called the **residual air**. This occupies the area inside the trachea,bronchi and bronchioles, often called the **dead space**.

The air which a person normally breathes in and out is called the **tidal air**. For the average man at rest, this is approximately half a litre, but during exercise it gradually increases until his vital capacity is reached.

Why do you breathe more quickly when you exercise?

When you exercise, your muscle cells need more energy and therefore more food and oxygen to produce it. This extra oxygen must be made available to your blood by your respiratory system and so you need to increase your breathing rate to achieve this.

At rest, the average adult person breathes 15 times per minute. During exercise, this breathing rate can more than double and tidal air can increase by up to eight times. The blood circulation also speeds up so that more blood goes through the lung capillaries, thereby picking up more oxygen to transport to the cells.

What else affects breathing rate?

Another important factor which affects your breathing rate is the concentration of carbon dioxide (hydrogen carbonate ions) in your blood. A rise in this results in an increase in breathing rate and a fall results in a decrease. Figure 2 summarises all the factors responsible for altering your breathing rate.

Questions

1 Copy out and fill in the missing words: Breathing involves the .... muscles and the .... The .... are found between the .... There are two sets. When the .... contract the ribcage pivots .... and .... When the .... contract, the ribcage pivots .... and inwards. The .... is a sheet of tendon surrounded by ....

2 a) Explain why your breathing rate changes during exercise.
b) State two more factors that will alter your breathing rate.
c) In addition to your breathing rate what else alters during exercise and why?

3 Explain the relationship between: tidal air, vital capacity, residual air, and dead space.

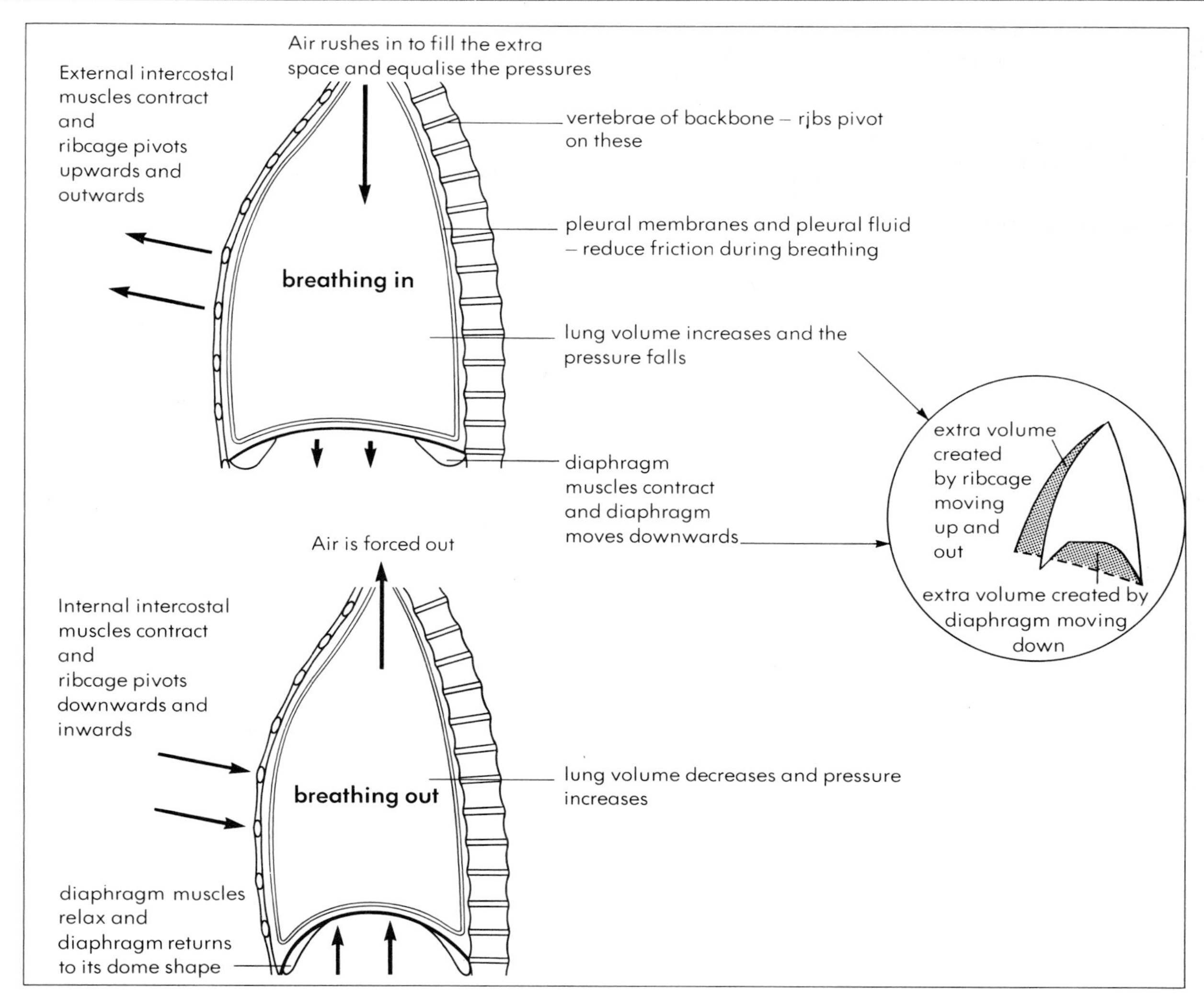

Figure 1 Breathing involves your diaphragm, intercostal muscles and ribcage

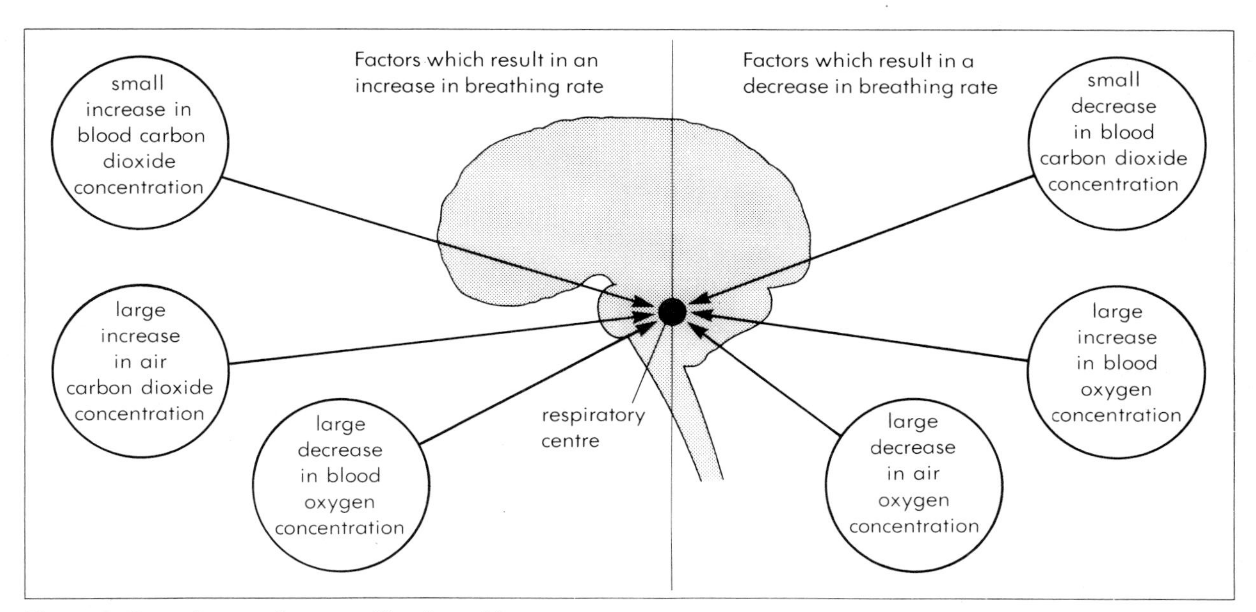

Figure 2 Some factors that can affect breathing rate

4.4 Smoking and lung diseases

Smoking can damage your health

You have probably seen advertisements for cigarettes which also say '**smoking can seriously damage your health**'. You may find this a bit puzzling but it was introduced by law to try and curb cigarette smoking. But a recent survey showed that 25 per cent of all 14-15 year olds, 30 per cent of older teenagers and 36 per cent of adults are still smoking.

How is smoking harmful?

Cigarette smoke contains over 1000 different chemicals, many of which can damage your body. Here are just a few examples of what it contains and the harm that can be done:

- **Nicotine**. Nicotine is a **drug** which in small doses stimulates your nervous system. One of its effects is to increase your blood pressure which makes your heart work harder. This extra work being done by your heart muscle requires more oxygen, but another substance in cigarette smoke – carbon monoxide – prevents it getting it.

  Nicotine also increases the levels of fatty acids in the blood and, therefore, the likelihood of a **blood clot** forming (see page 59).
- **Carbon monoxide**. Carbon monoxide is a poisonous gas which irreversibly combines with the haemoglobin in the red blood cells. This haemoglobin would normally carry oxygen.
- **Tar**. Tobacco tar is a brown sticky substance made up of several different chemicals, some of which are possible **carcinogens** (cancer producing). Others are **irritants**. Because it is sticky tar tends to accumulate and 'clog up the works'.
- **Irritants**. Some are dusts, such as soot, others are chemicals, like ammonia. All have the same effect in that they irritate the delicate mucous membranes lining your air passages, causing them to increase their production of mucus. This often leads to bronchitis and other serious diseases.

  The irritant dusts also settle in the alveoli, reducing the surface area available for gaseous exchange, with the result that you cannot get enough oxygen.

  Some of the irritant chemicals are known to be carcinogens in animals.
- **Heat**. Even the heat from a burning cigarette can add to the damage. Every time you draw on a cigarette, the heat burns away some of the cilia lining your trachea (figure 1). These cilia are normally covered by mucus and together they filter dust and germs out of inhaled air. The cilia then sweep the dirt and mucus mixture up to your throat, where it is swallowed. If the top few cilia are missing, the dirty mucus never actually reaches your throat and collects in the trachea. It is now called **phlegm** and must be coughed up every so often. If this coughing becomes persistent, it can damage the alveoli, resulting in **emphysema**, a serious and chronic condition.

Why is it so difficult to give up smoking?

Every cigarette contains nicotine which is a drug. Nicotine creates both **physical** and **psychological dependence** (see page 90). You soon begin to need it. Unfortunately, your body also builds up a **tolerance** to it so to get the same relaxing effect every time, you need to increase the dose. Eventually, even though you smoke 40 cigarettes a day, the dosage may still be too small. You are left feeling irritable and bad tempered all the time.

It is difficult to give up so the best thing you can do is not start smoking in the first place.

Questions

1. What percentage of people in the UK (all ages) still smoke? Does this represent an increase or decrease over the last decade?
2. a) Use the information in figure 3 to show that cigarette smoking is related to lung cancer.

   b) Which other cancers are you more likely to suffer if you smoke?
3. Cigarette, cigar and pipe tobacco smoke all contain the drug nicotine.

   a) Explain how nicotine can increase your chances of suffering from coronary heart disease.

   b) Why does nicotine make it so difficult to give up smoking?
4. State two reasons why people who smoke are often short of energy.
5. A person smoked on average 20 cigarettes a day for 25 years. Calculate how much shorter she can expect her life to be.

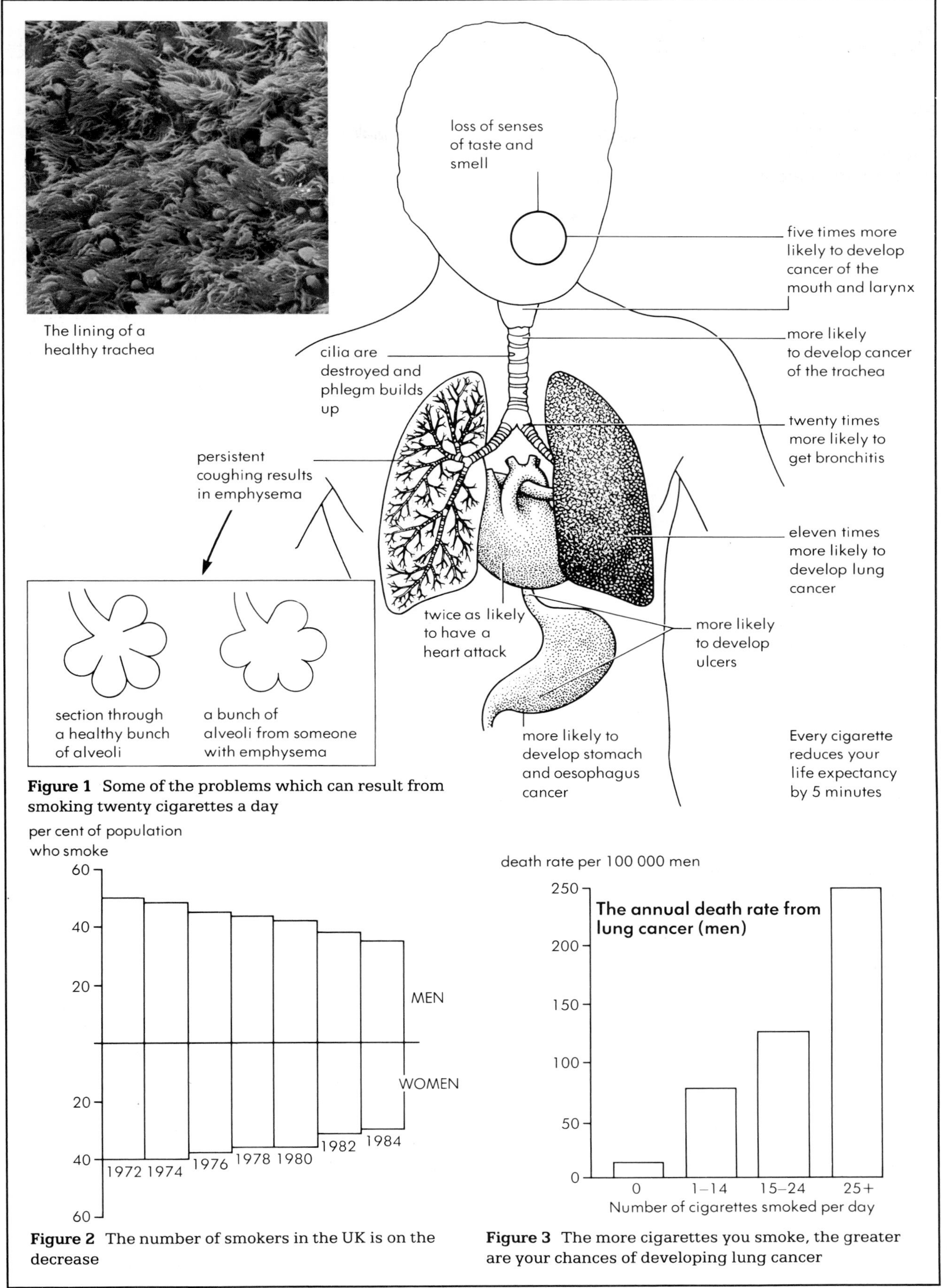

Figure 1 Some of the problems which can result from smoking twenty cigarettes a day

Figure 2 The number of smokers in the UK is on the decrease

Figure 3 The more cigarettes you smoke, the greater are your chances of developing lung cancer

Questions

Recall and understanding

1 Which of the following nutrients provides most energy per gram?
A carbohydrate
B fat
C protein
D water

2 The release of energy from food is called:
A photosynthesis
B burning
C phototropism
D respiration

3 Energy is released from food in the:
A nucleus
B chloroplasts
C mitochondria
D cytoplasm

4 Gaseous exchange takes place in the:
A alveoli
B lungs
C trachea
D nose

5 The most important factor affecting breathing rate is:
A the oxygen content of air
B the air pressure
C the size of your lungs
D the concentration of carbon dioxide in the blood

6 An oxygen debt will result from:
A aerobic respiration
B anaerobic respiration
C holding your breath
D not breathing deeply enough

7 Breathing movements are caused by the:
A ribs
B diaphragm
C diaphragm and intercostal muscles
D lungs

8 When the external intercostal muscles and diaphragm muscles contract air rushes into the lungs because:
A they need inflating
B the air pressure in the alveoli falls
C the pressure rises
D they are short of air

9 The vital capacity and tidal air volumes will be the same:
A during strenuous exercise
B when talking
C when exercising lightly
D when shouting

10 We do not breathe through our skin because:
A the surface area is not sufficient
B it is not moist
C it does not contain pores
D it is always covered by clothing

Application and interpretation of data

11 The pie diagram below shows how one day's energy is spent by the average man:

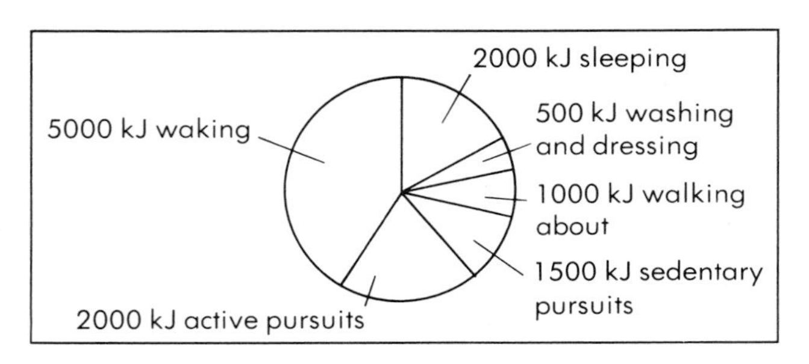

a) How much energy does the average man use in one day?
b) Suggest what pursuits fall under the headings of active and sedentary.
c) Define basal metabolic rate (BMR).
d) Consider the graph below:

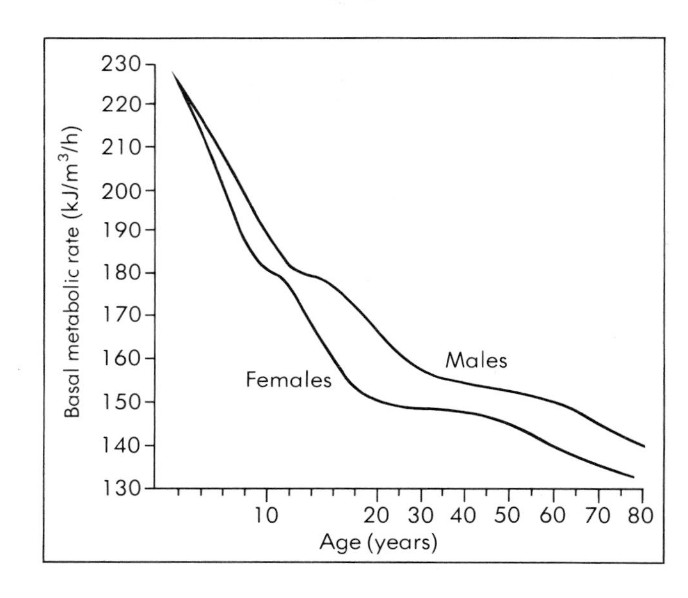

i) What does this show?
ii) At what age does the BMR for a male and female show the greatest difference?

12 The graph below shows what happens when we exercise vigorously.

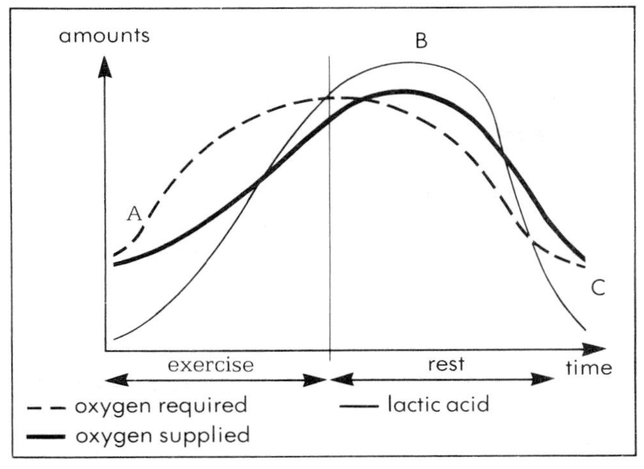

a) Why does lactic acid build up between A and B?

b) Why does the amount of oxygen taken in continue to increase once exercise has stopped?

c) From the graph, when is an oxygen debt
i) building up? ii) being paid back?

d) If you continued exercising you would probably end up with cramp. Why?

e) The bar chart below shows how athletes produce their energy whilst competing.

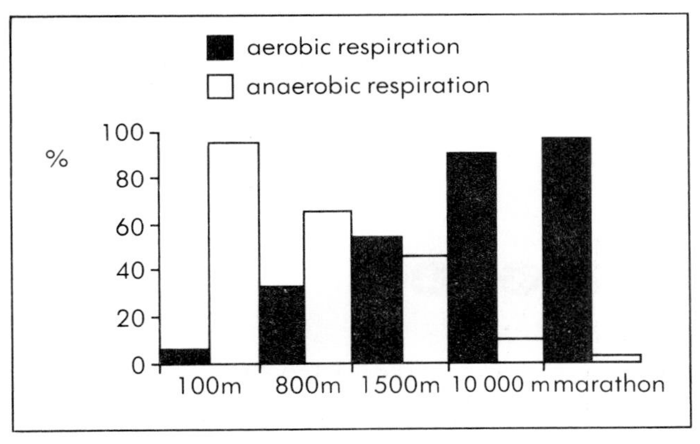

i) Suggest why a 1500 m runner could not possibly run a race at the same pace as a 100 m runner.

ii) If marathon runners start too quickly they are unlikely to finish. Why?

13 Copy the following table and fill in the spaces:

Cell	Length of each side	Surface area	Volume	Surface area to volume ratio
A	1 cm	$6 cm^2$	$1 cm^3$	6 : 1
B	2 cm			
C	5 cm			
D	10 cm			
E	100 cm			

For a cell to get the oxygen it needs by diffusion from the environment, it must not have a surface area:volume ratio of more than 6:1.

a) Which of the cells above would be able to get their oxygen from the environment by diffusion?

b) Which cells would not be able to?

c) Give reasons for your last two answers.

d) Calculate the surface area:volume ratio of the woman below. Measurements are in cms.

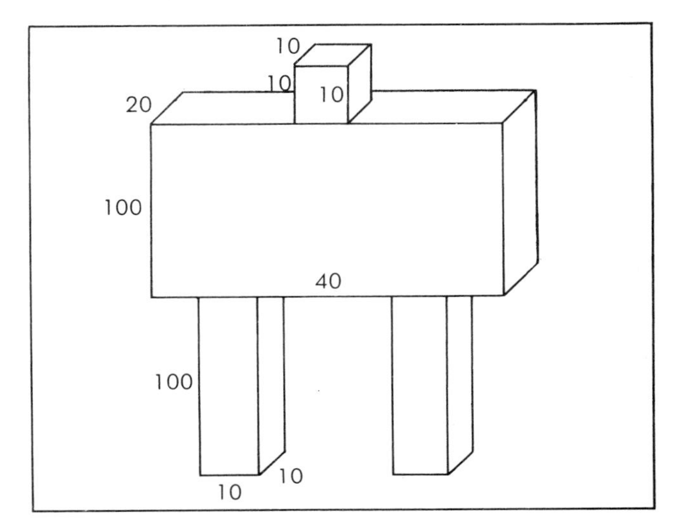

e) Suggest why it would be impossible for the woman to get the oxygen she needs by diffusion across her skin.

14

	Death rate per 100 000 people from lung cancer
Non smoker	7
Light smoker	47
Moderate smoker	86
Heavy smoker	166

a) Draw a bar chart using the data above.

b) Do the figures suggest there is or is not a relationship between smoking and lung cancer?

c) What would you say to someone who claims his grandad smoked 40 cigarettes per day yet lived to 95?

d) If cigarettes cost £1.20 for a packet of 20, how much would a person spend in a year if he/she smoked 20 a day?

e) Every cigarette smoked knocks 5 minutes off your life. How many hours would the person in (d) lose over the year?

CHAPTER 5: *SENSITIVITY AND CO-O*

Prick yourself with a needle and it hurts. Walk past a fresh bread counter and the smell makes your mouth water. If two cars crash behind you, you turn round to look. These may all seem quite simple responses, but in reality they involve a very complex system of communication within your body.

In this chapter you will learn about this communication system and how it helps you to respond and react to the changes in your environment.

The chapter also looks at what can go wrong and the effects of drugs on the system.

Dancers have to make very precise co-ordinated movements. This chapter explains how we manage such tricky manoeuvres.

How new drugs are developed

Where do drugs come from?

The first known drugs came from plants and 25% of those now used still do. Although many of the rest were first discovered in plants, they are now made artificially. Table 1 below shows the main sources of modern drugs.

Table 1

Source	Type of drug	Examples
Plants	Pain killer	Morphine
	Pain killer	Aspirin
Micro-organisms	Anti-biotic	Tetracycline
	Anti-viral	Interferon
Animals	Hormone	Insulin
	Hormone	Growth hormone
Synthesised	Barbiturate	Amylobarbitone
	Amphetamine	Methylphenidate

How are drugs discovered?

There are many companies in the world which develop and make new drugs. They are called pharmaceutical companies. Most new drugs are discovered through their intensive research programs. This is very expensive and only takes place if there is an urgent need for a drug. Unfortunately there are still many such needs.

The research may come up with as many as 1000 possible chemicals. Work now starts on screening these and eliminating those which are of no value. The full procedure is shown in figure 1.

Throughout the process, checks are made by various government agencies to make sure regulations are not being ignored.

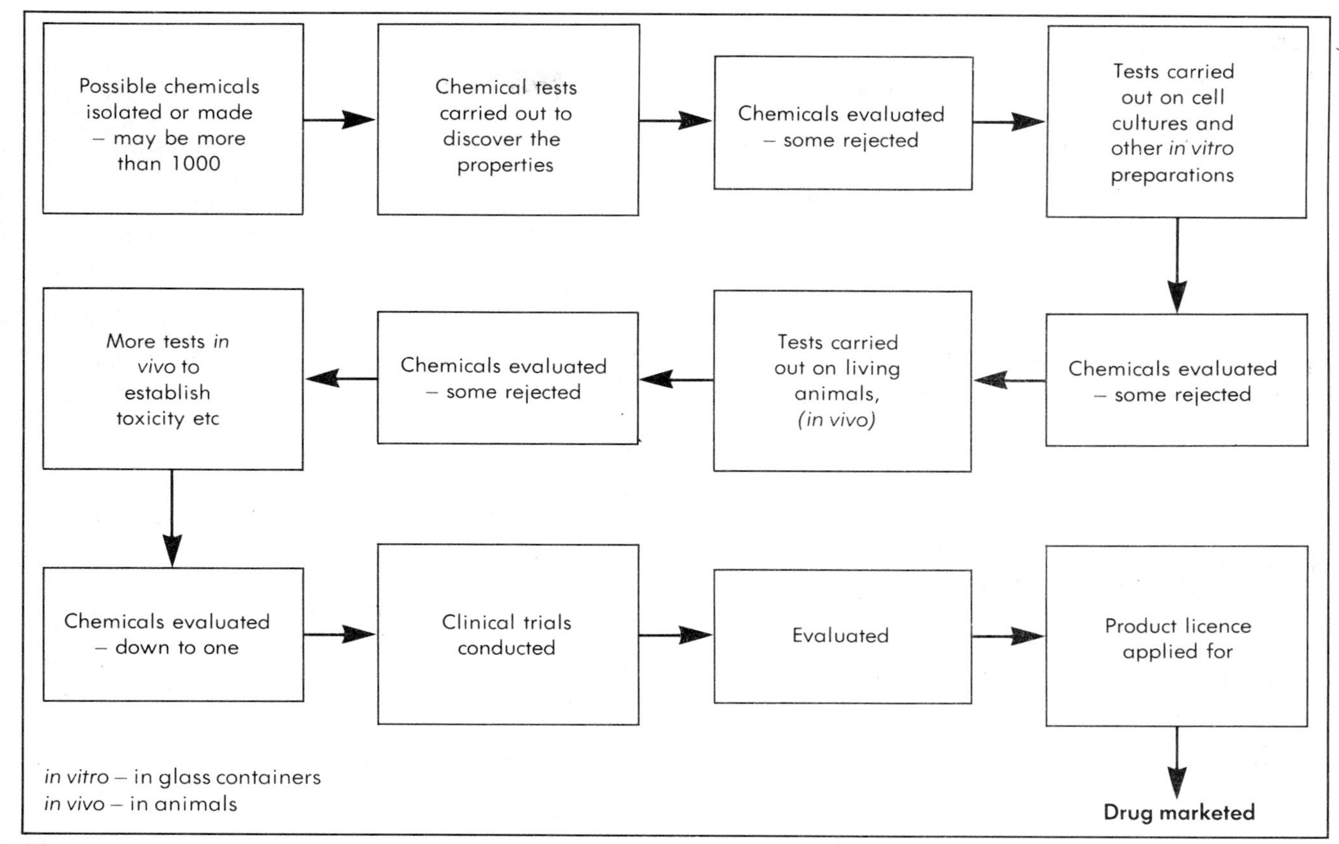

Figure 1 The development of a new drug involves many stages and can take a long time

How long does all this take?

Once chemicals have been suggested it can take ten years or more to get to the production stage. Most of this time is taken up by the biological testing. This is crucial to show up both long and short term effects of the drug.

The cost of all this is enormous, sometimes as high as £20 million. This is why drugs are so expensive.

What are all the tests looking for?

First and foremost they are trying to discover if the chemical does the job. Even if it does, they still need to know:

- Is it safe to use?
- Has it any side effects?
- How does it behave in the body?
- Can it be excreted?
- What is the best dose to use?
- What is the lethal dose?
- What is the best way to administer it?
- Can it be stored?... and so on.

What are clinical trials?

It is during the clinical trials that the drug company finds out if it has all been worthwhile because this is when they see if the drug works and is safe in humans. So far, all the tests have shown is that it works and is safe in animals.

Clinical trials involve:

- Trying the drug on healthy volunteers.
- Trying the drug on volunteers with the particular disease or disorder that it has been developed to treat.

All the tests are **controlled**, that is they use three identical groups of people. One group is given the drug, one group is given a fake dose containing no drug, and the last group is not given anything. Only by comparing results from all three can they assess the value of the drug.

Once this obstacle is out of the way, the company can apply for a licence to market the drug.

5.1 The senses

Responding to change

Things happen all around you all the time. It gets dark as night falls; there are noises in the street; it's cold outside and warm inside; your new shirt feels rough on your skin. Animals detect all these changes with special sensory cells, called **receptors**. In humans these receptors are grouped together forming **sense organs**. You respond to change through **effector organs** such as muscles. The connections between the receptors and effectors are formed by the **nervous system** (figure 1). This acts as the overall co-ordinator and is a vital part of your body.

What sort of things do we respond to?

Anything that produces a response is called a **stimulus**. The main changes in our environment which act as stimuli are due to light, sound, heat, cold, touch, pressure and chemicals. We have separate receptors to detect each of these.

The skin as a sense organ

Your skin is the largest of your sense organs. It contains a number of different receptors, each sensitive to a particular kind of stimulus. Some are sensitive to **touch**, some to **pressure**, some to **heat** and some to **cold**. Your skin even contains receptors which when stimulated give rise to the feeling of pain (figure 2).

Which parts of your body are sensitive to chemicals?

The receptors that are stimulated by chemicals are located in your **tongue** and **nasal cavity** (figure 3). They are permanently covered by a layer of moisture in which the chemicals must first dissolve if they are to have any effect.

There are four kinds of receptor in your tongue, each sensitive to a different chemical. These chemical stimuli are salt, sweet, bitter and sour. The flavour of a food is determined by how many of each of these receptors are stimulated, together with the smell. The **sensation of smell** is created when the chemical receptors in your nasal cavity are stimulated.

What are light sensitive receptors?

The receptor cells that are stimulated by light form the **retina** inside your eye (see page 79). There are two kinds: **rod** cells and **cone** cells. Rods can be stimulated by all wavelengths of light except red, but your brain only registers these as shades of black and white. They can, however, respond in low light.

There are three kinds of cone cell, each sensitive to one of the primary colours (red, yellow and blue). Your brain registers these colours, and when required, mixes them to produce other colours. Cones are only stimulated in bright light, such as daylight or a well-lit room.

You may have noticed that if you suddenly switch a light off you cannot see anything at all in the dark. But after a while you begin to be able to see a little more clearly. This is because your rods are now responding in the very low light, whereas before, when the electric light was on, your cones were responding.

A typical retina contains about 150 million rod cells and 7 million cone cells. The rods are more or less evenly spread out, whereas most of the cones occupy a small region of the retina called the **fovea**. Light striking this area produces the clearest images in the brain.

Where are your sound receptors?

The receptors that are stimulated by sound are housed in a structure called the **cochlea**. This forms part of your inner ear.

Questions

1 The following sentence illustrates what sort of response a change in the environment might produce: I turned my head in the direction of the loud noise. Write similar sentences to illustrate what sort of responses some of the other stimuli might produce.

2 Rewrite the following in the correct sequence: nerve cells, effector cell, stimulus, receptor cell, response.

3 When you burn yourself, the part of your skin affected loses all its sensitivity. Explain why.

4 Explain the following observations:
a) When your mouth is dry your sense of taste is poor.
b) Food tastes different when you have a cold.

5 Draw up a table of differences between rod and cone cells.

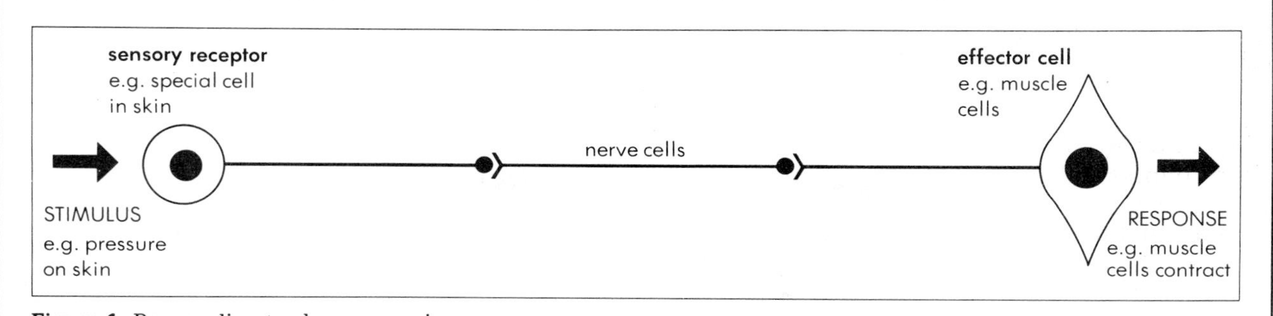

Figure 1 Responding to change requires many different structures all working together

heat receptor

cold receptor

touch receptor
These help us feel texture and shape

nerve endings sensitive to **pain**

These nerve endings are stimulated when the hair is moved. The brain will register **touch** or **pain** depending on how hard the hair is pulled.

pressure receptor
These help us control the force we use to hold things

Figure 2 The skin has several types of sensory receptors

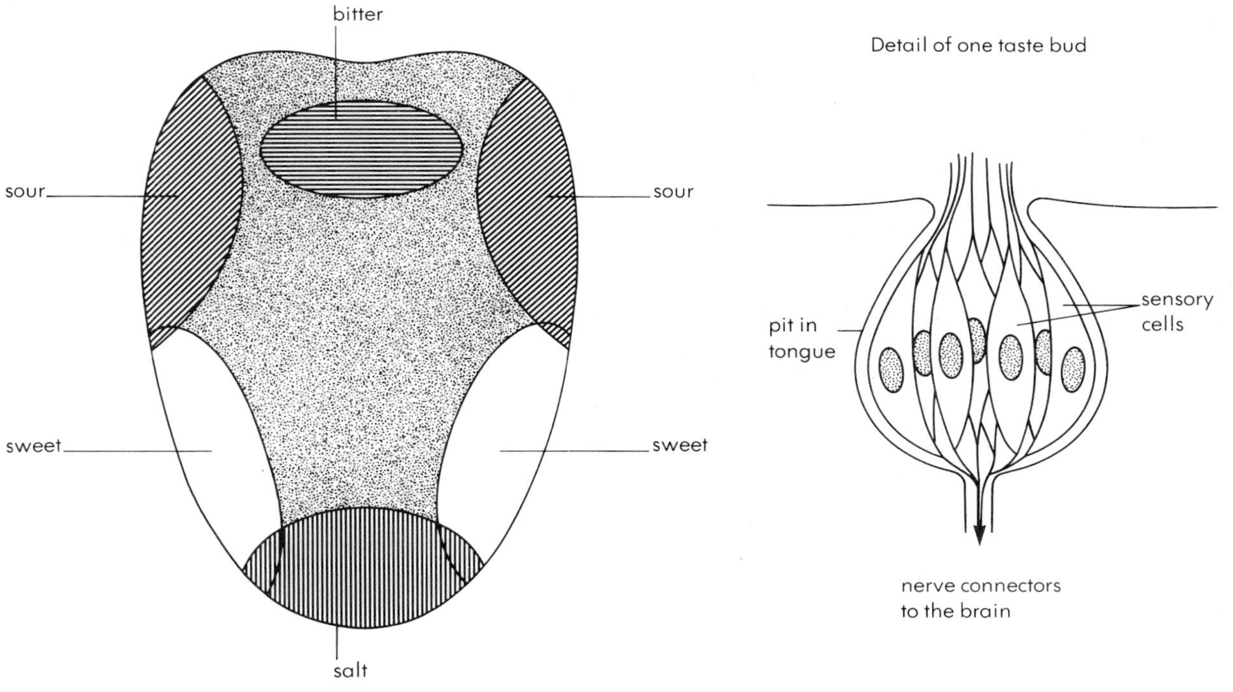

Figure 3 There are four different types of taste bud on different parts of your tongue

5.2 The eye

How do you see?

The eye is the sense organ responsible for detecting light. Your brain receives messages from this and uses them to 'form a picture'. This all happens very quickly and means that, when your eyes are open, you are constantly seeing things. This is the sensation we call **vision**.

The structure of the eye

The eye contains (figure 1):

- The light sensitive cells (receptors) which form the **retina**.
- Structures to focus the light onto the retina, i.e. the **cornea** and **lens**.
- Structures to protect and support the eye, i.e. the **sclera**, **choroid layer**, **iris** and **humours**.

Each eye is housed in a bony socket (orbit) in the front of the skull. It is held in position by six muscles which can move it up, down or sideways. This means you can move your eyes around to look at things in different places.

How does your eye work?

The eye works along the same lines as a camera. A controlled amount of light is allowed to enter and is focused onto a light sensitive layer (figure 2). An image is created by 'processing' the effects of this light.

In a camera, the amount of light entering is controlled by the diaphragm. In the eye, the same job is done by the **iris**, a membranous structure containing smooth muscle. The focusing in both is achieved by a **lens system**, but in the eye, this is helped by the **cornea**. If a camera was connected to a television set the processing would also be similar. In this case, the light sensitive layer sends a signal to the television, which produces a picture. Likewise in the eye, the retinal cells send a signal to the brain which also 'produces a picture'. But in reality the eye is much more complex than any camera yet devised.

Why do you blink?

The exposed surface of the eye is kept clean and free of germs by an antiseptic liquid produced by your tear glands. Blinking spreads this liquid over the surface and the excess liquid drains away near the corner of the eye, into your nose.

Why do you have two eyes?

There are at least three advantages to having two eyes. Firstly, with two eyes you can see over a larger area than with just one. Secondly, your two eyes help you to judge short distances better and thirdly, they enable you to see objects in three dimensions, that is, to see the depth of objects as well as the height and width. With 3D vision (sometimes called **binocular** vision), you can judge the size and shape of objects more accurately.

What is a squint?

A squint is caused by a lazy eye muscle. This makes it impossible for both eyes to look directly at the same spot or object at the same time.

Questions

1. In words, describe the structure of the eye.
2. The eye is held in the orbit by muscles. What is the other function of these muscles?
3. The way the eye works is often compared to a camera. Explain why you think this is a good or bad comparison.
4. A woman lost an eye in an accident. She was immediately banned from ever driving a car again. Explain why.
5. A man had the nerve connection from his eyes to his brain destroyed by a disease. Although his eyes were in perfect condition he could not see, and could not understand why. Explain it to him.

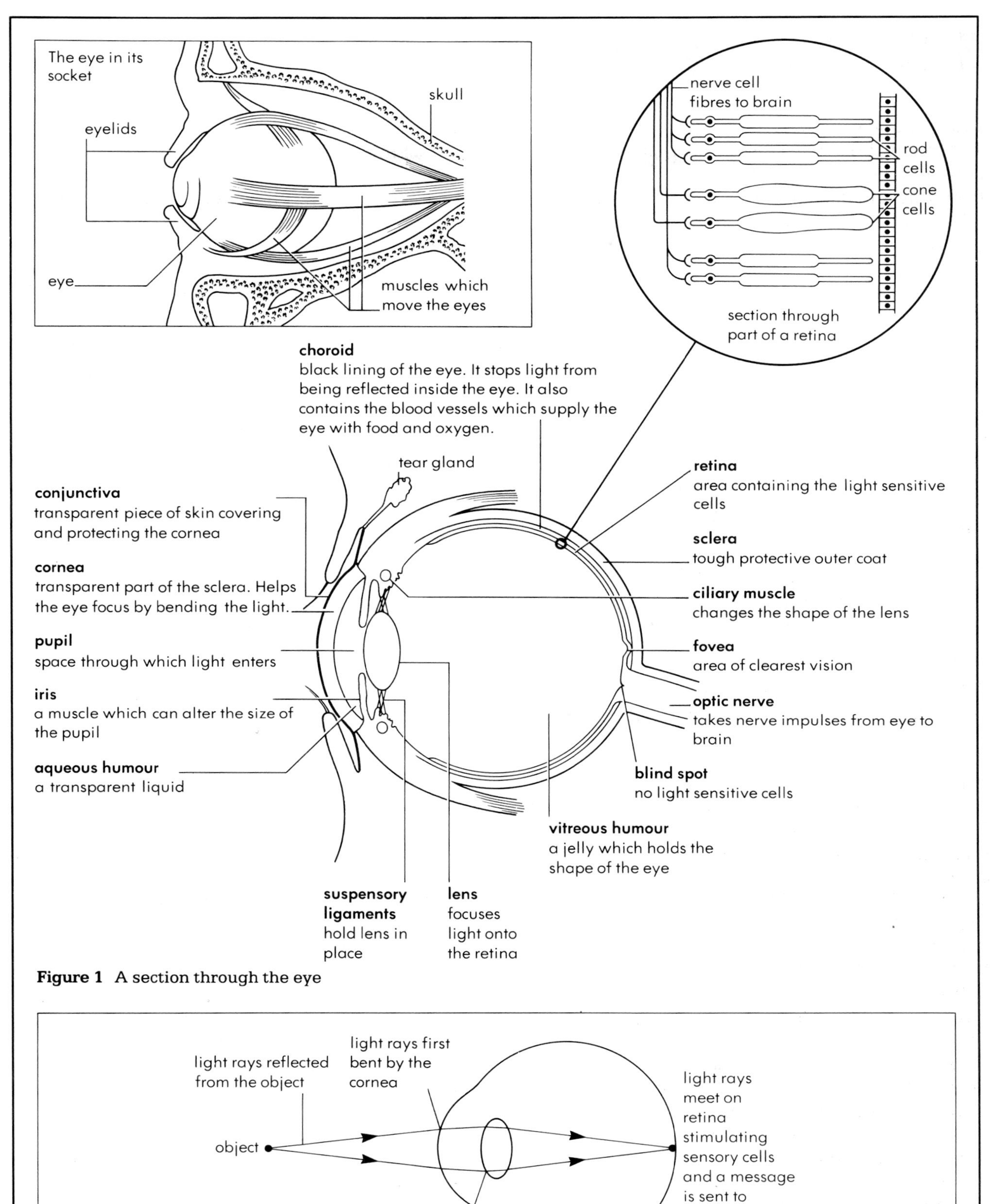

Figure 1 A section through the eye

Figure 2 The cornea and lens bend (refract) light rays so they come together on the retina

5.3 Focusing

How do your eyes focus the light?

You are able to see objects because every object **reflects** rays of light and some of these rays enter your eyes. To see the objects clearly (in **focus**), these light rays must be bent (**refracted**) inwards so that they meet exactly on the retina at the back of your eye. If they are bent too little, or too much, your image of the object will be blurred (out of focus). This rarely happens as the focusing of light rays is done automatically by the eye. It does this by altering the shape of its lens. In general, the fatter a lens is the more it bends the light rays. When your eye is focusing on a distant object it will have a thin lens. As the object moves closer, to keep it in focus, the lens gets fatter. This automatic adjustment of the lens to keep an object in focus is called **accommodation** (see figure 1).

How is the lens shape altered?

The lens is connected by **suspensory ligaments** to the **ciliary body**. The ciliary body contains the ciliary muscle. When this is relaxed the tension in the suspensory ligaments pulls the lens into the thin state needed for distant vision. When the ciliary muscle contracts, the change in tension in the suspensory ligaments allows the natural elasticity of the lens to change its shape and become fatter (figure 1).

As you get older, the elasticity of the lens often decreases, making accommodation impossible. This is one reason why so many older people wear glasses. Sometimes the lens becomes clouded as one gets older. This is known as a **cataract**.

Why do some people have to wear glasses?

The two main eye defects are **long** and **short sight**. A long sighted person can see distant objects clearly, but has difficulty focusing on near objects. This is usually due to the eyeball being too short, but can also be caused by a lens which is too weak (too thin). To correct long sight, the eye must be helped to bend the light rays more, so that they are brought into focus on the retina. This can be achieved by wearing glasses containing **converging (convex) lenses**. A converging lens will bend the light rays inwards a little before they enter the eye (figure 2).

A short sighted person can see near objects clearly, but has difficulty focusing on distant objects. This is usually due to the eyeball being too long, but can also be caused by a lens which is too strong (too fat). To correct short sight, the light rays must be bent outwards before they enter the eye. This can be achieved by wearing glasses that contain **diverging (concave) lenses** (figure 3).

What is astigmatism?

A person who suffers from **astigmatism** has blurred vision and sees distorted images. It is caused by having an uneven cornea. As light rays pass through this cornea, some of them are bent more than others and this makes it impossible to focus them at the same point on the retina.

Astigmatism can be corrected by wearing specially prepared lenses. In very bad cases a corneal graft may be needed. Lots of people have very mild astigmatism.

Questions

1. Explain why we can only see things when light is present.
2. Draw a diagram to show how the eye focuses light rays onto the retina.
3. Select from the list of statements below those which describe the state of various parts of the eye when it is focused on a distant object:

   ciliary muscle contracted, ciliary muscle relaxed, small pupil, large pupil, suspensory ligaments taut, suspensory ligaments slack, fat lens, thin lens.
4. Copy out and complete the table below:

Eye defect	Causes(s) of defect	How defect is corrected
Long sight Short sight Astigmatism		

5. When you move from a well lit room into a darker room, the size of your pupil alters. Design an experiment to test this hypothesis.

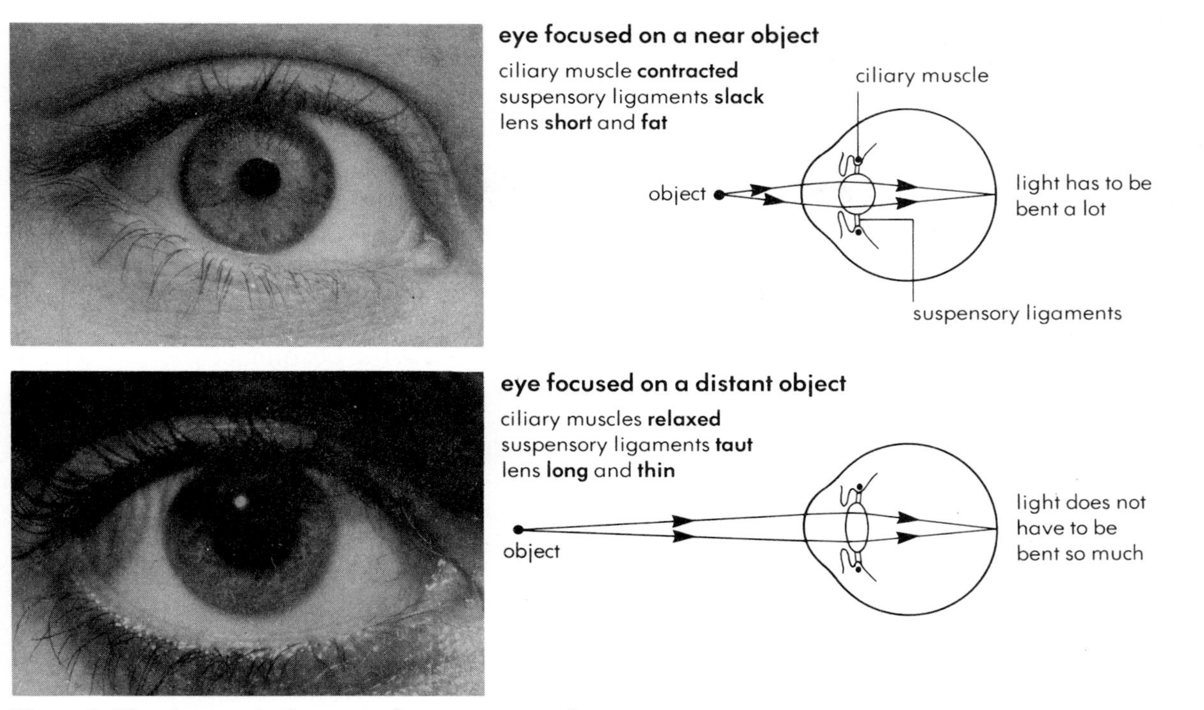

Figure 1 The changes in the eye to focus on near and distant objects are automatic

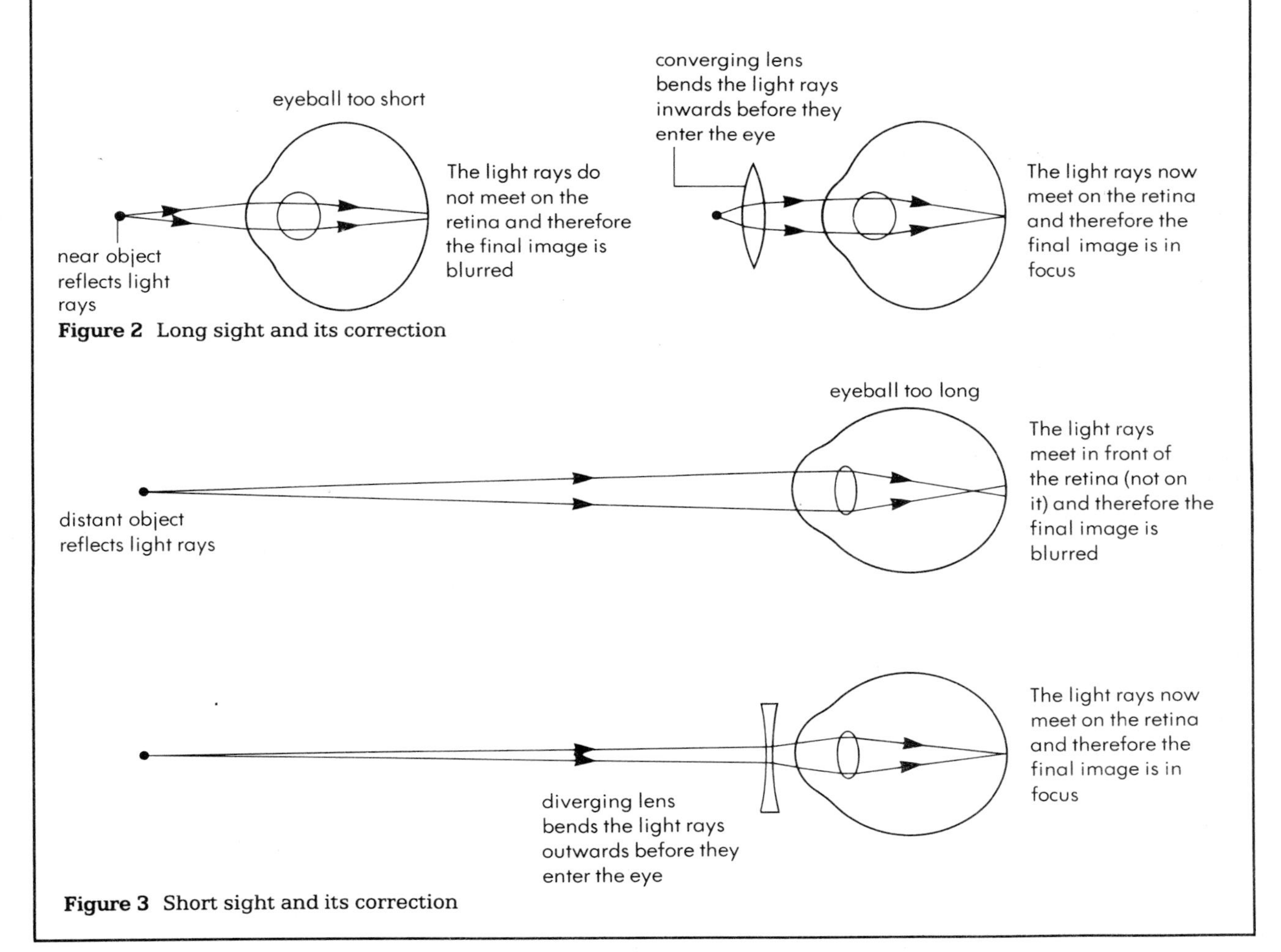

Figure 2 Long sight and its correction

Figure 3 Short sight and its correction

5.4 The ear

Your ear as a sense organ

The ear is the sense organ you use to hear sound. The receptors are housed in a part called the **cochlea**. They are not actually stimulated by sound but by vibrations caused by the sound.

The ear also contains the organ which gives you your sense of balance.

The structure of the ear

The ear contains (figure 1):

- The sound receptors. These form part of the cochlea within the inner ear.
- Structures to convert sound waves into mechanical vibrations, i.e. the **ear drum, ear ossicles** and **oval window**.
- Structures to direct sound waves into the ear, i.e. the **pinna** and **ear canal**.
- Receptors which are sensitive to movement of the head with respect to gravity. These are housed in the organs of balance, i.e. **semi-circular canals, utriculus** and **sacculus**.

Each ear is housed in a bony socket (auditory capsule) in the side of the skull.

How does the ear work?

Sound waves in the air are collected by the pinna and directed onto the **ear drum** which is made to vibrate. These vibrations are amplified by the ear ossicles in the middle ear before being passed onto the oval window and then into the fluid within the cochlea. The vibrations within the cochlea stimulate the sensory cells (receptors) which send a signal to the brain (figure 2). Each sound creates a different pattern of signals. These signals arriving at the brain are what we refer to as the sensation of **hearing**.

Loudness and pitch

A loud noise will cause large vibrations within the cochlea. Thus the brain gets the signal and assigns a loud sound. Noises of different pitch are picked up by different receptor cells in the ear. A high pitched noise stimulates the receptor cells near the oval window which then send the signal to the brain. A low pitched noise stimulates the receptors near the end of the cochlea.

Your brain can also tell you the direction the sound came from by using messages from both ears.

The sense of balance

Your body maintains its balance and posture in a very complex process. To do it properly, your brain requires continuous messages from the various parts of the body telling it where they are. Many of these messages come from special receptors in your joints and tendons. The messages about the position of your head come from the organs of balance in your ear. The utriculus and sacculus detect the sideways and up and down movements of the head, whereas the semi-circular canals detect its speed of movement.

Questions

1 Describe in words the structure of the ear.

2 Place the following statements in the correct sequence to hear a sound:
the oval window passes the vibrations into the cochlea,
the auditory nerve takes a message to the brain,
sound waves make the ear drum vibrate,
vibrations pass into the fluid in the upper chamber of the cochlea,
the ear ossicles pass the vibrations onto the oval window,
vibrations pass into the middle chamber of the cochlea and stimulate the sensory cells.

3 When ascending/descending in an aircraft your hearing is often affected. Explain why and suggest how the problem can be resolved.

4 Explain why you always seem to know where every part of your body is.

5 A woman suffered from a diseased auditory nerve. She eventually became deaf although the rest of her ear was in perfect condition. Explain why.

6 Explain why you can usually decide which direction a sound is coming from.

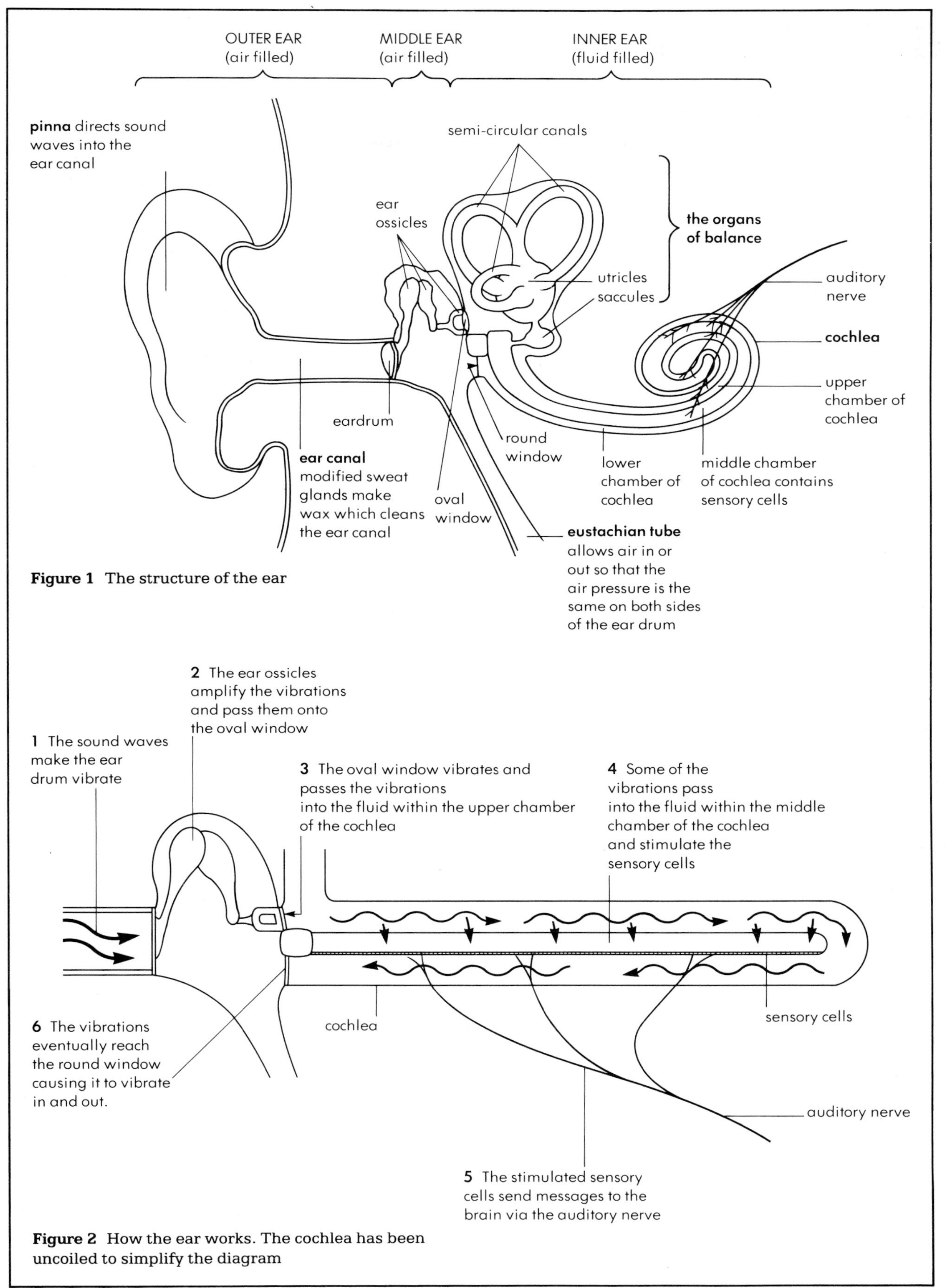

Figure 1 The structure of the ear

Figure 2 How the ear works. The cochlea has been uncoiled to simplify the diagram

5.5 The nervous system

The parts of the nervous system

The main job of the nervous system is to co-ordinate all your body actions and to carry messages from one part of your body to another to make it act as necessary.

The nervous system has two parts:

- The **central nervous system (CNS)** formed from the **brain** and **spinal cord**.
- The **peripheral nervous system (PNS)** formed from the **nerves**.

The basic unit of both is the **nerve cell**. The brain alone consists of over ten thousand million nerve cells bound together by supportive tissue.

Nerves are bundles of **nerve fibres** wrapped in connective tissue.

How does the nervous system work?

You can compare your nervous system to a telephone service. In this comparison the CNS is the exchange, receiving messages from all parts of the body, processing them and then passing them onto other parts of the body. The nerves connect the CNS with those other parts of the body. In our telephone service these would be the lines to and from the telephones.

Nerves which connect directly to the brain are called **cranial nerves**. Examples of these are the nerves to and from the eyes and ears. Nerves which connect to the spinal cord are called the **spinal nerves**. Examples of these are the nerves to and from the arms and legs (figure 1).

What form do these messages take?

The messages carried by the nervous system are in the form of both **electrical** and **chemical changes** within the nerve fibres. The correct name for these messages is **nerve impulses**.

What is the autonomic nervous system?

The **autonomic nervous system (ANS)** is made up of nerves which carry messages between your body's vital organs and the CNS. For example, when you start exercising, to meet the demand for extra oxygen, your brain sends out messages which result in an increase in your rate and depth of breathing. To speed up the delivery of this oxygen, a similar message is sent to your heart, resulting in an increase in the heart rate. These actions are completely involuntary, i.e. they happen automatically, as do all the actions involving the ANS. Some however can be influenced for short periods, for example, when you hold your breath.

Some of the actions involving the ANS are:

- breathing
- peristalsis
- accommodation
- emptying the bladder
- altering the heart rate
- production of digestive enzymes
- production of some hormones
- sweating
- sexual arousal

Questions

1 Copy out and fill in the missing words:
The nervous system consists of the ...., ... cord and nerves. The nerves are of two kinds; .... nerves and spinal nerves. The .... nerves connect directly to the brain and the spinal nerves connect to the ....

2 The working of the nervous system is often compared to a telephone service. Explain why you think this is a good or poor comparison.

3 Describe an action involving your autonomic nervous system.

4 Some actions which involve the ANS can be controlled voluntarily for short periods. Suggest several occasions when this could be an advantage.

Figure 1 The parts of the nervous system

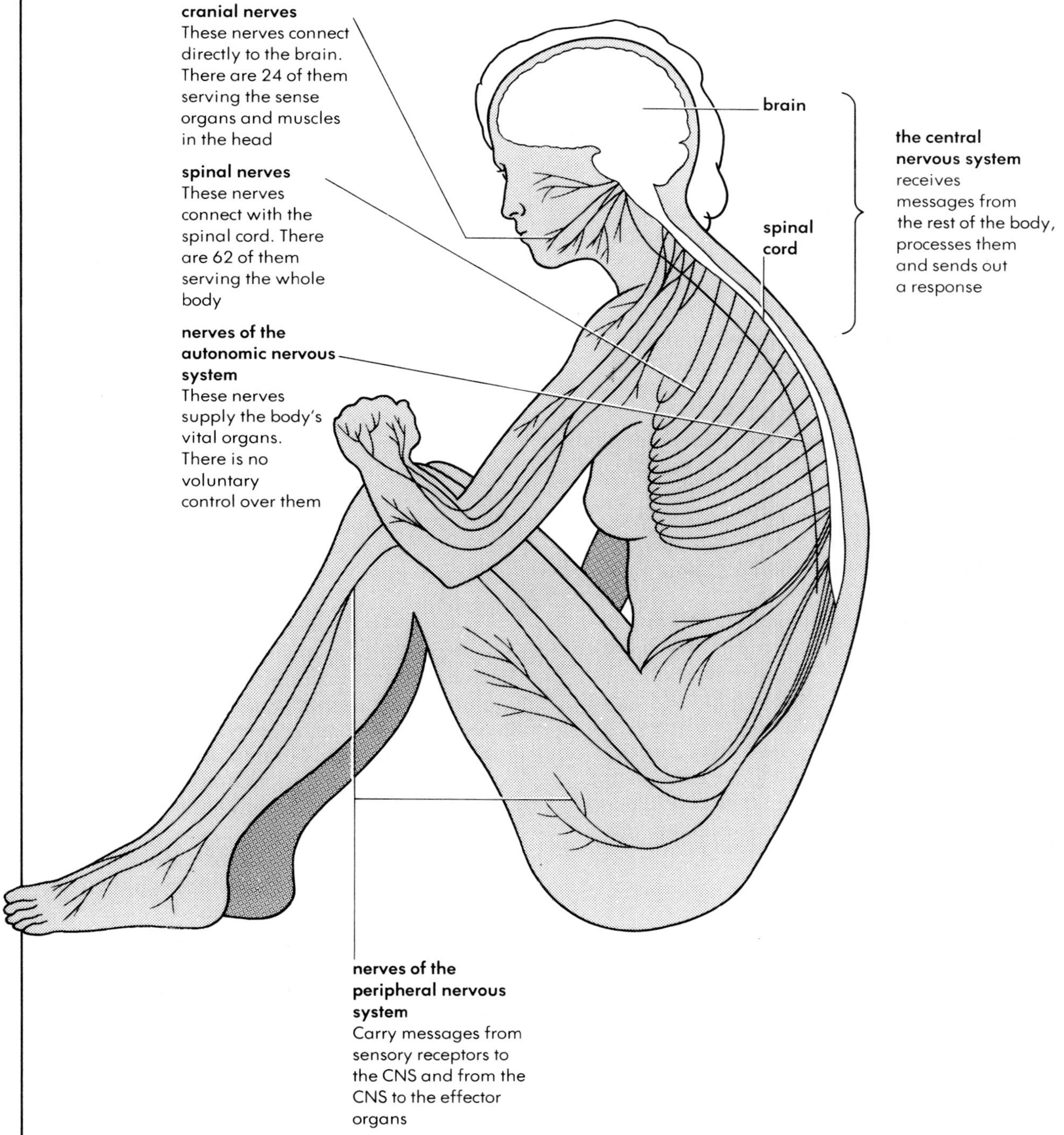

cranial nerves
These nerves connect directly to the brain. There are 24 of them serving the sense organs and muscles in the head

spinal nerves
These nerves connect with the spinal cord. There are 62 of them serving the whole body

nerves of the autonomic nervous system
These nerves supply the body's vital organs. There is no voluntary control over them

brain

spinal cord

the central nervous system
receives messages from the rest of the body, processes them and sends out a response

nerves of the peripheral nervous system
Carry messages from sensory receptors to the CNS and from the CNS to the effector organs

5.6 The central nervous system

What does the brain do?

Your brain is an immensely complex organ which will probably never be fully understood. From what is known, it is thought to function very much like a computer.

To start with, it is composed of circuits, but these circuits are not made from wires, transistors, etc., but from nerve cells. Some of these circuits form a memory to store information fed into it, whereas others are used to process the information and form a response.

The brain also acts as a monitoring device, making sure your body is functioning efficiently, and makes adjustments as necessary.

But there are some important ways in which the brain is different from a computer. You use your brain to make decisions; to make choices between good and bad; to register emotions and exercise the conscience for example. Figure 1 shows the general areas of the brain and which parts are thought to deal with certain activities.

What does the spinal cord do?

Your spinal cord runs from the base of your brain to the lower part of your back. It is also made from nerve cells. It acts as a switchboard, connecting incoming messages from the sense receptors to the appropriate nerves to the effector organs, as well as feeding messages to the brain.

A section through the spinal cord shows it to have two distinct regions. The outer area is white and is therefore called the **white matter**. This is made from nerve fibres entering and leaving the cord. The inner H shaped area is grey (**grey matter**), and is composed of the connecting nerve cells, including those taking messages to the brain (figure 2). A similar section through the brain would show that it also has white and grey matter.

What if the brain or spinal cord is damaged?

Unfortunately if human nerve cells are damaged or destroyed, they are rarely repaired and never replaced. Fortunately it is very difficult to physically damage the CNS because it is fairly well protected by a casing of bone in the form of the **skull** and **spinal column**. In between the bone and nervous tissue are two membranes called the **meninges** and in between these is a fluid called **cerebrospinal fluid**. This fluid not only protects the CNS, but also helps provide it with food.

Most damage is inflicted on the CNS by invading micro-organisms. Polio is one of the more serious infections. It is caused by a virus attacking the nerve cells at the base of the brain and top of the spinal cord. The damage can lead to paralysis of limbs and of the breathing muscles.

Questions

1. The way the brain works is often compared to a computer. Explain why you think this is a good or a poor comparison.
2. a) Draw a diagram of the brain and label the areas which are thought to be concerned with: vision, hearing, breathing, smell, thinking, memory and intelligence.
   b) Label the sensory and motor areas of the cerebrum. Explain what these two areas do.
3. Draw a diagram of a section through the spinal cord and label the grey and white matter. What do these areas contain?
4. It is very difficult to damage the CNS. Explain why.
5. a) Explain why polio often leaves the victim paralysed from the neck downwards.
   b) What do you think meningitis is?

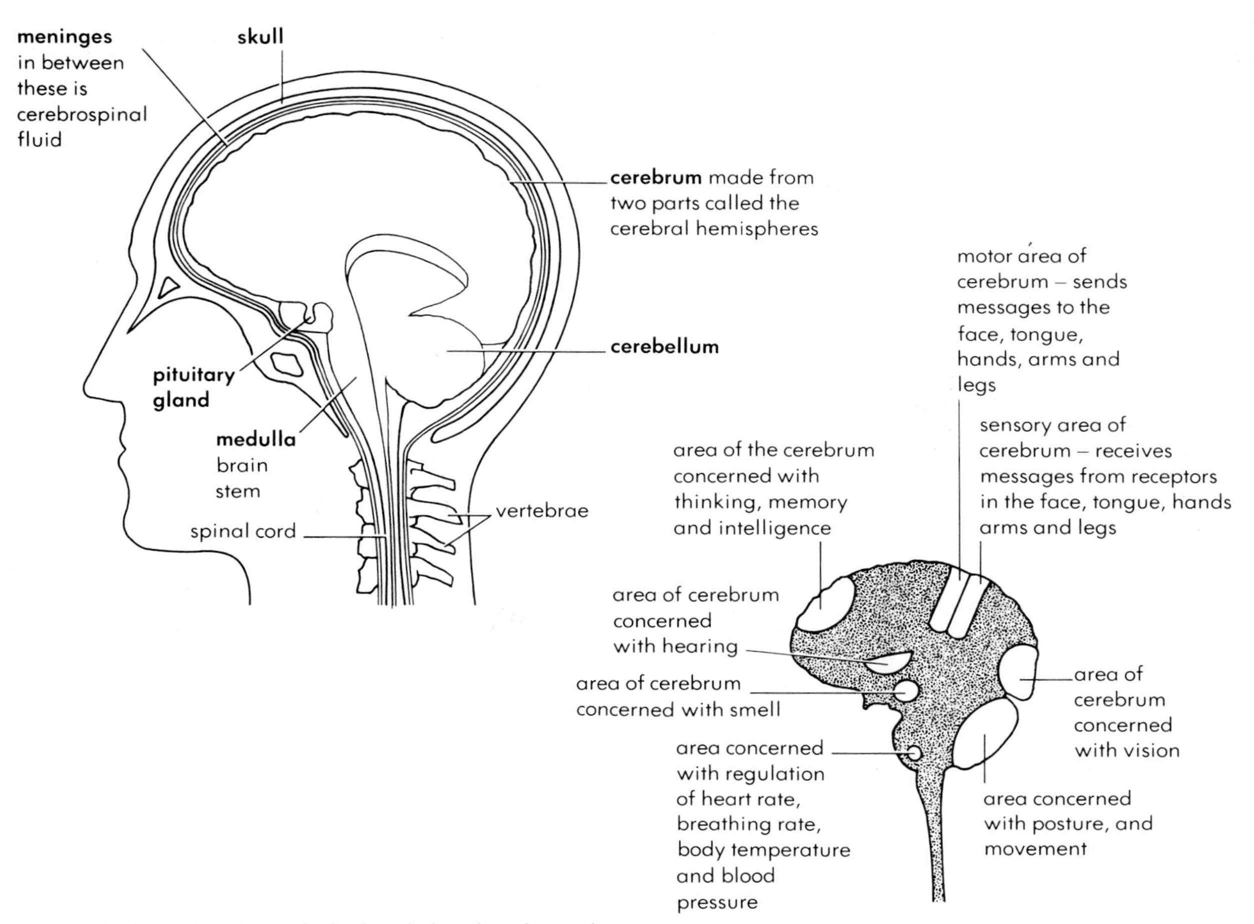

Figure 1 A section through the head showing the major areas of the brain and some of the areas with specific responsibilities

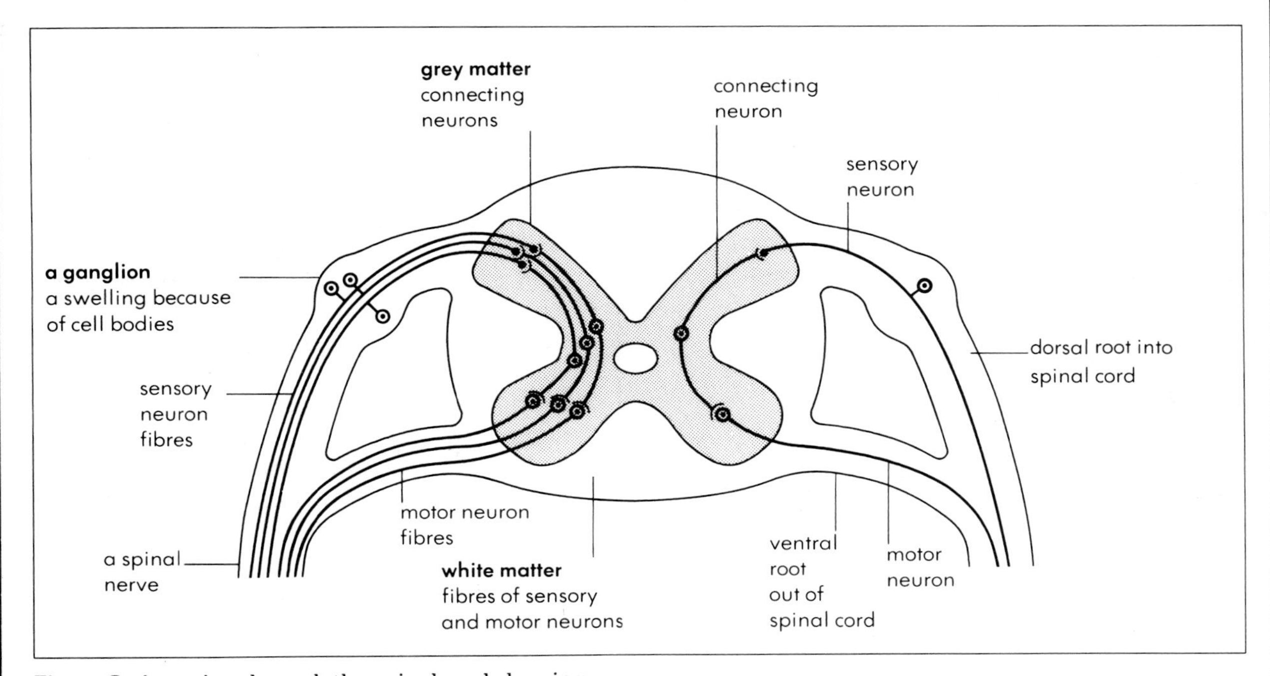

Figure 2 A section through the spinal cord showing how the neurons connect with one another

5.7 The nerves

What does a nerve cell look like?

A nerve cell is usually referred to as a **neuron**. There are three basic kinds of neurons:

- **Sensory neurons**. These carry messages from the sensory receptors to the CNS.
- **Motor neurons**. These carry messages from the CNS to the effectors (muscles and glands).
- **Connecting neurons**. These carry messages around inside the CNS and link sensory and motor neurons together.

All the neurons are structurally very similar. They have a nucleus surrounded by cytoplasm which forms what is known as the **cell body**. Part of the cytoplasm projects outwards from the cell body forming a long thin thread known as a **nerve fibre**. The nerve fibres of some sensory and motor neurons are over 1 metre long.

Nerve fibres outside the CNS are bundled together and enclosed in a tube. A fatty material called **myelin** is packed between the fibres to insulate them from one another. The whole structure is called a nerve (figure 1).

How are neurons linked together?

The simplest form of nervous activity is a **reflex action**. This is an involuntary response to a stimulus and involves a receptor, an effector and all three kinds of neuron. Together these form what is known as a **reflex arc** (figure 2).

The structures in a reflex arc are not physically joined together but each link is separated by a small gap called a **synapse**. For a message to travel from one link to another this gap must be temporarily filled by a substance which will enable the message to travel through the gap. The main one is a chemical called **acetylcholine**.

Synapses are extremely important to the working of the nervous system because they act like gates and in doing so

- make sure messages only travel in one direction;
- filter out weak or unwanted stimuli;
- enable information from different sources to interact.

A typical reflex action

A typical reflex action is the **knee jerk response**. This is illustrated in figure 3. All reflexes start with a **stimulus**, which in this case is a tap on the knee (the quadriceps muscle tendon). This stimulates a stretch receptor which in turn sets up a nerve impulse in the sensory neuron. This nerve impulse travels to the spinal cord where it is passed onto the appropriate motor neuron. (In some reflex arcs this would involve a connecting neuron but not in this case.)

The motor neuron takes the nerve impulse to the effector organ, which also happens to be the quadriceps muscle. The muscle responds by contracting, causing the lower part of the leg to jerk upwards.

Can you control your reflexes?

Most reflex actions are inbuilt survival mechanisms. You can see this in the actions of a newborn baby (see page 44). These kind of reflexes can often be controlled. Other reflex actions are necessary to maintain stability in the body – they operate as part of a **homeostatic** mechanism (see page 114). These cannot be controlled, but can often be influenced for short periods. For example, emptying a full bladder is a reflex action. Young children do this automatically, whereas older children and adults learn to delay it until they reach a toilet.

Learning involves setting up what are often called **conditioned reflexes**. Learning to walk for example, may be difficult to start with, but then you do it without even thinking about it, i.e. it becomes a reflex action. It is a conditioned reflex.

Questions

1 Describe the three kinds of neurons forming a reflex arc.

2 a) In the knee jerk reflex state what each of the following are: the stimulus, the response, the effector organ, the sensory receptor.
b) Place these in the correct sequence.
c) What is a conditioned reflex? Give an example.

3 When you accidently touch something hot you immediately withdraw your hand. Explain fully what has happened to produce this response.

4 Why are synapses so important to the working of the nervous system?

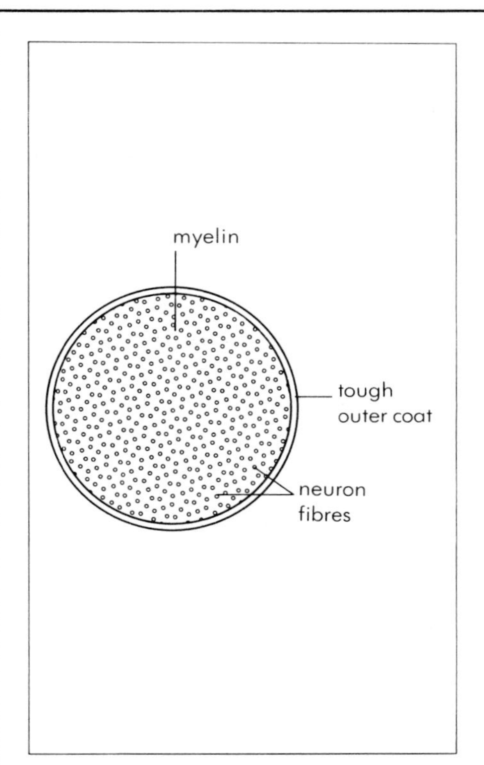

Figure 1 A section through a nerve

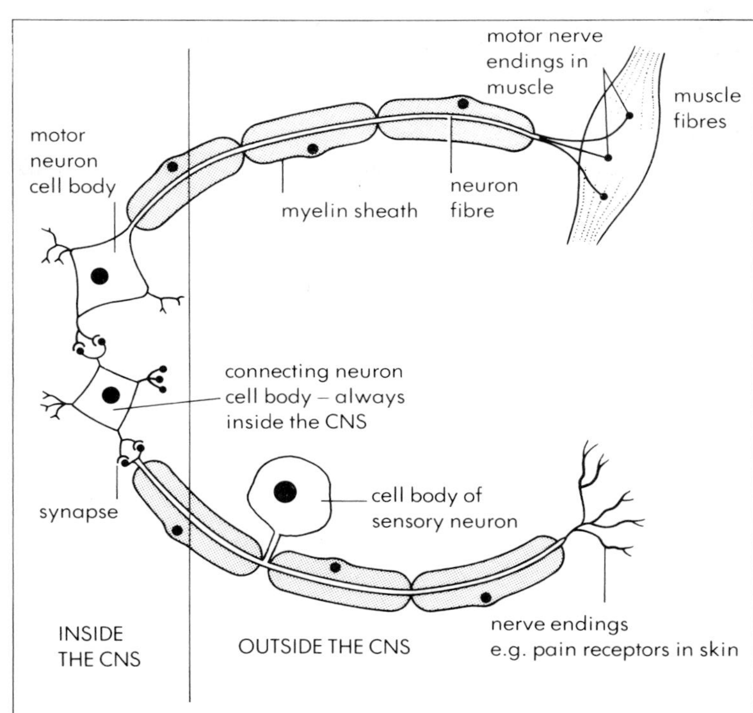

Figure 2 A reflex arc

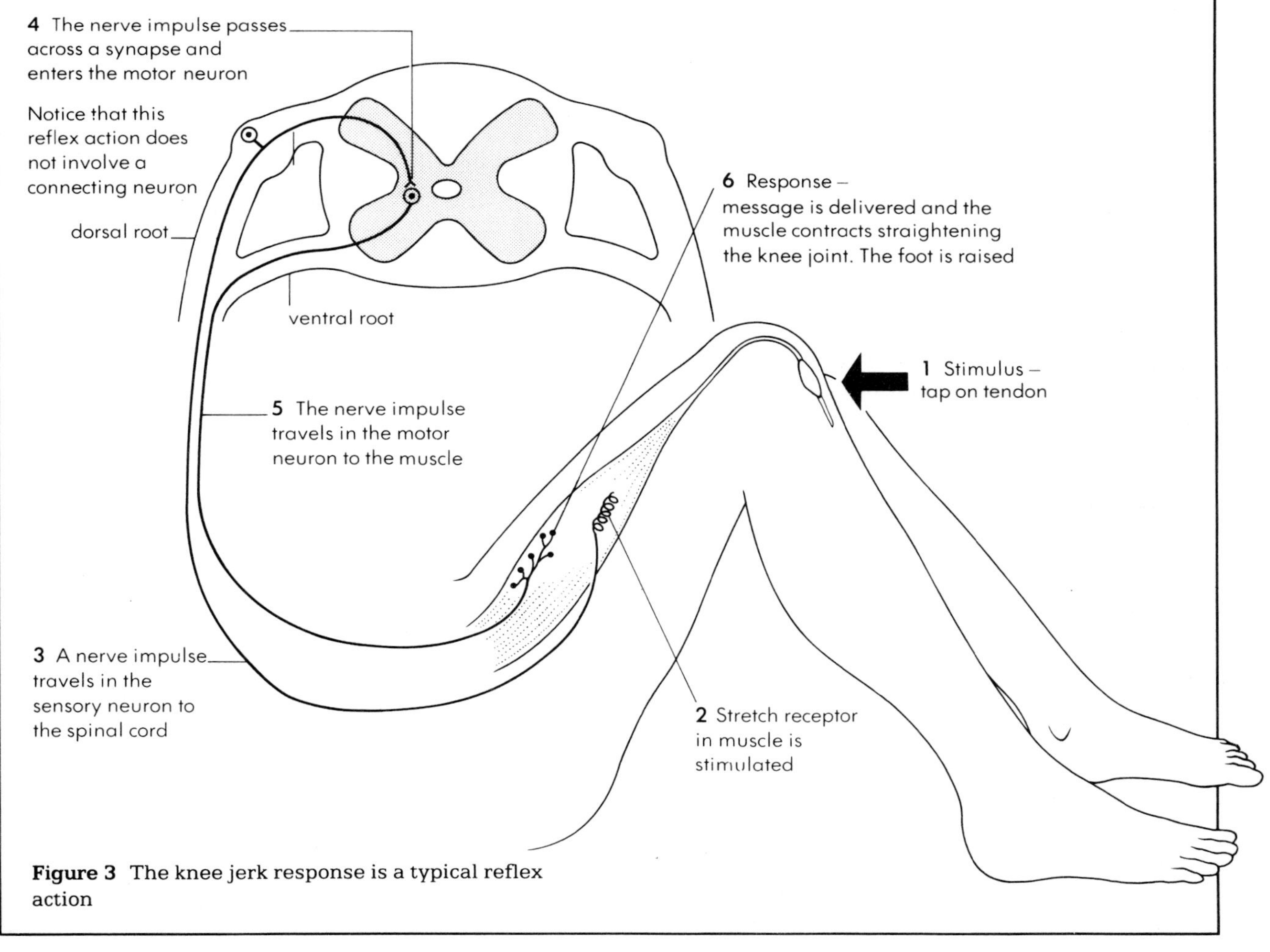

Figure 3 The knee jerk response is a typical reflex action

5.8 Drugs and the nervous system

What is a drug?

A drug is a chemical substance that can alter the way your mind or body works. Many do this by acting directly on the nervous system. The following are examples of these:

- **Stimulants**. These speed up the actions of your nervous system, making you feel more alert and confident. There are two main kinds:
  Amphetamines The main effects of amphetamines are that they temporarily relieve depression and increase mental alertness and energy. They are sometimes called pep pills. Amphetamines, like many other drugs, can cause **drug dependence**; that is, your body reaches a stage when it cannot work properly without the drug. Your body also builds up a **tolerance** to amphetamines so you have to increase the dose each time in order to get the same results.
  Cocaine Cocaine is a powerful stimulant to which you can quickly become dependent. Unfortunately its use as an illegal drug is quite widespread in many countries. Cocaine is used medically as a local anaesthetic to combat pain.

- **Depressants** These slow down the actions of your nervous system, often relieving anxiety and producing a sleepy feeling. There are many different kinds and they all induce drug dependence.
  Barbiturates Barbiturates were once prescribed as sleeping pills because they cause drowsiness. This practice was stopped when it was discovered just how dependent people could become on them and just how easy they are to overdose.
  Tranquilisers Tranquilisers were developed to replace barbiturates. They work in a similar way, producing similar effects but do not induce drug dependence so easily.

- **Narcotics** These are all **pain killers** and are therefore sometimes called **analgesics**. Their action is similar to depressants except that they work mainly on the part of your brain which registers pain.
  Opiates e.g. heroin All the drugs in this group come from the opium poppy. Their main use was originally as painkillers. However heroin and other opiates are widely used by people who take them for pleasure because at first they produce a feeling of great happiness, warmth and contentment. But opiates, especially heroin, are the worst drugs for inducing drug dependence and only bring misery and great distress to users. Such dependency often leads to early death through an accidental overdose, infection or suicide.
  Aspirin and Paracetamol These are mild painkillers used for headaches and fever. Aspirin also reduces inflammation.

- **Hallucinogens** These drugs alter your senses, producing mental illusions. There is no evidence to suggest that they induce drug dependence.
  LSD LSD is made from a fungus called ergot. People who take it often have vivid hallucinations. They talk of the experience as 'tripping'. The nature of the 'trip' depends upon the person's mood or frame of mind. It can be terrifying, or sometimes very joyous and mystical.
  Cannabis Cannabis is obtained from a weed called hemp. It comes in two forms: grass (dried flower) and hashish (resin from the flower). Cannabis produces a relaxed feeling and sometimes the importance of everyday objects seems exaggerated. LSD and cannabis are not used medically.

Questions

1 Below is a list of drugs. Arrange them into groups based on things they have in common. There are several possibilities.

opium	methedrine	Valium
Librium	LSD	cannabis
cocaine	aspirin	heroin
paracetamol	morphine	seconal

2 You are offered some cocaine at a party. Write out a balance sheet of possible benefits and drawbacks should you accept.

3 A friend of yours has started taking drugs to help him get over a girlfriend problem. Explain to him why this is foolish and can only lead to more serious problems.

4 Many people who take drugs illegally share needles. This is foolish, particularly in view of the AIDS problem. Design a poster telling of the dangers of sharing needles.

Name of drug	Slang name	Main effect on nervous system	Cause drug dependence (physical)	Medical use	Why the drug is abused	Effects of the drug when taken over a long period
AMPHETAMINES Benzedrine Dexadrine Methedrine	bennies dexies speed } uppers	stimulants	✓	once prescribed as slimming pills	taken to help the user stay awake and increase mental alertness	user may develop paranoia and start to hallucinate, impurities can be lethal
BARBITURATES Nembutal Seconal Pentathal Phenobarbitone	downers goof balls sleepers	depressants	✓	once used as sleeping pills and tranquillisers Phenobarbitone is used to treat epilepsy	cause a feeling similar to 'drunkenness' if the user deliberately stays awake	user becomes sluggish, irritable and finds it hard to co-ordinate movements, very easy to overdose
TRANQUILLISERS Librium Valium		depressants	✓	reduce tension and anxiety. Used as sleeping pills	taken as a 'pick-me-up' because they relieve depression	user becomes confused and finds it difficult to concentrate
OPIATES opium morphine heroin	 M horse, junk	depressants	✓	pain-killers	when first taken, these produce a warm, contented, cheerful feeling	the user's mind and body deteroriate and he may suffer mental illness, impurities can be lethal
codeine	schoolboy	depressant	✓	ease pain and coughing	as opiates	can cause constipation
HALLUCINOGENS LSD	acid	hallucinogen	✗	none	some 'trips' are extremely pleasurable	can alter the users personality causing depression and other mental illnesses
cannabis	marijuana hashish pot grass	hallucinogen	✗	none	can produce a very relaxed state of mind	the user's motivation decreases depression may get worse
OTHERS cocaine	coke	stimulant	✓	local-anaesthetic	when first taken it produces a feeling of happiness and alertness	user becomes forgetful, irritable and cannot sleep, he starts to suffer from depression, mania, paranoia and may hallucinate.
aspirin		depressant	✗	eases pain, reduces fever and inflammation,		can result in stomach bleeding
paracetamol		depressant	✗	pain-killer		can damage kidneys and liver

5.9 Social drugs

What kind of drug is nicotine?

Cigarette smoking is common all over the world. It is socially acceptable to smoke (although there are signs that it is being discouraged through no smoking cinemas, cafes, tubes and buses). Yet cigarettes contain **nicotine** which is a drug. In small doses nicotine is a stimulant in that it makes you feel more relaxed and alert. Larger doses however can cause heart or respiratory failure leading to death. Nicotine can also induce drug dependency, which is the reason why people find it hard to give up cigarettes.

Do tea and coffee contain drugs?

Tea, coffee, cocoa and cola all contain a drug called **caffeine**. Caffeine is a mild stimulant and therefore, reduces tension and increases mental alertness. It is possible to become dependent upon it.

Is alcohol a drug?

Alcohol is a drug: it belongs to the depressant group of drugs. You can develop a tolerance to it and it can lead to dependence, yet over 90 per cent of the adult population regularly drink it. It is particularly easy to become dependent on alcohol given the lifestyle and attitudes in the UK. The average man drinks 1 pint of beer (or equivalent) per day and the average woman drinks half a pint of beer (or equivalent) a day. The problems with alcohol come when it is regularly taken in excess. For a man, this means over 4 pints a day and for a woman 3 pints a day (or equivalent). One in 20 people exceed these quantities and become alcoholics.

The effects of alcohol depend on the quantities consumed and how tolerant your body has become. In small quantities, it produces a feeling of well being, mainly because it increases the blood flow to the skin (which makes you feel warm). This in itself can cause serious loss of body heat. The effects of drinking larger quantities are shown in figure 1. At worst alcoholism can kill.

What about glue sniffing?

The sniffing of glue and other solvents produces similar effects to alcohol and like alcohol, it can kill. It nearly always induces nausea and vomiting. Many of the deaths are from people choking on their own vomit while in a stupor. Glue sniffing is particularly worrying because glue and other solvents are easy and cheap to buy. Thus young people have taken up glue sniffing in the UK in recent years with tragic results.

What is meant by drug abuse?

A person who abuses drugs takes them on a regular basis, purely for the pleasurable effects they have on the body. Such people are **addicts** and sooner or later their health deteriorates.

What is meant by drug addiction?

The term addiction is no longer used because it has no medical applications. **Drug dependence** has replaced it and this describes the compulsion to continue taking a drug as a result of taking it in the past. Drugs can induce two kinds of dependence:

- **Physical dependence**. A person has become physically dependent on drugs if, when he or she stops taking them he or she suffers **withdrawal symptoms**. When drug dependence is used in this book, it refers to physical dependence unless specifically stated.
- **Psychological dependence**. A person has become psychologically dependent on drugs when he or she constantly has a craving for them because of the pleasure or stimulation they give. *Nearly all drugs* are capable of inducing this kind of dependence.

Questions

1 Your friend started smoking one year ago and has gone from five a day to 25 cigarettes a day. Suggest to her a reason for this.

2 A friend goes to the pub and drinks 3 pints of beer. While he is in the pub the weather turns very cold outside. He has not brought a coat and intends to walk the two miles home. You know this is a foolish thing to do. How would you convince him he would be better getting a taxi?

3 Your father drinks 2 pints of beer, three glasses of wine and two brandies during a business lunch. He arrives home two hours later and offers to take you in his car to the cinema. Would you accept or decline? Use the information in figures 2 and 3 to justify your answer.

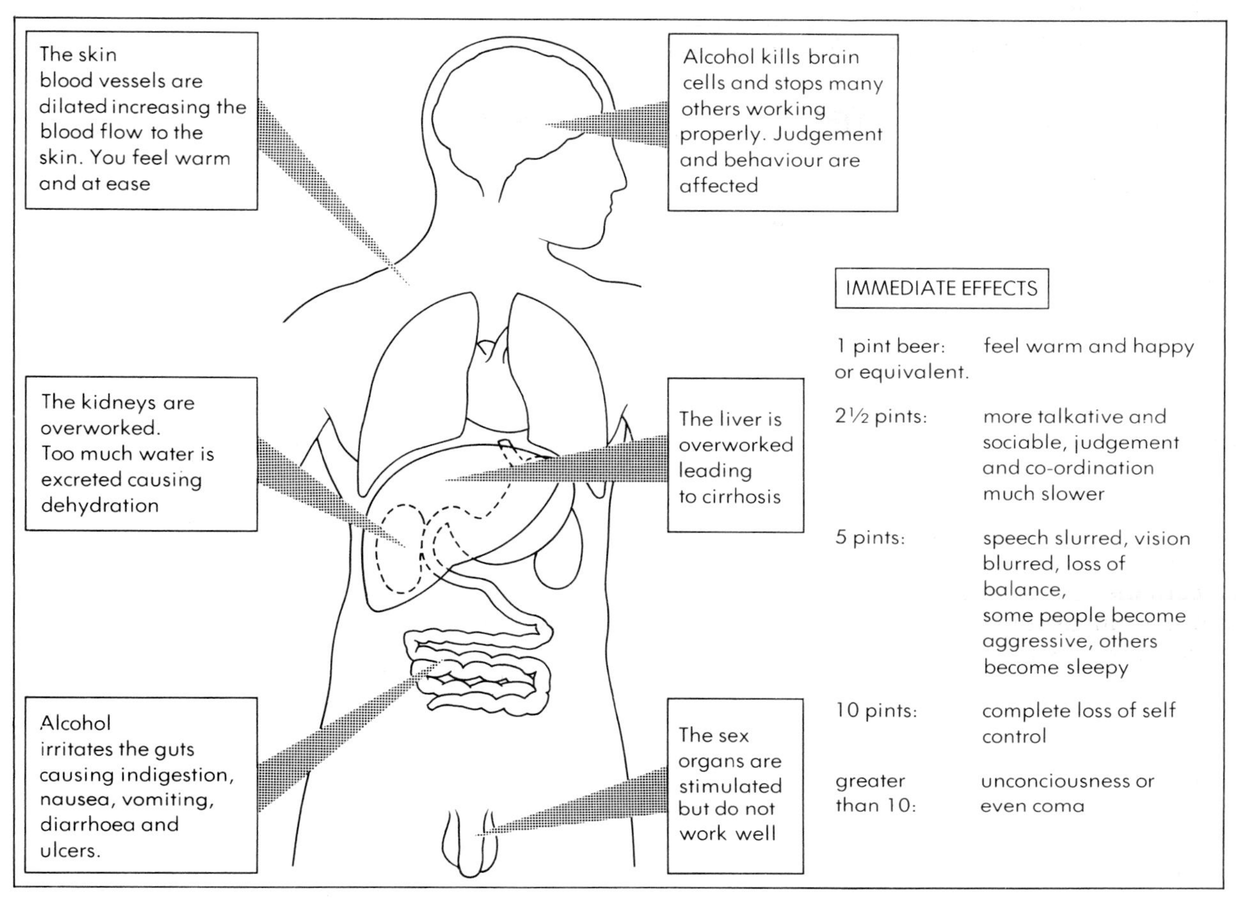

Figure 1 Alcohol can affect the human body in a number of ways. Some effects are only caused by regular drinking, often in excess

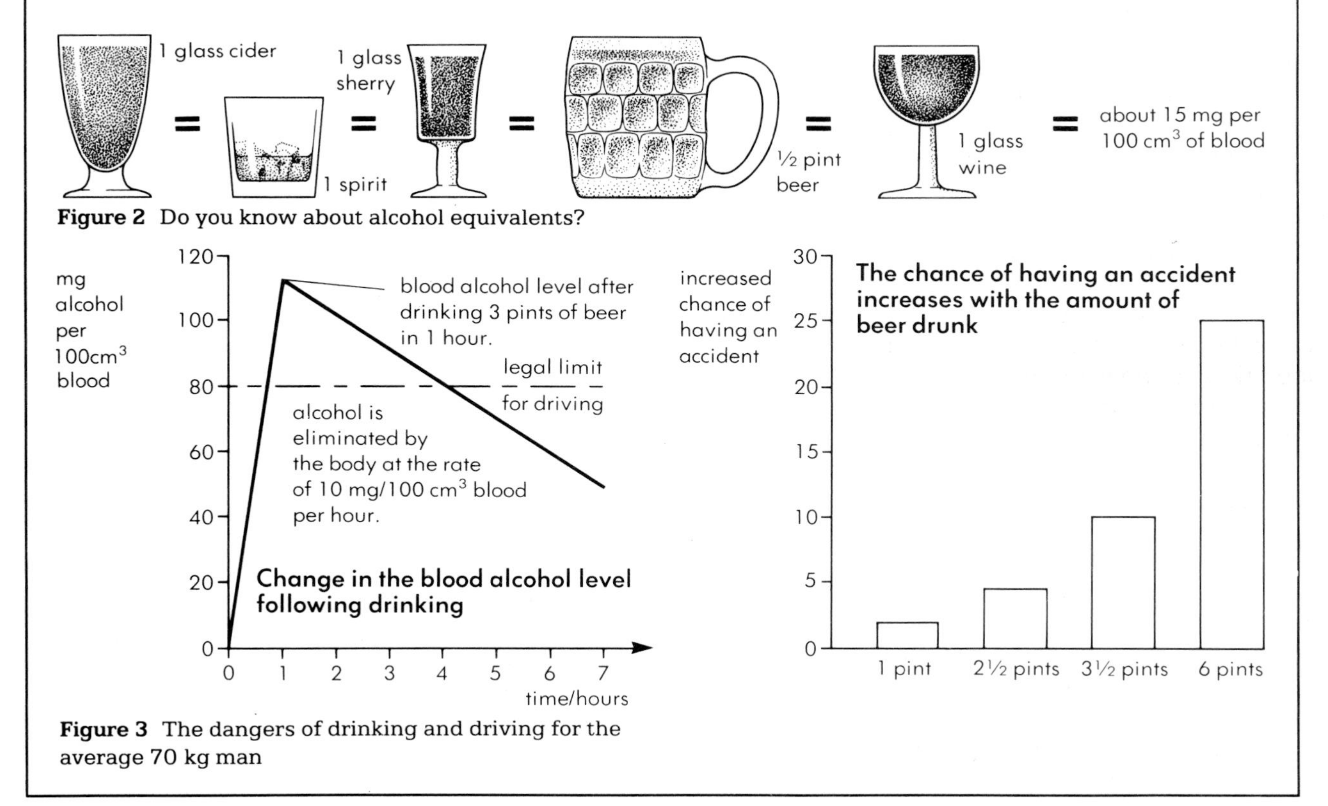

Figure 2 Do you know about alcohol equivalents?

Figure 3 The dangers of drinking and driving for the average 70 kg man

5.10 Mental illness

What is mental illness?

Mental illness is an illness of the mind. About one in ten of the UK population will be affected by such an illness in their lifetime. Most of these people will make a full recovery.

Mental illness should not be confused with mental handicap. Someone with a **mental handicap** has a particular condition which is usually present from birth, and although it can never be fully cured, special education may help the person overcome it.

Mental illnesses are usually divided into two groups: the **neuroses** and the **psychoses** (figure 1). Neuroses are illnesses which usually have emotional or social causes. Psychoses usually, but not always, result from loss of brain function due to some form of damage. A psychotic person loses touch with reality and his or her behaviour often becomes anti-social.

What are the causes of mental illness?

It is difficult to pin down any one thing as being the cause of a mental illness. Many of us feel unhappy or disturbed at times. Sometimes this is because something has happened; you've failed an exam or split up with your boyfriend. Sometimes it just feels like a mood that you can't get out of. Depression such as this is quite common.

Some of the things which can be associated with depression and other mental illnesses are: puberty, menopause, post-operation blues, other illnesses, accidents, family and relationship problems, death of loved ones.

What should you do if you are feeling depressed?

If you are feeling anxious, depressed or disturbed in any other way, it is best to go and see your doctor. She may be able to help you herself or refer you to a psychotherapist or counsellor who will help. These people are trained to talk through your problems with you and offer support over a period of time.

What is stress?

Stress is very difficult to define, yet everybody knows what it is because they have probably felt stressed at some time. It is really the feeling of being overburdened, under pressure and anxious so that you cannot cope with everything in your life. Major changes or conflicts such as those listed in figure 2 can cause stress. However, for many people everyday life is stressful.

Who is most prone to stress?

Research has shown that people who demand a lot from themselves, such as very ambitious and competitive people, are most likely to have a high stress rating. Figure 3 shows what kind of jobs are the most stressful.

How can you avoid or reduce stress?

There are drugs available, but these should only be taken as a short term measure. It is far better to try and work out why you are feeling under stress and try to change your circumstances or your attitude to them. You could get help in doing this from a counsellor or psychotherapist.

Questions

1 Copy out and complete the table:

Mental illness	Neurosis or Psychosis	Signs of the illness
Anxiety and Depression		
Schizophrenia		
Dementia		
Obsessions and Compulsions		
Manic Depression		
Hysteria		

2 What do you think a nervous breakdown is?

3 What things in your life cause you stress? What can you do to reduce or avoid this stress?

4 The main phases of your life are childhood, adolescence, adulthood and old age. For each write a list of the most likely causes of stress.

5 Look at figure 3. Decide which member of your family (including uncles, aunts etc) is most likely to be under stress.

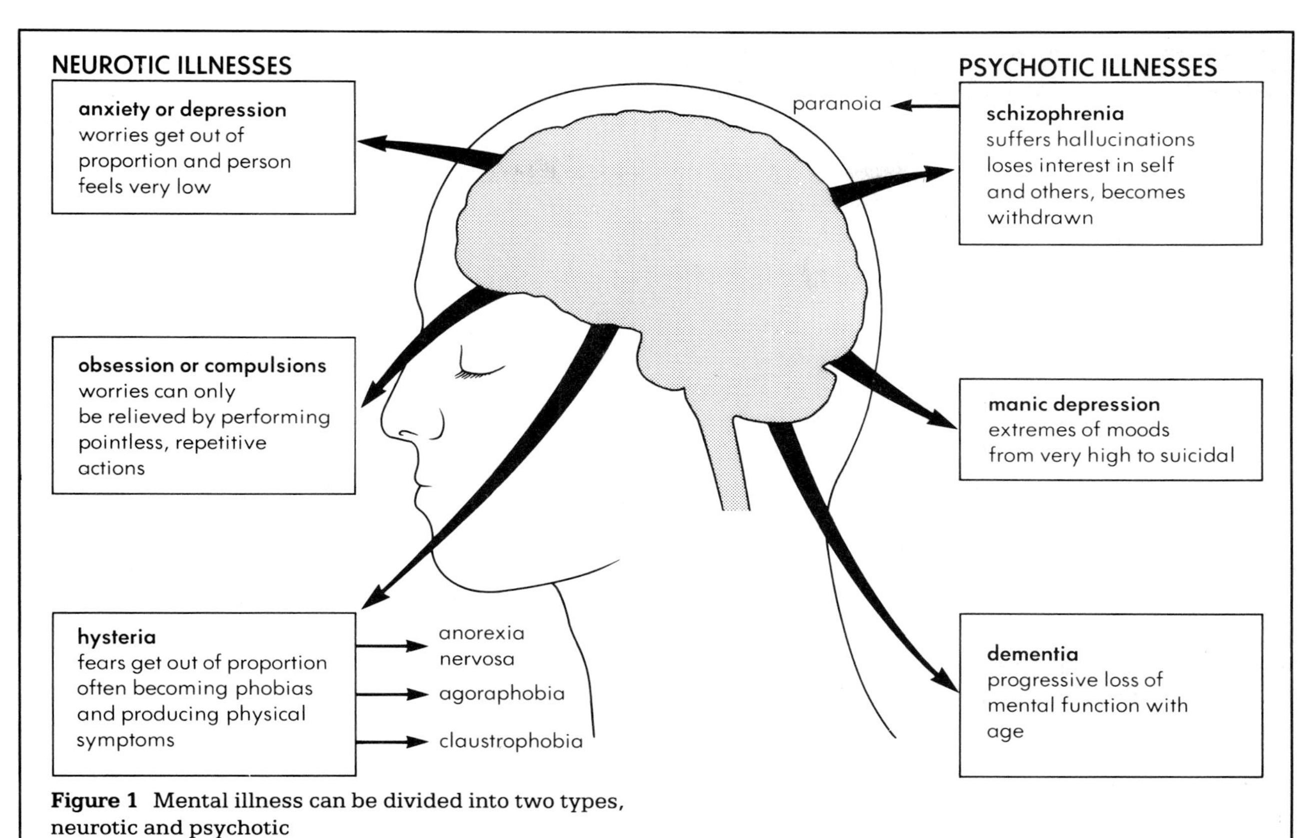

Figure 1 Mental illness can be divided into two types, neurotic and psychotic

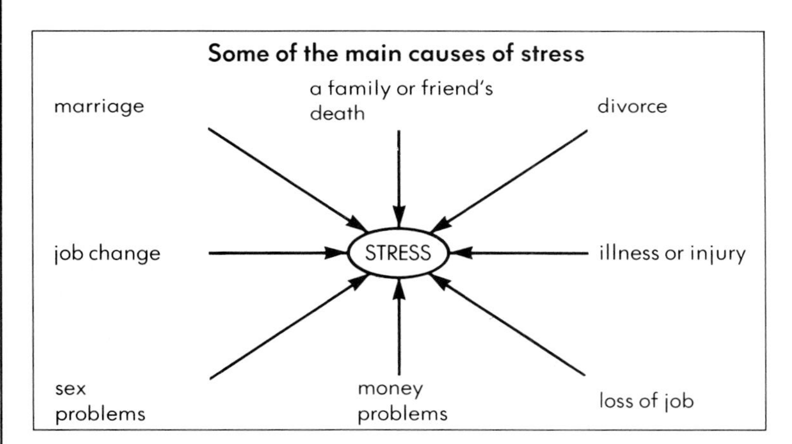

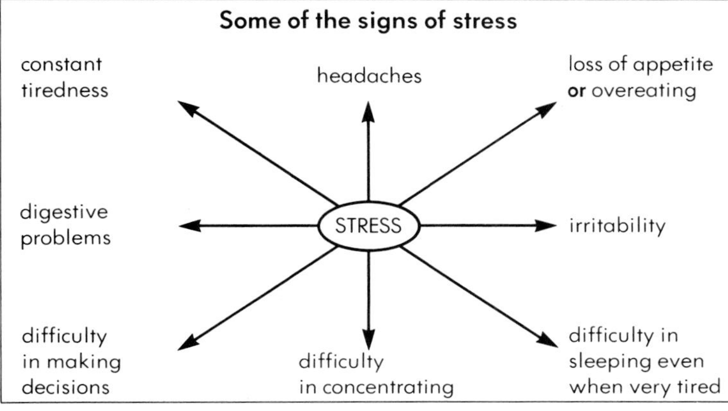

Figure 2 Stress is very common and can be caused by one major factor or several factors together

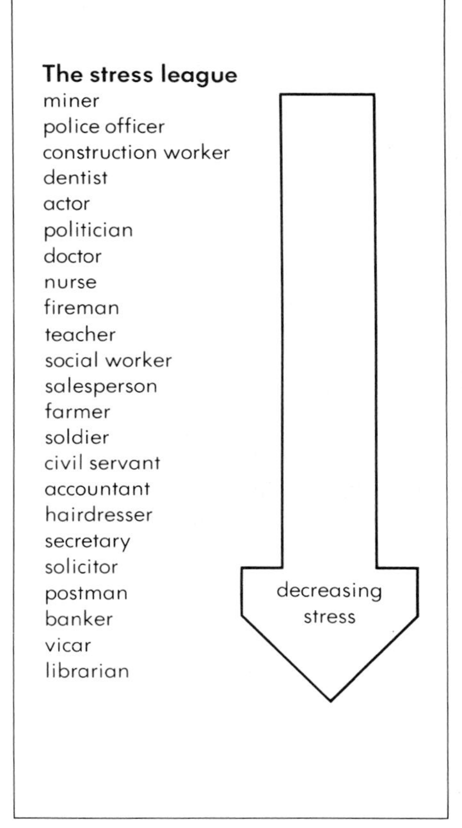

Figure 3 Stress partly depends on what job you do

5.11 The endocrine system

The nervous system is not the only way your body uses to send messages from one part to another. It also has the **endocrine system** which produces **chemical messengers** called **hormones**.

What are hormones?

Hormones are chemicals produced by special structures called glands. They diffuse from these **glands** into the blood and circulate with this until they reach their place of action (i.e. target organ). Here they diffuse out of the blood and deliver their message. This message usually has an important modifying effect on the structure or function of that part of the body.

There are about 30 known hormones produced by eight glands.

What are these glands?

The most important glands in your body are the **pituitary**, **thyroid**, **pancreas**, **adrenals**, **testes** (male), and **ovaries** (female). The positions of these and the main hormones they produce are shown in figure 1.

The pituitary gland

The pituitary gland is often called the **master gland** because it produces special hormones which control all the other glands. These hormones are called **trophic hormones**. For example, the pituitary gland produces **thyrotrophic** hormone which delivers its messages to the thyroid gland telling it to release its own hormone, **thyroxine.**

Adrenalin – the fight/flight hormone

Adrenalin is the hormone which is released in times of stress – for example when you are in a race or attending an interview. Its overall effect is to prepare your body for action. This is why it is called the fight/flight hormone. Some of the ways in which it does this are:

- By making sure that some of the stored glycogen in your liver and muscles is converted back into glucose. This temporarily raises the blood sugar level enabling your body cells to release more energy.
- By increasing your rate and depth of breathing so that more oxygen is available to the blood.
- By constricting the blood vessels in the skin and gut and diverting the blood from these regions to your muscles and brain.
- By increasing your heart rate, thereby making your blood move more quickly.

Some of the symptoms of adrenalin's effects are: a dry mouth, a pale face, a pounding heart and a sinking sensation in the stomach.

Exocrine glands

The tear and salivary glands are different kinds of gland to those of the endocrine system. Both produce chemicals, but these are not hormones and do not travel in the blood. They get to their sites of action via tubes called ducts. Glands like these are called **exocrine glands**.

The **pancreas** is sometimes called a dual organ because it functions both as an endocrine and exocrine gland.

Questions

1 Copy out and fill in the table:

Endocrine gland	Hormone(s) produced	Effects of hormones
Pituitary Thyroid Adrenal Pancreas Testis Ovary		

2 Why is the pancreas often called a dual organ?

3 Separate into exocrine and endocrine glands; tear glands, salivary glands, sweat glands, pituitary gland, pancreas, thyroid gland.

4 You are getting ready to meet a date for the first time. Your mouth is very dry your face is pale and your heart is beating fast. What is causing these symptoms?

5 Why is the pituitary gland often called the master gland?

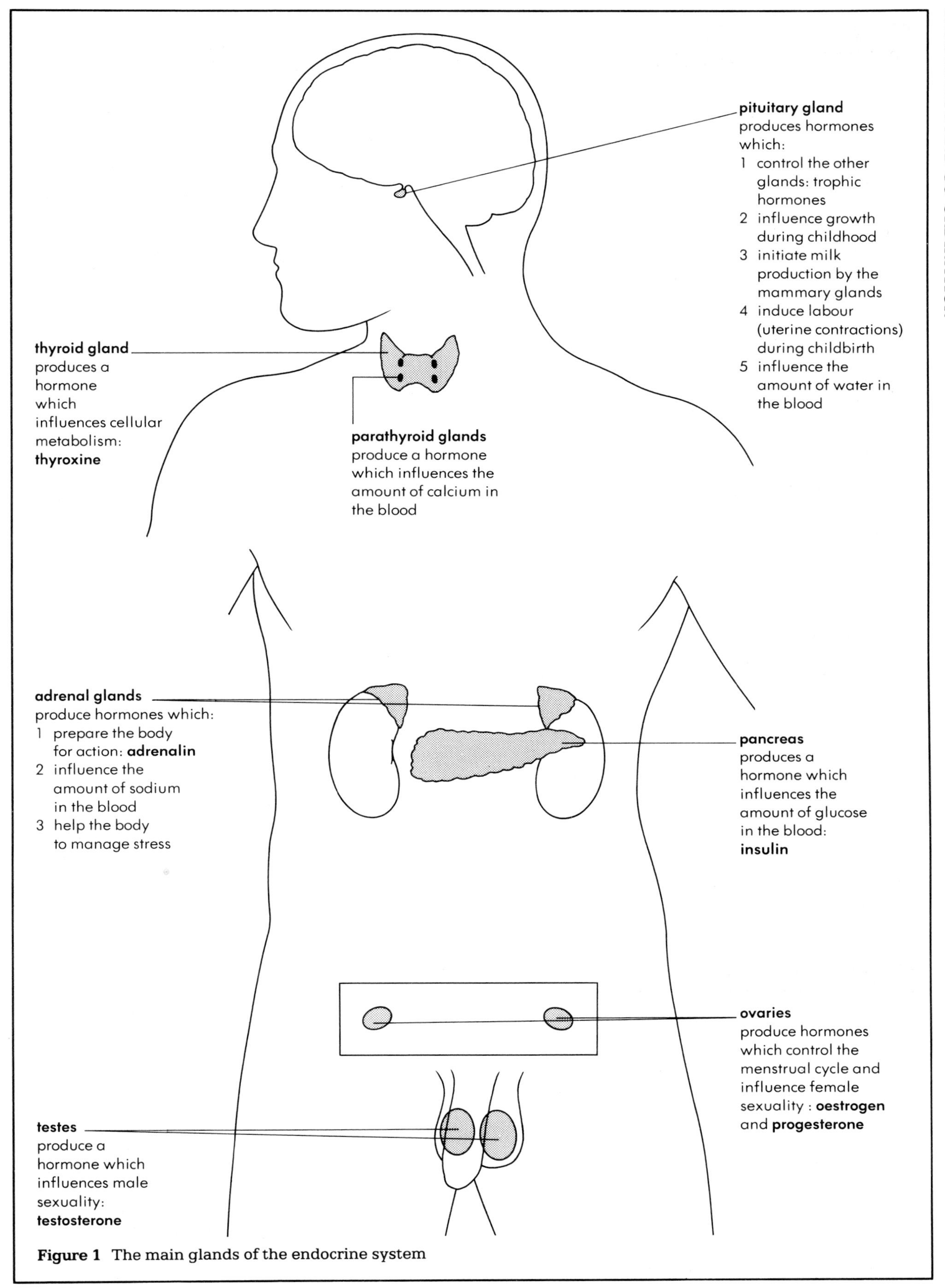

Figure 1 The main glands of the endocrine system

Questions

Recall and understanding

1 Which of the following is not a sense organ?
A eye C liver
B tongue D skin

2 The part of your eye which contains the light sensitive cells is the:
A front C lens
B choroid D retina

3 To see with binocular vision you require:
A binoculars C spectacles
B two eyes D a binocular microscope

4 Messages from the CNS to effectors travel along:
A sensory neurons C motor neurons
B the spinal cord D cranial nerves

5 Learning is an example of a:
A skill C conditioned reflex
B good memory D simple reflex

6 Alcohol is a:
A hallucinogen drug C stimulant drug
B aphrodisiac D depressant drug

7 Correcting long sight requires a:
A diverging lens C operation
B converging lens D lot of vitamin A

8 Alcohol is eliminated from your body at the rate of 10 mg per hour. How long would it take to eliminate the alcohol contained in 2 pints of beer?
A 6 hours C 8 hours
B 4 hours D 2 hours

9 Which hormone is released in times of stress?
A insulin C thyroxin
B cortisone D adrenalin

10 The part of the human brain which is more highly developed than other animals is the:
A cerebrum C medulla
B cerebellum D hypothalamus

Application and interpretation of data

11 An experiment was carried out to find the range of binocular vision of each pupil in a class of 12 year olds.

Results

Angle	Angle	Angle	Angle
132	120	119	109
144	109	127	111
145	112	131	126
132	146	138	151
143	152	104	141
151	143	147	152
144	147	149	132
107	106	153	125
130	119	156	121
147	157	159	141

a) What is binocular vision?
b) Use the data to fill in the table below:

Angle of binocular vision	Number of children
100 – 120 °	
121 – 140 °	
141 – 160 °	

c) Predict which group of children will be best at ball games. Explain why.
d) What proportion of the class is this?
e) When a ball comes towards you, the size of your pupil alters. Why?

12 The following graph shows the range of sound frequencies audible to percentages of the population.

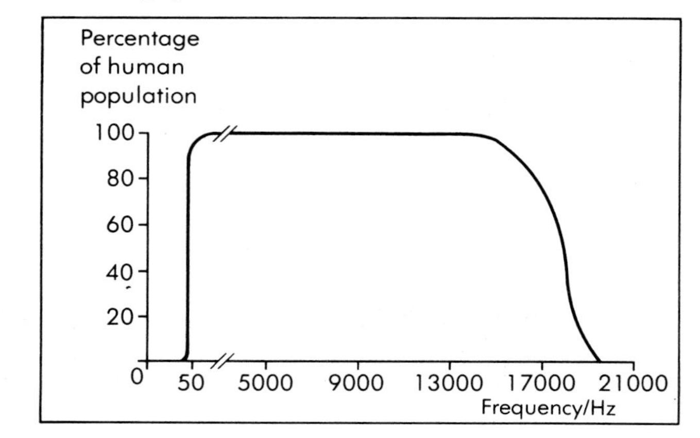

a) What is the full range of human hearing?
b) What percentage of the population can hear above 17 000 Hz?

c) The graph below shows the threshold of hearing of someone after spending 4 hours in a noisy disco.

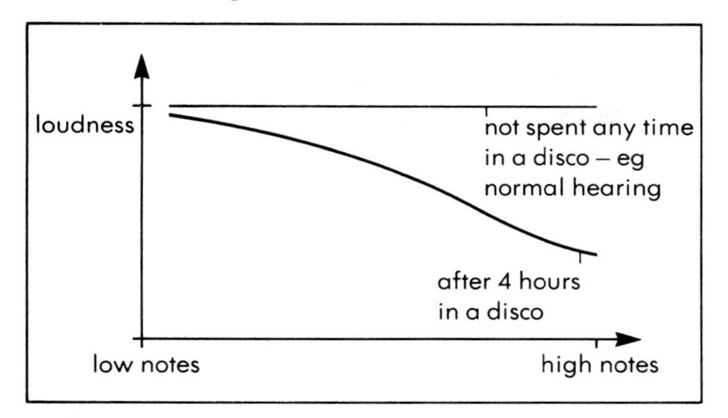

i) How does spending 4 hours in a disco affect a person's hearing?
ii) Why will someone who works 4 hours a night, 5 nights a week in a disco probably have poor hearing?

13 The figures below refer to the number of new cases of drug dependence over a 10 year period in the UK.

Year	1973	1976	1981	1982	1983	1984
Male	644	745	1607	1976	2979	3840
Female	163	239	641	817	1207	1575

a) Plot these figures on a graph.
b) What does the graph show about the incidence of drug dependence between 1973 and 1984?
c) Calculate the average number of new cases per year for men for
i) 1973 – 1979   ii) 1979 – 1984
d) Suggest two reason for the differences.

14 The diagram at the bottom of the page shows road accident casualties at different times of day in the UK, 1984.

a) At what time of day do most accidents happen?
i) Monday – Friday
ii) Weekends
b) At what time of day are there most fatalities?
c) Suggest reasons for your last two answers.
d) The graph below shows stopping distances with and without alcohol.

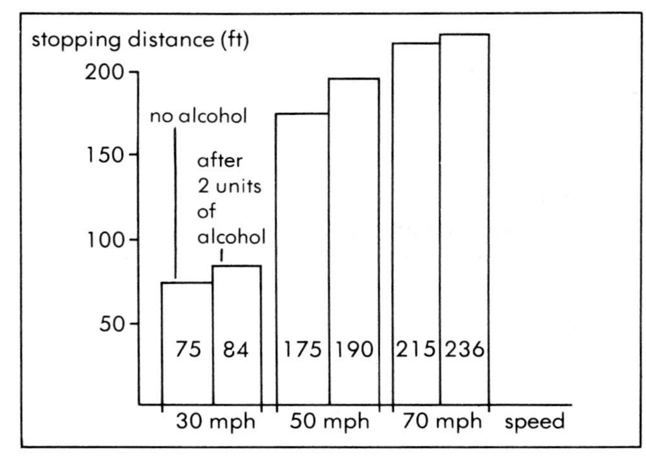

i) How does alcohol affect reaction speeds?
ii) For each speed, what is the increase in stopping distance?
iii) The average car length is 15 feet. How many car lengths does it take to stop at 50 mph?

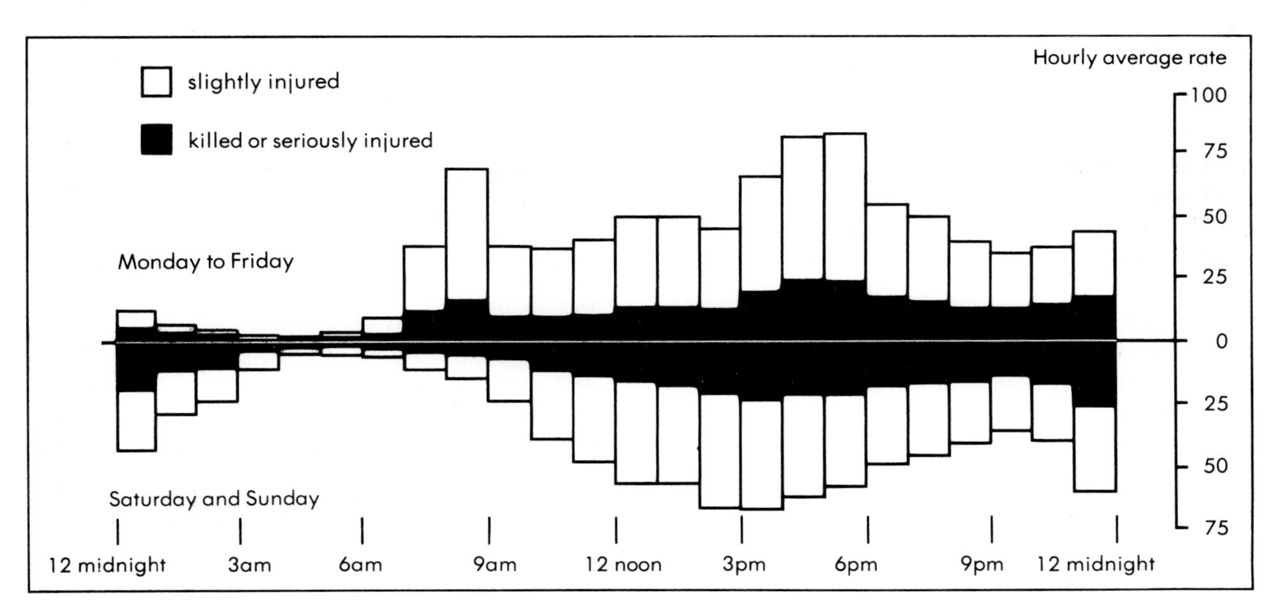

CHAPTER 6: *SUPPORT AND LOCOM*

Karate experts can chop through bricks with their bare hands. Pianists can perform delicate precise movements with their fingers. A weightlifter can lift up to three times his or her own weight, and a top 100 m runner can reach speeds of over 25 mph.

Not one of these actions would be possible without muscles and bones. Your body contains over 600 muscles and 200 or so bones. This chapter explains how these work together to produce movement, provide support and maintain posture.

Muscle power

This section looks at the power produced by muscles and how it can be increased.

How much power do muscles produce?

The strength of a muscle is related to its size. In general, the larger a muscle is the stronger it is. Strength is usually measured by the force that the muscle can apply to an object. The power generated by a muscle depends on how quickly it can apply this force. Forces of 30 newtons are not uncommon, giving a power of more than 60 watts.

The maximum power a human can produce using all of his or her muscles is about 80 watts. To show what sort of exertion this would require, an olympic sprinter generates a power of about 35 watts on take off.

The amount of power a muscle can generate is determined by the amount of energy it can obtain. This depends on the amount of food and oxygen available. Getting the food is not usually a problem but the oxygen can be.

How do weight lifters increase their strength?

Muscular strength can be increased by training. There is however a limit to the extra strength a person can gain and no amount of extra training will push you past this.

The type of training weight lifters do is designed to build up all of their muscles, not just those in their arms. It must also improve their general fitness especially the functioning of the cardiovascular system. Large muscles are no good if your blood and heart cannot bring food and oxygen there quickly enough.

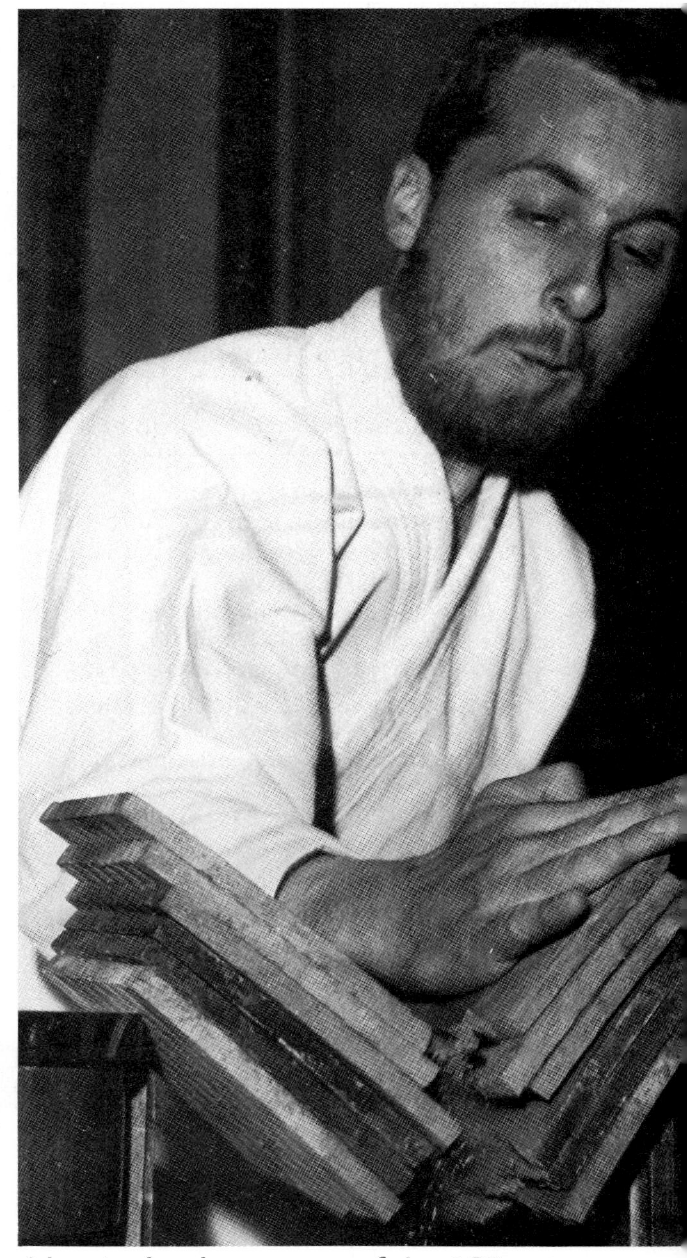

A karate chop has a power of about 25 watts

A weight lifter uses most of his body muscles, not just those in his arms

Why do other athletes train so much?

This can be answered by a series of graphs.

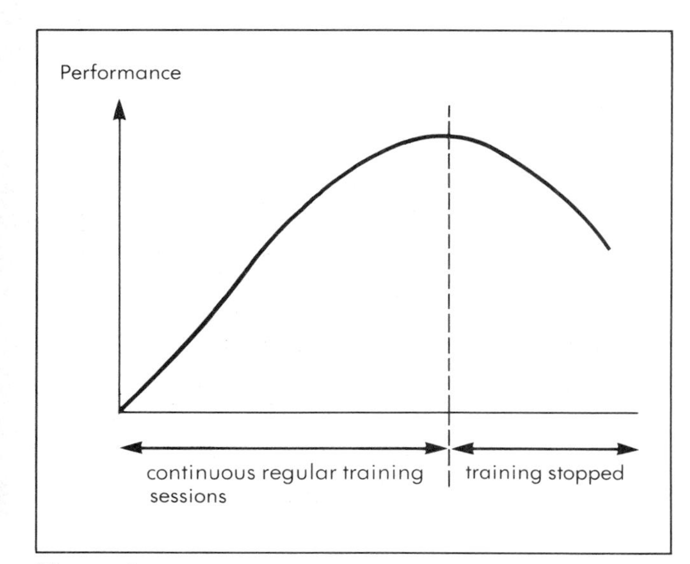

Figure 1

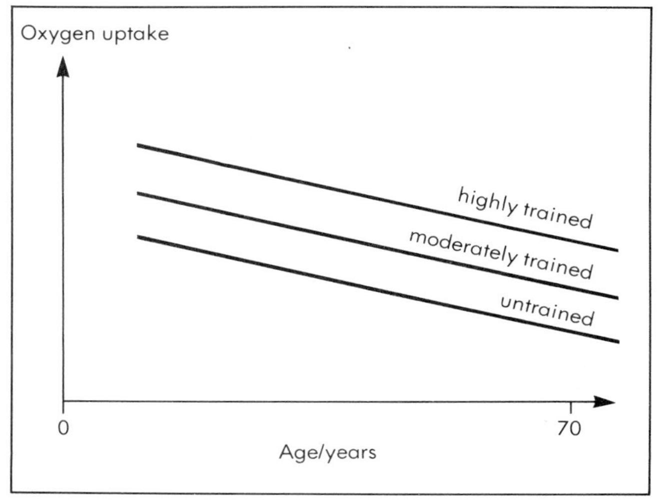

Figure 2

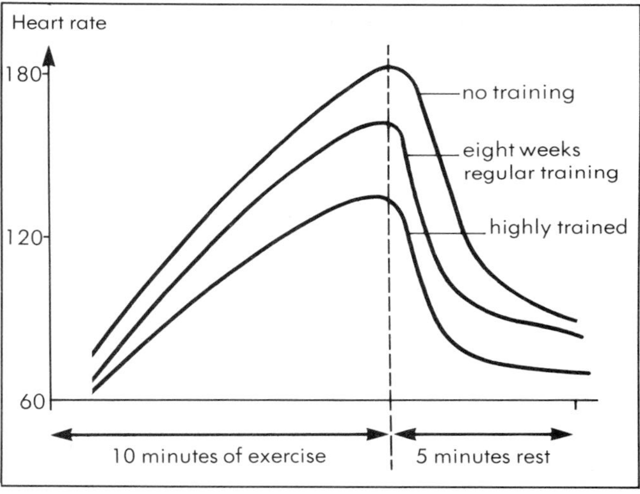

Figure 3

Figure 1 clearly shows that training improves performance. Figures 2 and 3 show why; training increases a person's capacity to take in oxygen and strengthens the cardiovascular system so that the oxygen has more chance of being delivered to the muscles.

Why do athletes have random drugs tests?

Many drugs and other substances can improve a person's performance making the competition unfair. For example, **steroid** drugs such as the male hormones can improve strength by increasing muscle size.

Amphetamines are the drugs most commonly taken by athletes. These increase alertness. Bicarbonate and caffeine increase endurance. These are all banned.

6.1 The skeleton

The framework of your body

You've probably got some idea about your skeleton just by knowing the shape of your body and how it moves. Your skeleton is the basic framework inside your body - without it you would collapse!

What does your skeleton do?

- It brings about movement or locomotion by working with the many muscles which are attached to it.
- It supports your body and gives it a shape.
- Parts of it provide protection for vital body organs, such as the brain.
- Some of the bones themselves produce new blood cells. Others can store calcium.

What is your skeleton made from?

The human skeleton is made from structures called bones (see page 104). There are 206 of these, some very large and some very small. The table below shows where these are.

Part of skeleton	Number of bones
Skull	22
Ears	6
Throat	1
Backbone	26 (33★)
Ribcage	25
Shoulder girdles	4
Hip girdle	2
Arms	60
Legs	60

★ Some of the bones are fused together

The parts of the skeleton

The skeleton can be divided into two parts:

1 The **axial** skeleton which consists of the **skull**, **ribcage** and **backbone**. All these parts provide protection for vital organs:
   - The skull protects the brain, eyes and ears (figure 1A).
   - The backbone protects the spinal cord and is also very important for support and movement (figure 1B).
   - The ribcage protects the heart and lungs and is involved in breathing.

2 The **appendicular** skeleton which consists of the **pectoral** and **pelvic girdles**, the arms and the legs. These parts are mainly involved in movement.
   - The girdles connect the limbs with the backbone and are important areas for muscle attachment.
     There are two of them:
     i) The pectoral (shoulder) girdle which connects the arms to the backbone (figure 1F);
     ii) The pelvic (hip) girdle which connects the legs to the backbone and protects the organs in the lower part of the abdomen (figure 1E).
   - The limbs act as levers when creating movement. The pattern of bones in them is very similar (figure 1D) and this same pattern is found in the limbs of birds, bats, horses and whales suggesting a common ancestry (see page 182).

Questions

1 The skeleton is sometimes divided into two functional parts; the axial and the appendicular skeleton.
a) What is the basis behind this division?
b) Divide the following into those parts which form the axial skeleton and those which form the appendicular skeleton:
Skull, girdles, arms, legs, ribcage, backbone.

2 Make a paper skeleton. Use different colours for the axial and appendicular parts.

3 How many bones are there in the skeleton? Find as many of these as you can in your own body.

4 Why are the vertebrae larger near the base of your back?

5 The same pattern of bones is found in the limbs of several different groups of animals. What does this imply about the origin of these animals?

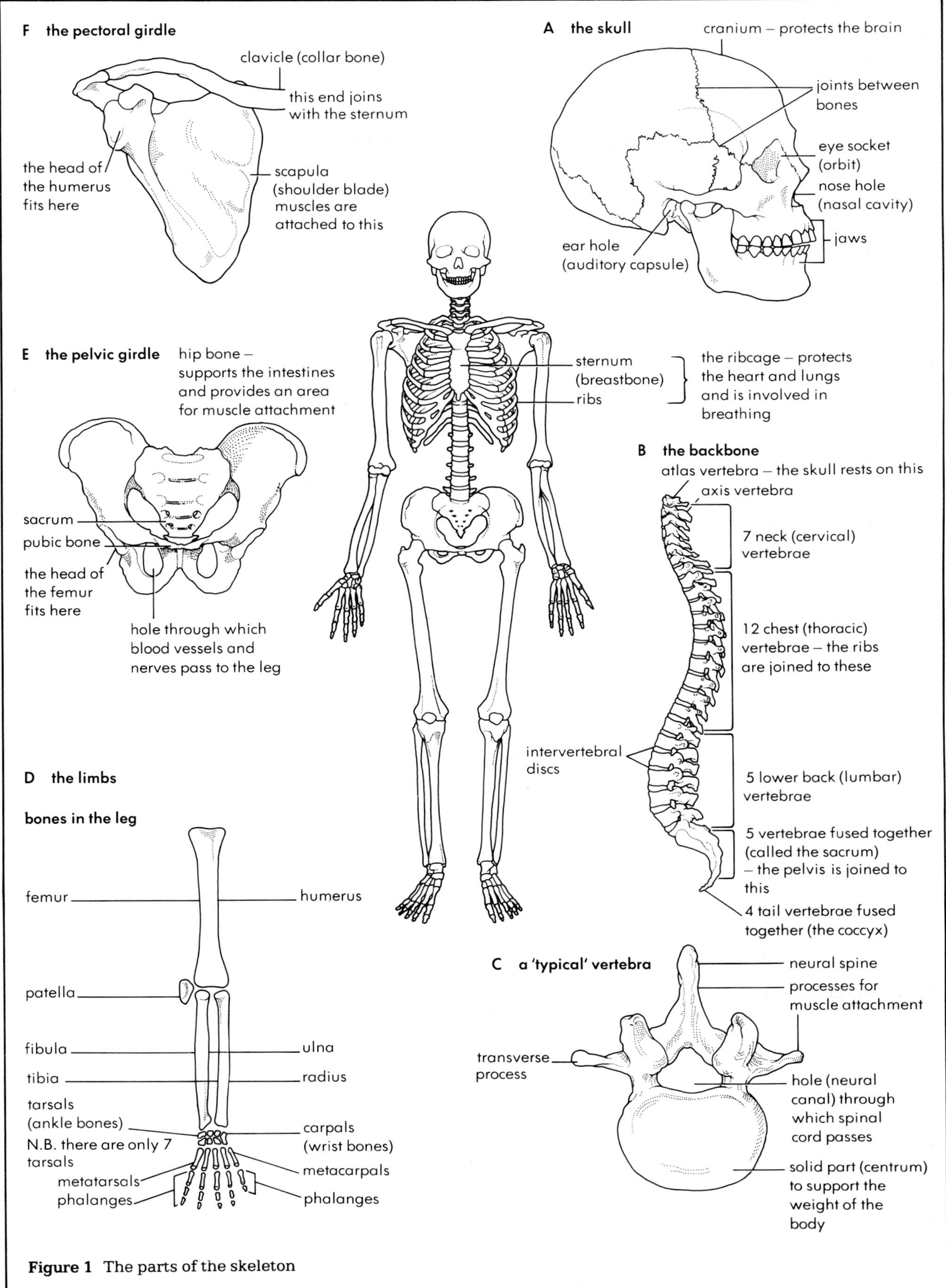

Figure 1 The parts of the skeleton

6.2 Bone

What are bones made from?

Bones are made from bone tissue (figure 1). There are two kinds:

- **Compact bone tissue** on the outside of the bone;
- **Spongy bone tissue** on the inside of the bone. This contains lots of spaces which, in some bones, are filled with **red bone marrow**. New blood cells are made here.

Some of the larger bones in your body are hollow to make them lighter. The cavity is filled with **yellow bone marrow**, which is mainly fat. All bones are enclosed in a protective skin made from fibrous connective tissue. The tendons of muscles are attached to this.

The ends of the bones which form movable joints are covered by **cartilage** which is continually replaced as it wears.

What is cartilage?

Cartilage is a tough, yet flexible material made from carbohydrates and proteins. There are several kinds, suitable for different jobs. The cartilage in the pinna of the ear is very flexible, but the cartilage covering the heads of bones is fairly rigid and very smooth.

Cartilage can be turned into bone tissue by the process of **ossification**.

Can bones grow?

All bones are made from cartilage at first. When a baby is born, most of this cartilage has already been turned into bone (**ossified**), but not all of it. The remaining areas of cartilage can be clearly seen on X-rays of the child's skeleton (figure 2). They appear as gaps in the bones.

These areas of cartilage are the growth areas of the bone. These get bigger as the bone grows (figure 3). As the skeleton reaches full size, all the cartilage becomes ossified and the bone can no longer grow, although it can still repair itself when damaged.

Why is bone so strong?

Bone tissue is made from bone cells embedded in a firm matrix. This matrix is made from a mesh of flexible protein fibres embedded in a hard mixture of calcium salts (mainly calcium phosphate).

The properties of each of these materials can be clearly seen in a simple experiment. If you place the bone from a chicken's leg into dilute acid and leave it there for several hours the calcium salts are removed. The bone still has the same shape but is now very flexible.

If you take another bone and heat it very strongly, you can burn away the protein leaving behind the calcium salts. In this state, the bone is hard but brittle.

It is the combination of this flexibility and hardness that gives bone tissue its great strength and durability.

Why do children need to drink lots of milk?

Milk contains calcium and vitamin D. Calcium is one of the substances that makes bone hard and rigid. Vitamin D is needed for the absorption of calcium from the food we eat.

If you do not get enough calcium your bones are soft and will bend easily (see page 21).

Questions

1. Explain why bone tissue is so strong and durable. Describe how it is arranged in a typical long bone.
2. What are the differences between red and yellow bone marrow?
3. Children who do not drink enough milk or eat enough dairy products grow up with bow legs. Pregnant women who do not drink enough milk or eat enough dairy products suffer from fragile bones and tooth caries. Explain why.
4. An X-ray of a child's hand gives the impression that the bones are too short and have gaps in them. Explain why.
5. Explain how bones grow.

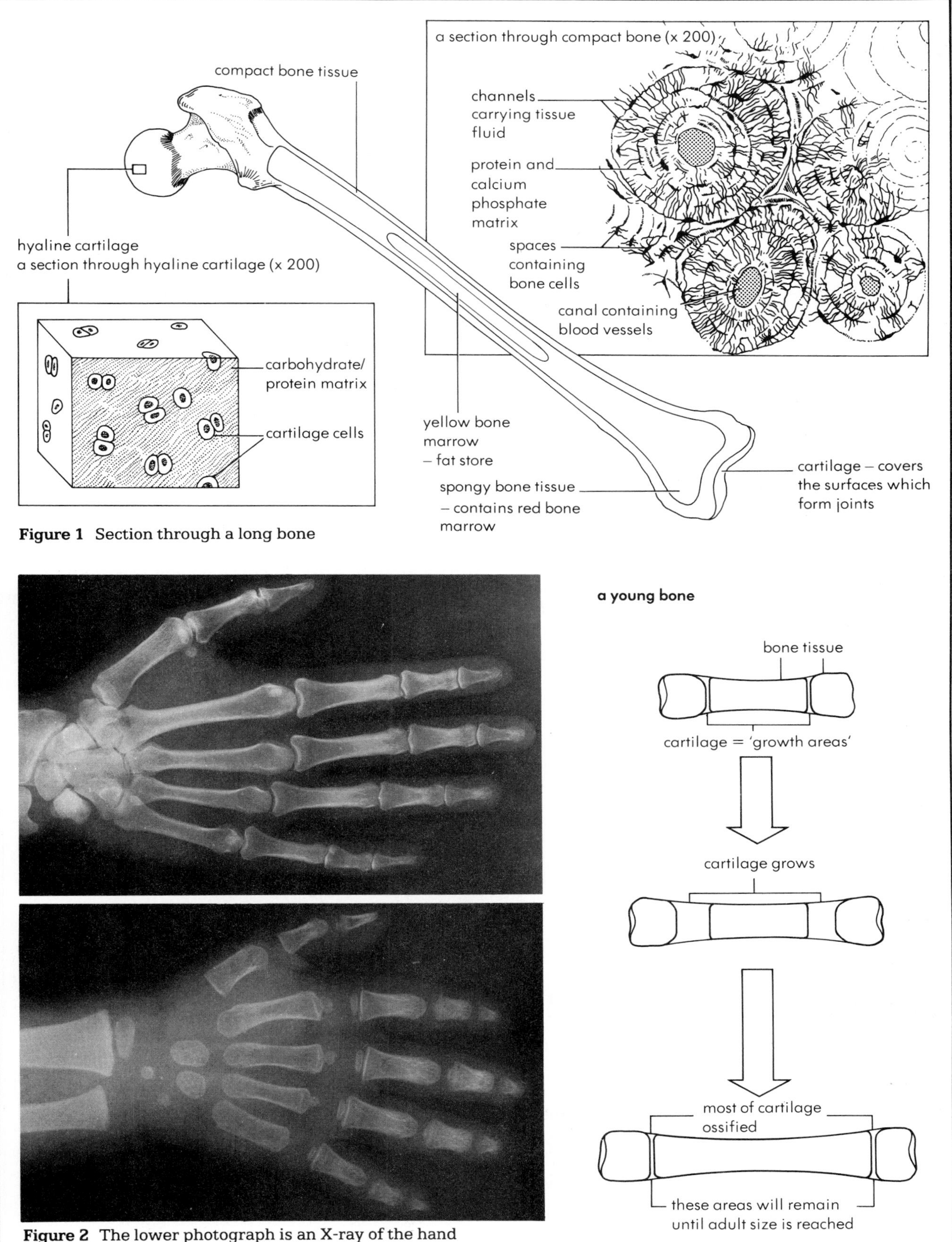

Figure 1 Section through a long bone

Figure 2 The lower photograph is an X-ray of the hand of a 2½ year old child. It shows the areas of cartilage where the bone grows. Compare it with the adult's hand above

Figure 3 How a bone gets bigger

6.3 Joints

Where are your joints?

Wherever two bones in your body meet you have a **joint**. There are over a hundred joints in your skeleton. The ones you probably know about are in your shoulder, elbow, hip, knee, hands and feet. Your hand alone has over twenty joints.

The joints that allow movement are called **movable joints**. Some joints, however, e.g. those in the cranium, do not allow movement and are therefore called **fixed joints** (figure 1).

Are all movable joints the same?

There are several types of movable joint:

- **Ball and socket joints** which are found in the hips and shoulders. They are formed by the head of one bone fitting into a cup-shaped cavity of another bone (figure 2). Ball and socket joints allow movement in all directions.
- **Hinge joints** which are found in the elbow, hand, knee and foot. The bones are joined together in such a way that movement is only possible in one direction, just like a hinge on a door (figure 2).
- **Slightly movable joints** which are found in the backbone. Sandwiched between the bones which form these joints is a pad of cartilage (figure 3). The bones can move a little against this cartilage.
- **Pivot joints** such as that found between the atlas and the axis vertebrae. In these joints one bone pivots on another (figure 4).
- **Gliding joints** which are found in the wrist and ankle. The flat surfaces of the bones glide across each other during movement.

How do bones stay together when we move?

Bones are held together at joints by pieces of strong fibrous connective tissue called **ligaments**. Ligaments are fairly elastic, but can be torn. This happens when joints are **dislocated**. You may have heard of someone 'tearing a ligament' in a sports injury. If this happens to you see a doctor as soon as possible. Meanwhile, reduce the swelling by using a cold compress.

What is arthritis?

Arthritis is a disease of joints that usually affects older people. There are two main forms:

- **Rheumatoid arthritis** which seems to be caused by our own immune system attacking the synovial membrane. This membrane first swells up and then starts to disintegrate. Once it is gone, no synovial fluid is made so that the heads of the bones wear more quickly, and the bones may even stick together.
- **Osteoarthritis** is brought on by old age. Athletes who use their joints a lot often suffer from this. Gradually, as we age, the parts of the joint wear away more quickly than they can be repaired. Movement then becomes very painful.

What is a torn cartilage?

Footballers often get torn cartilages in the knee. The knee has two extra pieces of cartilage to strengthen and protect the joint. If these tear and pieces lodge between the bones, the joint locks. Usually, an operation is needed to remove the damaged piece of cartilage.

Questions

1. What are the five kinds of movable joint? Name two places in your body where each type of joint can be found.
2. Draw a ball and socket joint. Label the articular cartilage, synovial membrane, synovial fluid and capsule. Under each label write the function of the part.
3. A hinge joint works very much like the hinge on a door. Suggest similar comparisons for the other types of movable joints.
4. Explain why rugby players often have to give up playing when they tear the ligaments in their shoulders.
5. Explain why footballers often have the cartilage removed from their knee joint.

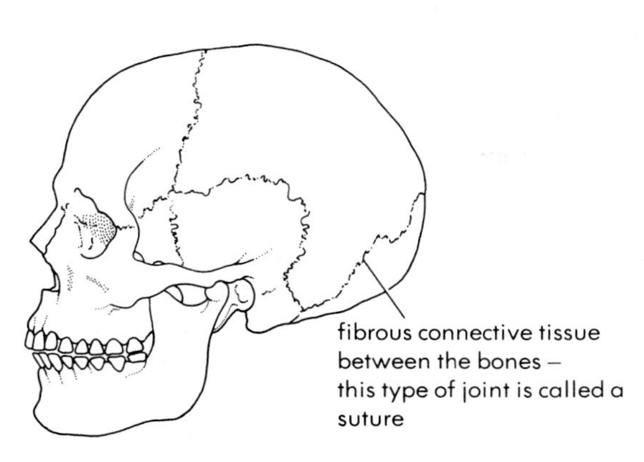

Figure 1 The joints of the skull do not allow movement

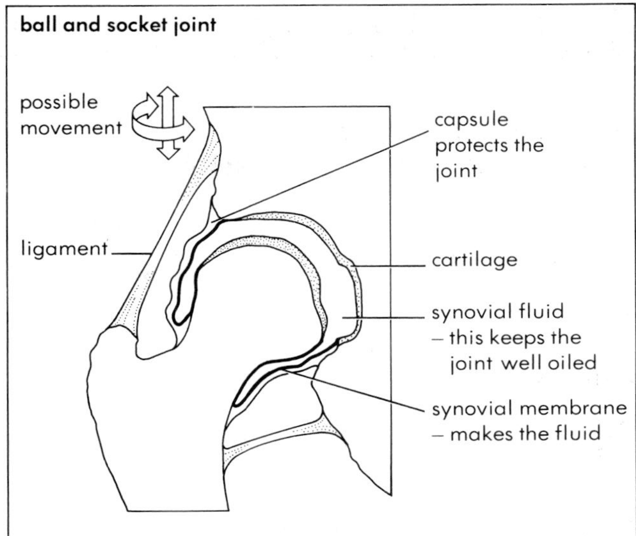

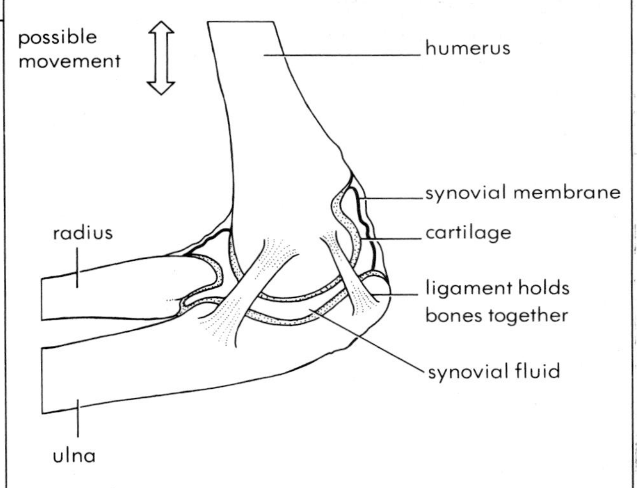

Figure 2 These two types of joint allow the greatest amount of movement. They are found in our limbs

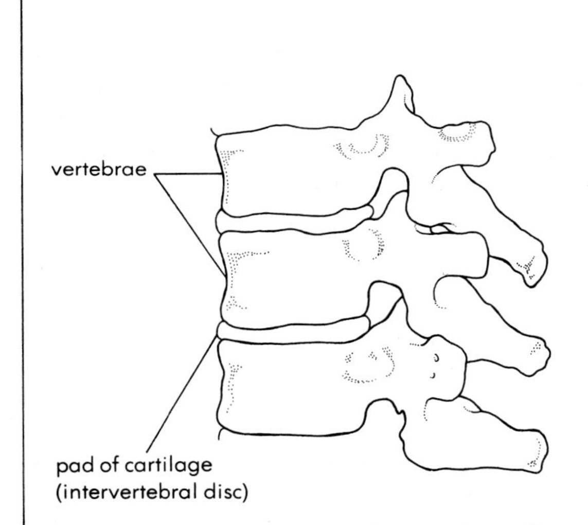

Figure 3 The joints between the vertebrae allow a little movement

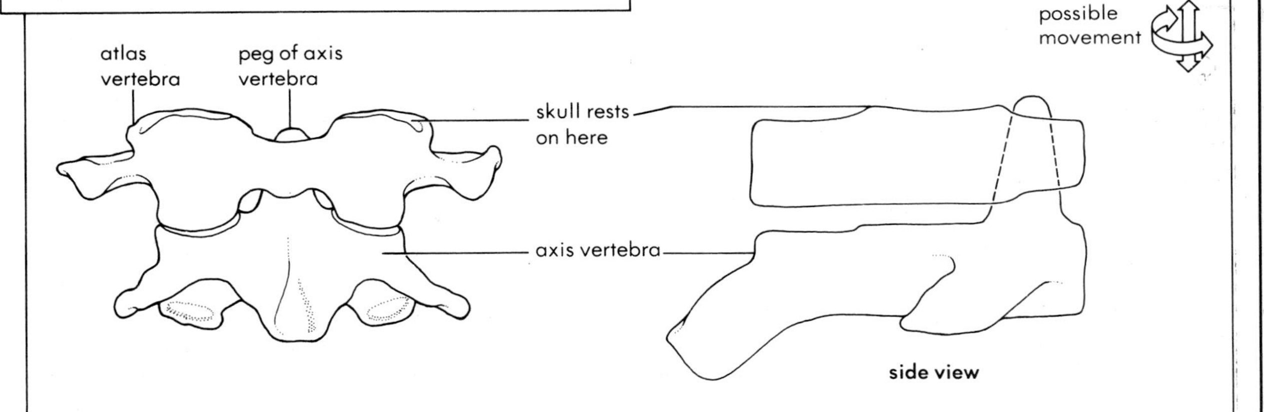

Figure 4 A pivot joint allows the head to move up and down and from side to side

6.4 Muscles and movement

What is muscle tissue?

Muscle tissue makes up about 50 per cent of your body weight. There are three kinds, each having specific functions, but all are made from cells containing **contractile proteins**. When energy is available, these proteins use it to contract, thereby shortening the cells in the muscle tissue which gives the muscle its ability to move – to contract and expand.

The types of muscle tissue

- **Skeletal muscle tissue** The cells of skeletal muscle are joined together to form long fibres. These fibres are then bundled together and several of these bundles form a 'muscle' (figure 2a).

  Skeletal muscles get their name from the fact that they are all attached to the skeleton. Their job is to pull on the bones to produce movement and to keep the body in the right position. To do this they have to be large and powerful and capable of very fast or very slow sustained contraction. Skeletal muscles work when told to do so by the brain, i.e. movement is voluntary. Unfortunately they also tire quickly.
- **Smooth muscle tissue** Smooth muscle is made from long thin cells held together by a fibrous connective tissue (figure 2b). It is found in the walls of hollow tubes such as the gut, ureters and blood vessels. The muscle enables these to move things through them by peristalsis.

  Peristalsis is often a continuous process and requires a kind of muscle that does not tire easily, can produce smooth, sustained contractions and does not have to be told to do so (i.e. movement is involuntary).
- **Cardiac muscle tissue** Cardiac muscle is very similar in structure to skeletal muscle except that the parallel fibres are cross-linked together at intervals (figure 2c). Its properties however are very different. Cardiac muscle is only found in the heart. It contracts on average 70 times a minute for 70 years, without ever tiring and without ever being told to, even by the autonomic nervous system. It is completely involuntary. The ANS can, however, tell it to speed up or slow down.

How do muscles move bones?

The skeletal muscles are attached to bones by pieces of non-stretching fibrous connective tissue called **tendons**. One end of a muscle (the **origin** end) is always attached to a bone which will not move and the other end (the **insertion** end) is attached to a bone near a joint. When the muscle contracts, it pulls on the bones and the one near the joint moves in the direction of the pull. Movement is therefore produced by lever systems, the joint acting as the fulcrum, the bone as the lever and the muscle providing the effort force to overcome any resistance (load). Figure 3 illustrates how the biceps muscle moves the bones in the lower part of your arm. Try and work out what provides the load and effort forces in each case.

How is the arm straightened again?

To straighten your arm, another muscle is required to pull on the bones at the opposite side of the elbow joint (see figure 3). All muscles work in pairs like this. The muscle that bends the joint is called the **flexor muscle** and the muscle that straightens it is called the **extensor muscle**. The flexor and extensor muscles are known as an **antagonistic pair of muscles**.

Muscles and posture

Skeletal muscles are responsible for maintaining your posture. They do this by always being slightly contracted, the ever present tension being called **muscle tone**. If you don't look after your muscles and exercise them regularly the muscle tone gradually disappears and your posture deteriorates.

Questions

1 Copy and complete the table below:

Muscle type	Where found	Properties
Skeletal Smooth Cardiac		

2 Explain how tendons and ligaments are different.

3 Look at figure 3.
a) What will happen to the arm if the flexor muscle contracts? Explain why.
b) Explain how the arm can be straightened again.
c) When you bend your arm, which is the bigger force; the effort or the load?

4 How can posture show the state of someone's muscles?

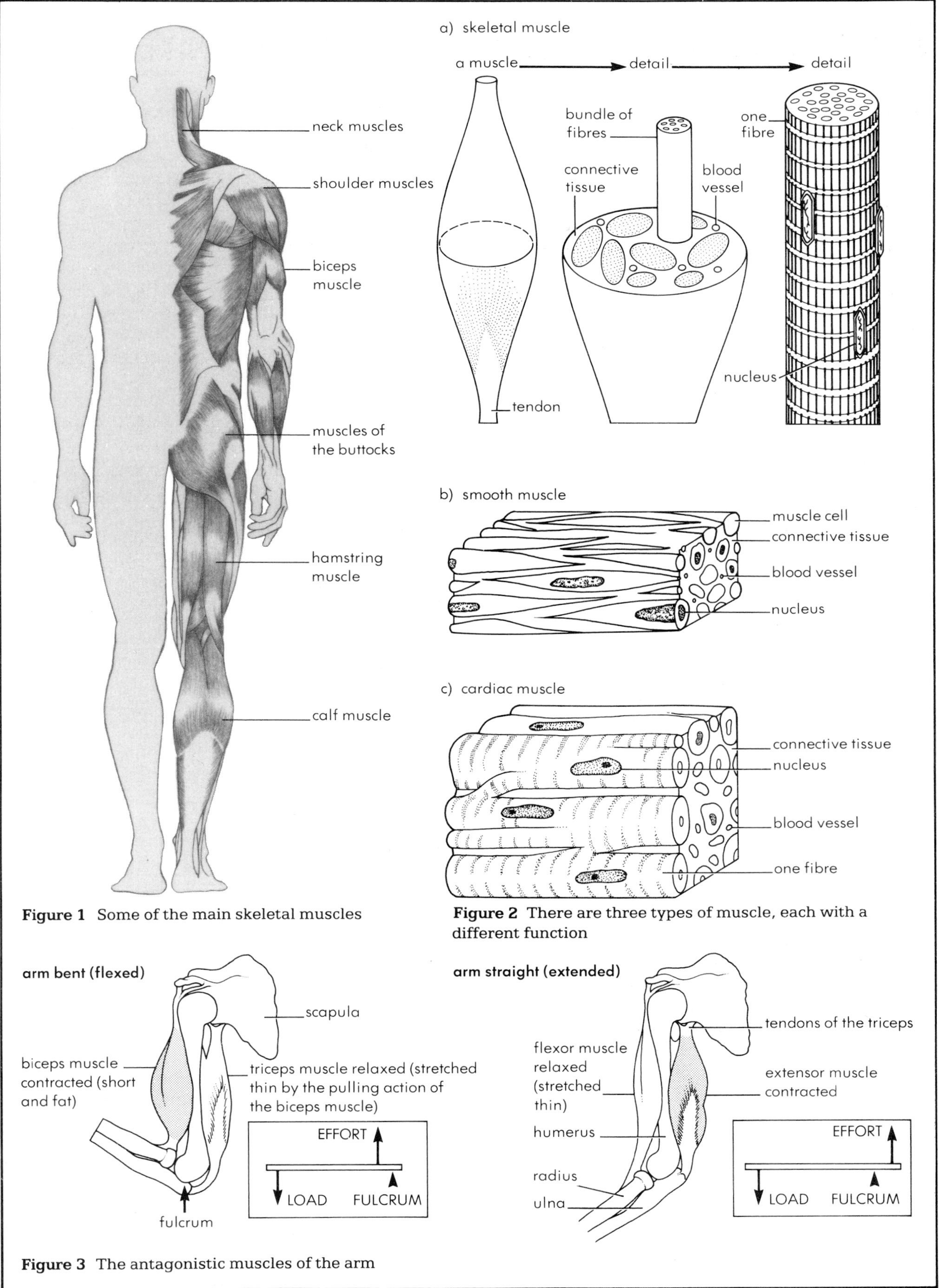

Figure 1 Some of the main skeletal muscles

Figure 2 There are three types of muscle, each with a different function

Figure 3 The antagonistic muscles of the arm

Questions

Recall and understanding

1 The spaces in spongy bone tissue are filled with:

A fat C blood
B marrow D red bone marrow

2 The number of vertebrae in the backbone is:

A 34 C 33
B 29 D 26

3 The bones of a movable joint are held together by:

A ligaments C muscles
B tendons D protein threads

4 A ball and socket joint can be found in the:

A fingers C wrist
B ankle D hip

5 The antagonistic muscle to the triceps is the:

A biceps C temporal
B quadriceps D masseter

6 Cardiac muscle is found in the:

A leg C brain
B heart D eye

7 The cranium protects the:

A eyes C teeth
B ears D brain

8 Muscle tone is important for:

A swallowing C rapid movement
B peristalsis D maintenance of posture

9 Muscles are involved in each of the following except:

A flexing the arm C eating
B running D secretion of hormones

10 When a piece of bone is put into dilute hydrochloric acid for a few days:

A protein is removed
B mineral salts are removed
C it completely dissolves
D it strengthens it

Application and interpretation of data

11 The graph shows the changes in bone density as you age.

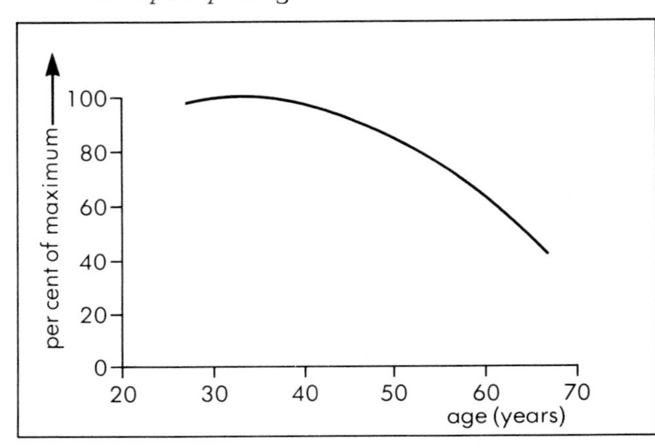

a) At what age is the density of your bones at its greatest?
b) Calculate the rate of loss of bone density per year from age 30 to age 70.
c) What is most of the loss in bone density a result of?
d) As you get older your bones get more brittle. How does this tie in with the information above and your last answer?

12 The table below shows the number of deaths from rheumatoid and osteoarthritis in 1984.

Age	under 15	15–24	25–34	35–44	45–64	65+
Rheumatoid arthritis	0	0	0	7	246	1246
Osteoarthritis	0	0	0	1	16	948

a) What are rheumatoid and osteoarthritis?
b) Which age range do they affect mostly?
c) At what age do these diseases start appearing?
d) Identify the main risk groups, e.g. sports persons.

13 The table below gives the breaking tensions of some of your body materials and steel.

	Breaking force (N per cm^2)
Bone	12600
Muscle	56
Tendon	7000
Steel	90000

a) Approximately how many times stronger is steel than bone?
b) Tendons are almost as strong as bone. What do these do?
c) The table below gives the number of deaths from fractures in 1984.

Age	under 1	1–4	5–14	15–24	25–34	35–44	45–54	55–64	65–74	over 75
male	8	50	166	674	374	288	231	281	379	824
female	2	27	75	170	80	72	78	135	372	212

i) Plot these on a graph.
ii) Which age ranges have the most fractures resulting in death? Suggest why.
iii) How do the figures for males and females differ?

14 Leukaemia is the term given to a group of cancers which affect the white blood cells and tissues producing these. The data below is for 1984. The figures are rates per 100 000.

Age	under 1	1–4	5–9	10–14	15–19	20–24	25–29	30–34	35–39
male	2.5	8.2	3.7	2.7	2.4	1.9	1.9	2.2	2.7
female	2.3	6.7	3.2	2.1	1.5	1.3	2.0	2.4	2.0

40–44	45–49	50–54	55–59	60–64	65–69	70–74	75–79	80–84	over 85
3.7	4.3	7.1	10.6	16.8	25.8	36.8	59.5	70.6	84.4
2.8	3.2	4.3	7.4	9.8	15.8	20.2	29.6	36.7	52.5

a) Leukaemia is often referred to as a young person's disease. Do the figures agree or disagree with this?
b) At what age does the incidence of leukaemia start to increase rapidly?
c) Calculate the average rate for
i) ages 0 - 50,
ii) ages 50 upwards.
d) Comment on the relationship between age and leukaemia.

CHAPTER 7: *HOMEOSTASIS*

Your body stays more or less the same inside even though your surroundings outside may change.

In this chapter you will learn how you keep your body temperature constant, even in a very cold or hot place. You will also find out what happens to the waste materials that your body produces every day.

Your blood is the main transport system in your body. One of the most important jobs it does is to take glucose (sugar) to your body cells. It is vital that there is always enough glucose in your blood. It is just as important that there isn't too much.

Genetic engineering solves the insulin problem

This section looks at what happens to people whose body cannot maintain the correct blood sugar (glucose) level.

What is insulin?

Insulin is one of several hormones which regulate your blood sugar level (see page 122). It enables your body cells to make use of the glucose in your blood. Without insulin, this glucose would build up in the blood and cause the disease **diabetes**. If not treated this can cause death.

This girl is having to inject herself with insulin before she eats a meal.

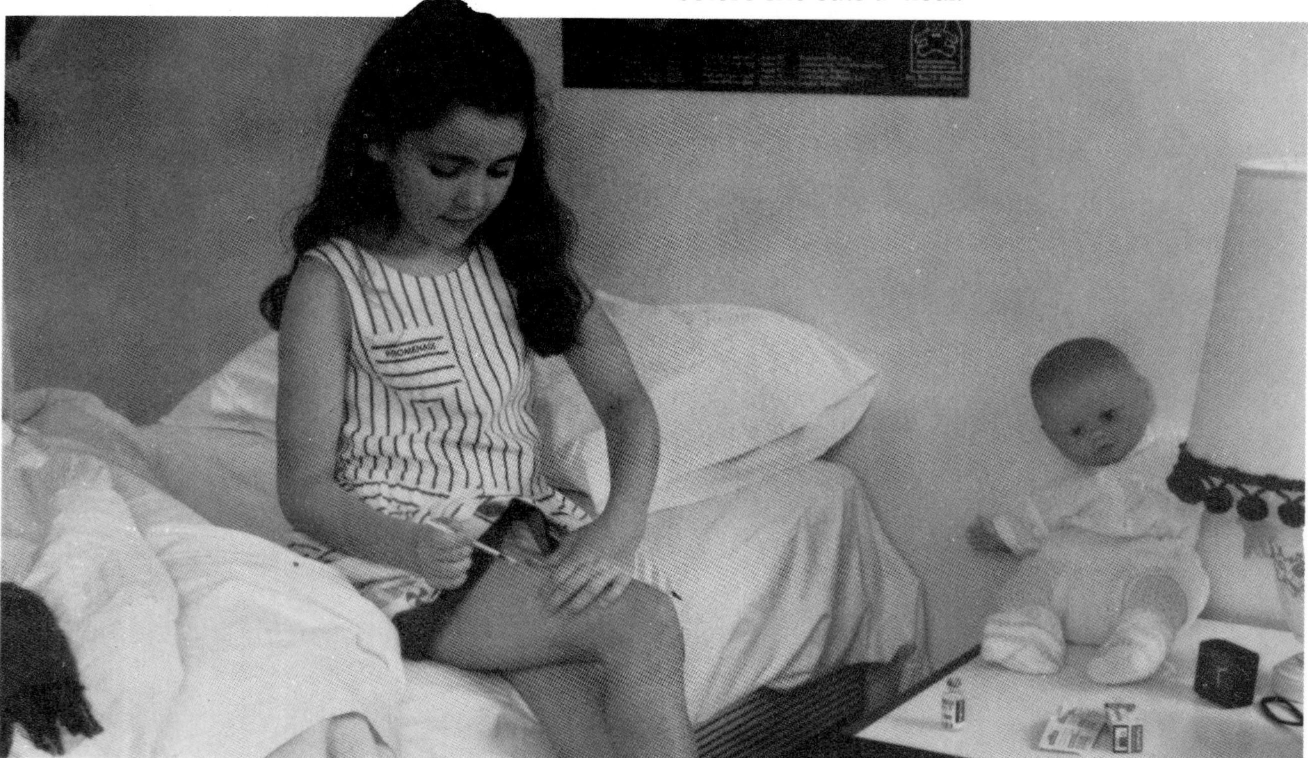

Why do we need to manufacture insulin?

Approximately 1 – 2% of the population in the developed world are known to suffer from diabetes. For about 20% of these people, the symptoms can be relieved by twice daily injections of insulin. This is about 4 million people, so a lot of insulin is needed.

How do we get insulin for these people?

Insulin is extracted from the pancreases of pigs and cattle killed for their meat. The amount collected therefore depends on the demand for meat. Sometimes there is not enough.

Animal insulin is also slightly different from human insulin. This sometimes causes problems.

What problems does animal insulin cause?

The animal insulin is treated by the body as a foreign invader and antibodies are produced to destroy it. This makes it less effective. To counter this a larger dose must be taken.

Can we use human insulin?

Until recently human insulin was not available for obvious reasons! However we can now make it thanks to two recent developments in biotechnology.

- Pig insulin can be converted into human insulin using special enzymes.
- Bacteria can be persuaded to make human insulin by **genetic engineering** (page 171).

  Briefly, a human insulin gene is inserted into a bacterium. That bacterium can then make human insulin. Bacteria cells grow and reproduce very quickly so that one bacterium soon gives rise to millions (figure 1). All these will have the human insulin gene. If these are grown in a large container, the insulin they make can be collected.

Enough human insulin is currently being made for it to be used medically alongside the animal insulin.

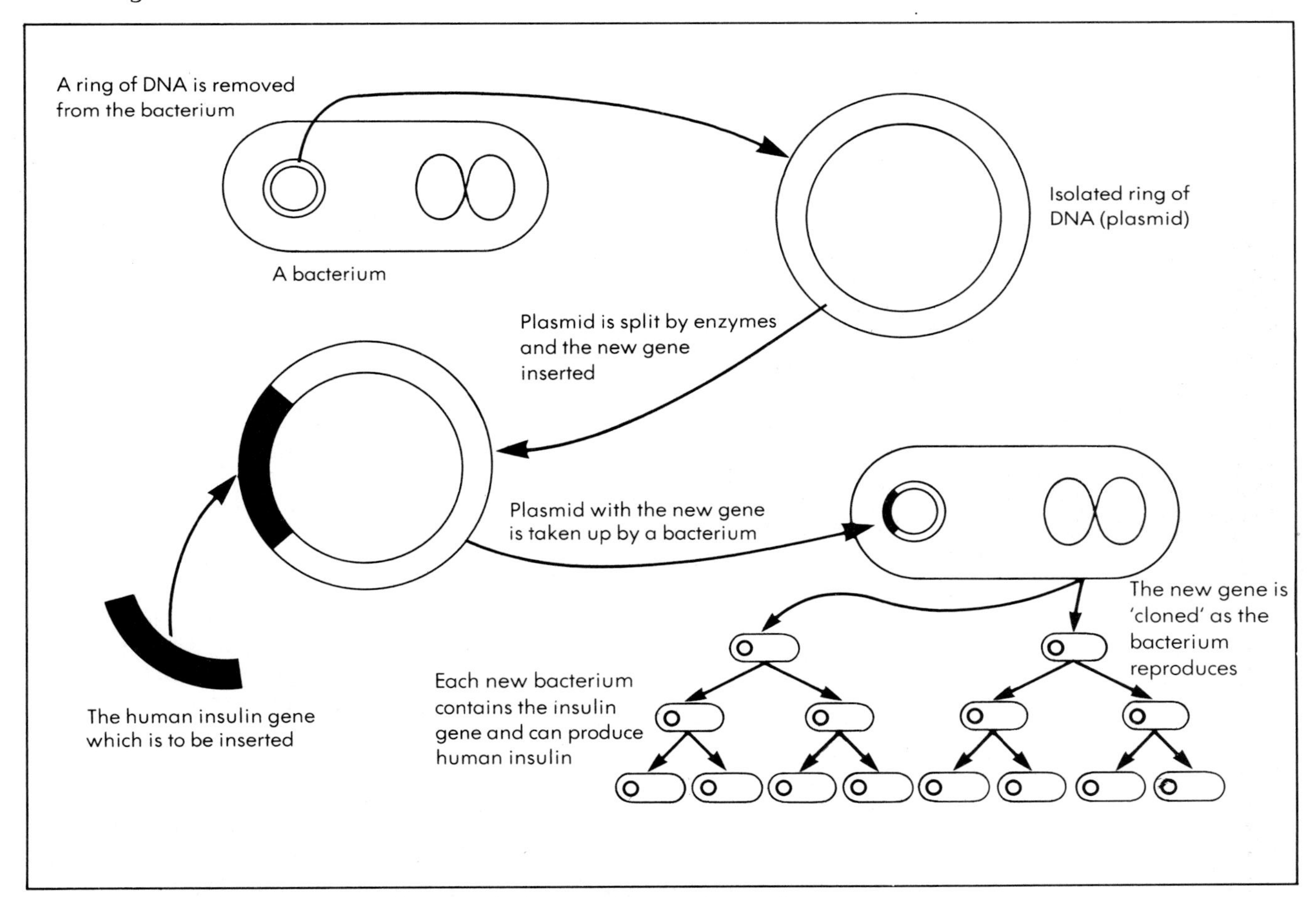

Figure 1 How a human insulin gene is inserted into a bacterium.

7.1 Temperature control

What is homeostasis?

Your cells are very delicate structures and will only function properly in the correct conditions. Large changes in these conditions will kill them. Keeping constant conditions in the tissue fluid around the cells is called **homeostasis** (figure 1).

How is this done?

Homeostasis is achieved by a number of different organs:

- The **skin** (figure 2) and **liver** help to maintain your body at a suitable temperature.
- The **kidneys** excrete metabolic waste products and maintain suitable salt and water levels in the blood.
- The **liver** and **pancreas** work on maintaining a suitable glucose level in your body.
- The **lungs** are involved in controlling the oxygen and carbon dioxide content of your blood.

The work of these organs is co-ordinated by your brain which must be continually informed of the conditions so that it can direct changes if necessary. The information is fed to it from sensory receptors located all around your body.

For example, if your body temperature rises, temperature receptors in the brain itself are stimulated as the blood passes through it. The brain acts on the information from this and gets all the mechanisms going which will lower your body temperature. The same temperature receptors will tell your brain when the temperature is back to normal. Without this continual feedback of information, homeostasis would be impossible (see figure 1).

What is a suitable body temperature?

In general, animals fall into two groups with respect to body temperature:

- Those which maintain a constant body temperature whatever the outside temperature. These are referred to as **endotherms**.
- Those whose body temperature falls or rises with the outside temperature. These are referred to as **ectotherms**.

The human body behaves as an endotherm. It tries to maintain an internal temperature as near to 37 °C as it can despite large fluctuations in the outside air temperature.

Why 37 °C?

The optimum temperature for your body's enzymes is 37 °C. At temperatures below this they still work, but much more slowly. At 27 °C the enzymes are only working half as fast and this is not good enough to keep you alive.

If your body temperature rises a few degrees above 37 °C, the enzymes speed up which itself causes problems. At 41 °C and above they start to be destroyed by the heat and stop working altogether. This eventually results in death.

What causes hypothermia and hyperthermia?

These are both forms of **heat stress**. **Hypothermia** (exposure) is the condition where your body temperature falls below 30 °C because too little heat is being generated. Many old people in the UK suffer from hypothermia because they live in unheated surroundings. They may not eat enough food to generate enough heat to replace all that is lost in cold living conditions.

Hyperthermia is caused by your body being unable to shed enough heat to stop its temperature rising. The main way this comes about is by not replacing the water (by drinking water) which is lost in sweat. When this happens, sweating stops and you overheat. **Heat exhaustion** is a mild form of heat stress, caused by the failure to replace the salt lost in sweat. This usually results in cramps and dizziness.

Questions

1. Animals are either ectotherms or endotherms. List the advantages and disadvantages of each.
2. With reference to body temperature, explain how homeostasis operates.
3. Explain why if your body temperature falls below 27°C or rises above 41°C and stays there, you will almost certainly die.
4. Design a poster which will make people think more about hypothermia or hyperthermia.

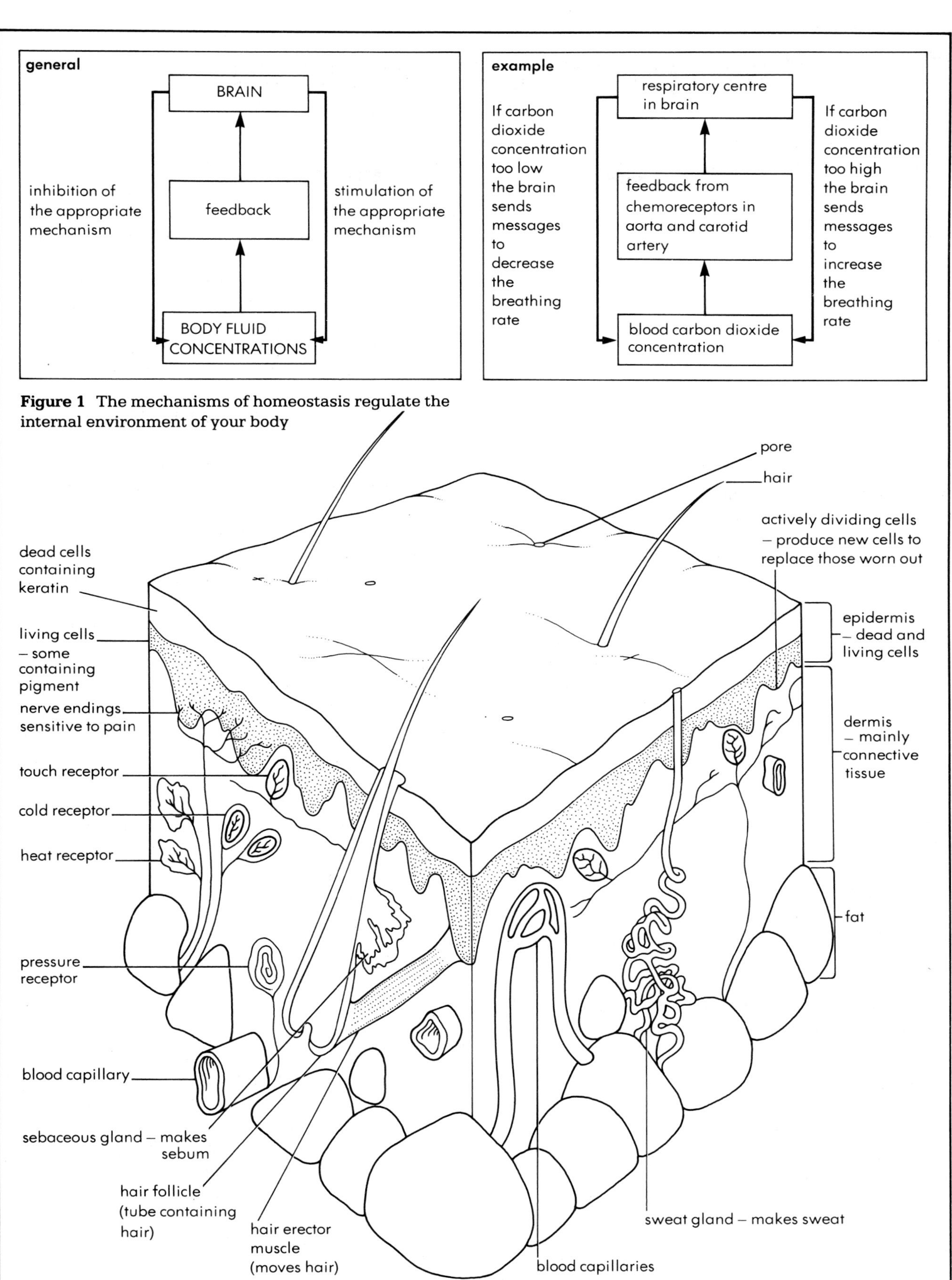

Figure 1 The mechanisms of homeostasis regulate the internal environment of your body

Figure 2 The skin is your body's largest organ. It contains many different structures all with specific functions

7.2 Overheating and overcooling

How do you monitor your body temperature?

Your blood temperature is checked as it passes through the **hypothalamus** of the brain. If it rises or falls then your body tries to make adjustments.

Overheating

The mechanisms involved in preventing overheating are (figure 1):

- The blood vessels in your skin get wider (**dilate**) and more blood is directed through them.
- Your sweat glands make sweat and deposit it on the skin surface. The heat from the blood is used to evaporate this.
- The hair erector muscles relax and your body hairs lie flat against your skin, allowing air currents nearer to it.
- Cellular metabolism slows down so less heat is produced.
- Your breathing rate increases so more heat is lost with expired air.
- Behaviour changes. Clothing is removed and cold drinks are taken.

These mechanisms are only effective between 37 °C and 41 °C. Above 41 °C, they start to fail and your body goes into shock, followed by death.

Overcooling

The mechanisms involved in preventing overcooling are (figure 2):

- The blood vessels in your skin become narrower (**constrict**) and less blood is directed to them.
- Sweat production stops.
- The hair erector muscles contract (creating goose pimples) raising your body hairs. This prevents air currents getting near your skin surface. (In humans this is only really important for preventing heat loss from the head).
- Cellular metabolism increases, producing more heat for the blood to distribute around your body.
- Your skeletal muscles start contracting regularly which causes shivering. This movement generates heat.
- Behaviour changes. Warm clothing is put on and heating is turned up. Hot food and drinks are taken.
- If a person is in a cold environment for many months, fat is deposited under the skin to insulate the body.

These mechanisms start to fail at 29 °C and your body goes into a coma, followed by death.

What other jobs does the skin do?

Your skin is the boundary between the harsh outside world in which you live and the delicate, precise world inside your body. As well as helping to maintain a constant body temperature, it also:

- **protects the tissues** beneath it from injury.
- **stops the entry into your body of poisonous chemicals and germs**.
- **acts as a waterproof barrier** stopping water getting in and if necessary out.
- **protects** the body from the effects of ultraviolet radiation in sunlight.
- **contains sense organs** which help us detect changes in the outside world.
- **makes vitamin D**.
- **excretes excess salt and water**.
- **stores food**.
- **can be modified to form teeth, nails and mammary glands**.
- **makes hair**.

Questions

1 Explain each of the following:
   a) You feel less hungry in warm weather.
   b) Your skin turns pink when you are warm and white when you are cold.
   c) It is easier to keep cool on a hot dry day than it is on a hot humid day.
   d) babies and small children in particular must be well wrapped up in cold weather and given plenty of food.

2 Imagine you accidently get locked in a cold room. Describe the changes which take place in your body over the next hour.

3 You are at an airport in Spain waiting for a flight back to England. You know it will be cold in England so you have dressed in warm clothing. An announcement is made saying that your flight has been delayed for four hours! It is 10 a.m. and already very warm. Describe the changes which take place in your body as you sit out the next few hours.

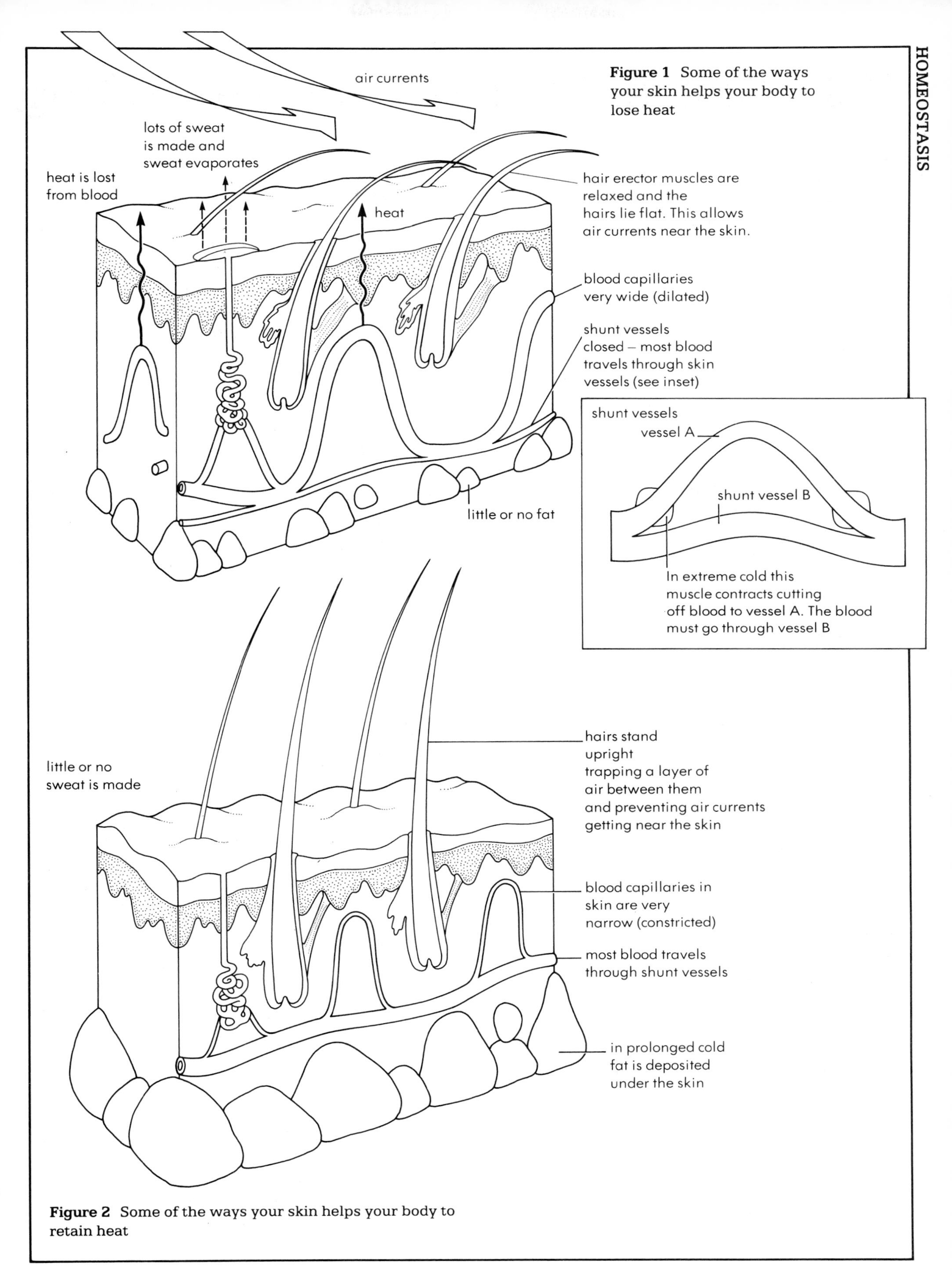

Figure 1 Some of the ways your skin helps your body to lose heat

Figure 2 Some of the ways your skin helps your body to retain heat

7.3 Excretion of waste materials

What is excretion?

Excretion is the way your cells get rid of waste substances made during their normal metabolic processes. Removal of undigested waste from your intestines should not really be called excretion because this waste is not made by the cells. Getting rid of this waste is called **defaecation**.

What sort of substances do we excrete?

Our main body excretions are:

- **Carbon dioxide** produced by the cells during respiration. This is excreted mainly by your lungs.
- **Water** produced during respiration and also extracted from food during digestion. A lot of this water is retained, but the rest is excreted by your kidneys or lost by evaporation from your lungs and skin (in sweat).
- **Urea** made in the liver from surplus amino acids and carbon dioxide. This is excreted by your kidneys.

Excretion by the kidneys

The kidneys are your specialist excretory organs. You have two, which together filter your blood at the rate of 1200 cm^3 every minute, removing waste and harmful substances, together with a lot of water. This is the mixture we call **urine**.

Your kidneys form part of the **urinary system** (figure 1). The other parts are:

- Your **bladder** to store the urine. When this is full it holds about half a litre.
- Two muscular tubes which take the urine from the kidneys to the bladder. These are called the **ureters**.
- A tube through which your bladder can be emptied. This is called the **urethra**.

What exactly does urine contain?

Urine is a pale yellow watery fluid. The table below shows roughly what it is made up of in a healthy body:

Waste substance	Percent by volume
Water	96%
Urea (nitrogenous waste)	2%
Salt	1%
Other waste substances	1%

How much you have of certain substances in your urine is a good indication of the state of health of your body. Urine tests are often used to diagnose disorders. For example, if the urine contains glucose, this could mean the person has diabetes, and protein in the urine could be an indication that the kidneys are failing.

Urine can even be used to confirm a pregnancy. A fertilised egg releases a hormone which the mother excretes in her urine. This can usually be detected from about 14 days after fertilisation and is the basis of most pregnancy tests.

How much urine is made every day?

In the average adult the kidneys filter about 1500 litres of blood every day. From this, they produce about 1.5 litres of urine, but this does depend on how much fluid you drink and how much you sweat etc.

Questions

1 Copy out and complete the table below:

Excretion	How produced	Organs which excrete it
Carbon dioxide Water Urea		

2 With the help of a diagram, describe the structure of the urinary system.

3 Under each of the labels you have put on your diagram, write one sentence to describe the function of that part.

4 A friend of yours has just started work in a pathology laboratory. Today he has been testing urine samples and found one of them to contain glucose and another to contain protein. Explain to him what these are signs of. Write your explanation down.

5 Try and explain why in the morning your urine is usually very yellow compared to later on during the day.

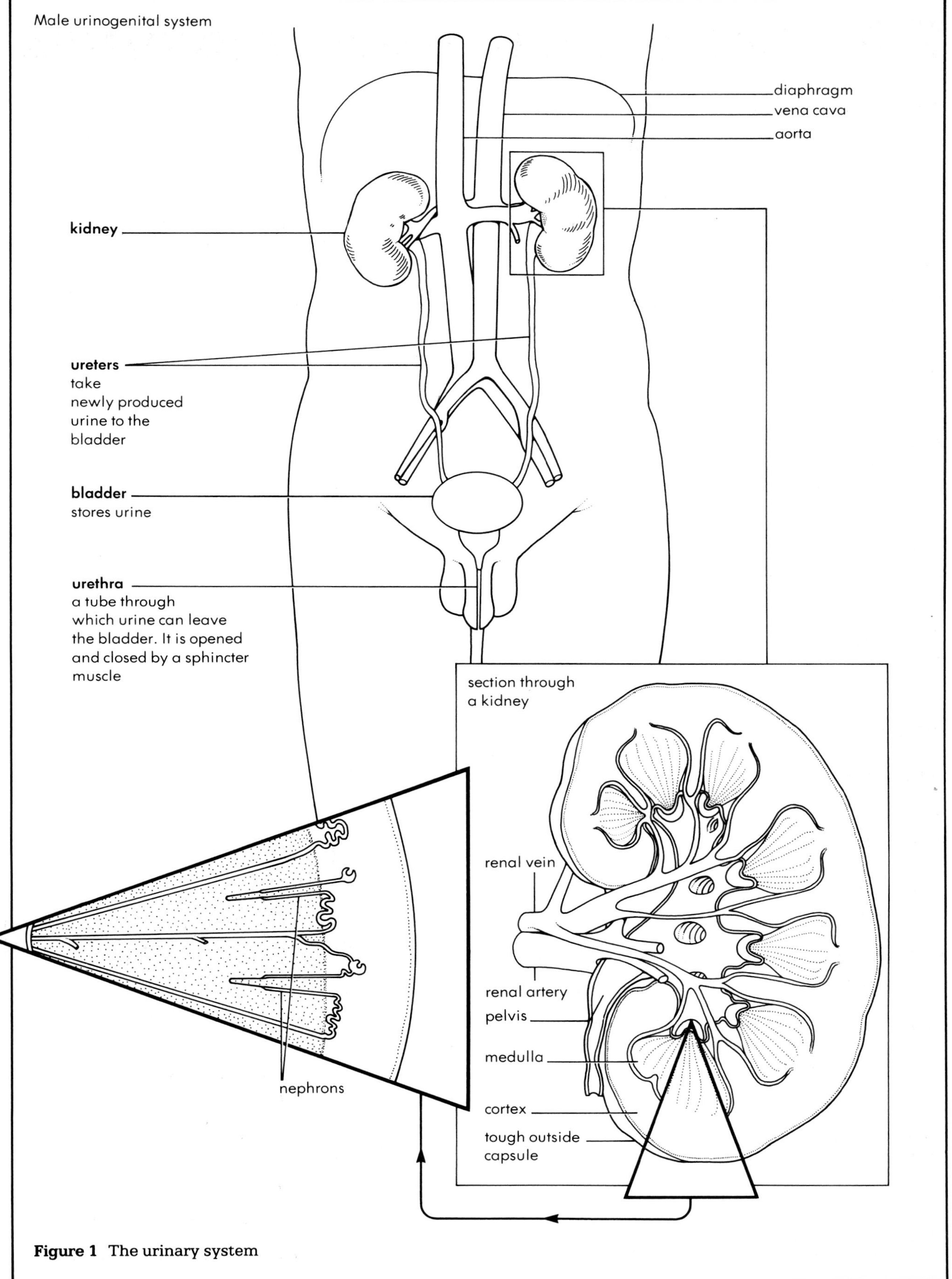

Figure 1 The urinary system

7.4 Excretion and osmoregulation

What happens inside the kidneys?

Each of your kidneys contains over a million tiny units called **nephrons**. Each nephron consists of a cup-like end capsule, called the **Bowman's capsule**. This is connected to a narrow tube which twists and turns and leads to a collecting duct (figure 1). Urine is produced in the nephrons by two processes:

- **Ultrafiltration** Blood entering the **glomerulus** finds itself suddenly having to enter a narrower blood vessel. This raises the pressure inside the vessel to such an extent that some of the liquid in the blood passes out between the cells of the vessel wall. This **filtrate**, as it is called, drains into the **Bowman's capsule**.
- **Selective reabsorption** As the filtrate passes along the nephron, some of the more valuable substances are reabsorbed by the blood. By the time it arrives at the **pelvis**, the filtrate has the composition of urine.

Osmoregulation

The kidney is also responsible for regulating the amount of water in the blood. For example, if you drink too much liquid, your blood is diluted, lowering its **osmotic potential**, i.e. its concentration (see page 12). This change is detected as the blood travels through the **hypothalamus** and a message is sent to your kidneys. This message tells the nephrons to stop reabsorbing so much water. This results in the production of more dilute urine, but more importantly it reduces the amount of water in the blood and thereby raises its osmotic potential back to an acceptable value.

If your blood is too concentrated the message sent to the kidneys tells them to increase the amount of water reabsorbed, thereby increasing the blood water content and lowering its osmotic potential. Full details are shown in figure 3.

The kidneys also maintain the correct **blood sodium level** and **pH**.

What happens if your kidneys stop working?

If only one kidney fails, the other can usually cope quite well with the extra burden. If both kidneys stop working the waste substances they normally filter out of your blood start to build up and if nothing is quickly done about this, you will die.

What does a kidney machine do?

Fortunately, people whose kidneys have both failed to work can be connected to a kidney machine. This machine will remove the waste substances from their blood and adjust its salt and water content just as your kidneys would. It does this by a process called **dialysis**. In dialysis blood is removed from an artery, usually in the arm, and passed over a sheet of cellophane. On the other side of the cellophane is a watery solution called **dialysing fluid**. This fluid contains the normal concentrations of glucose and salts that are found in your body. As the blood passes over the cellophane, waste substances diffuse out of it and into the dialysing fluid. The larger components of the blood such as the red and white blood cells are held back. The blood is eventually returned to the body through a vein in the arm or leg. This is continued until the blood has a normal acceptable concentration which could take up to 12 hours, and must be done two or three times a week. This is an awful discomfort for a person to go through every few days.

Some people are able to have kidney transplants, where they receive a kidney which belonged to another person. But these are difficult and costly to arrange – the kidneys have to be perfectly matched (see page 143).

Questions

1 Explain fully what the kidneys do in homeostasis.

2 a) Draw a typical nephron and label the Bowman's capsule, glomerulus and collecting duct.

b) Under each of these labels write a sentence saying what happens in the structure.

c) Mark on your diagram the place where ultrafiltration takes place and two places where reabsorption occurs. Explain how the nephron is involved in the production of urine.

3 Kidney machines are expensive to buy and cost on average £10,000 per year to run. Not surprisingly there are not enough of them to treat all the patients. Some doctors think the solution is to buy kidneys from healthy people for £5,000. How do you think the problem should be solved?

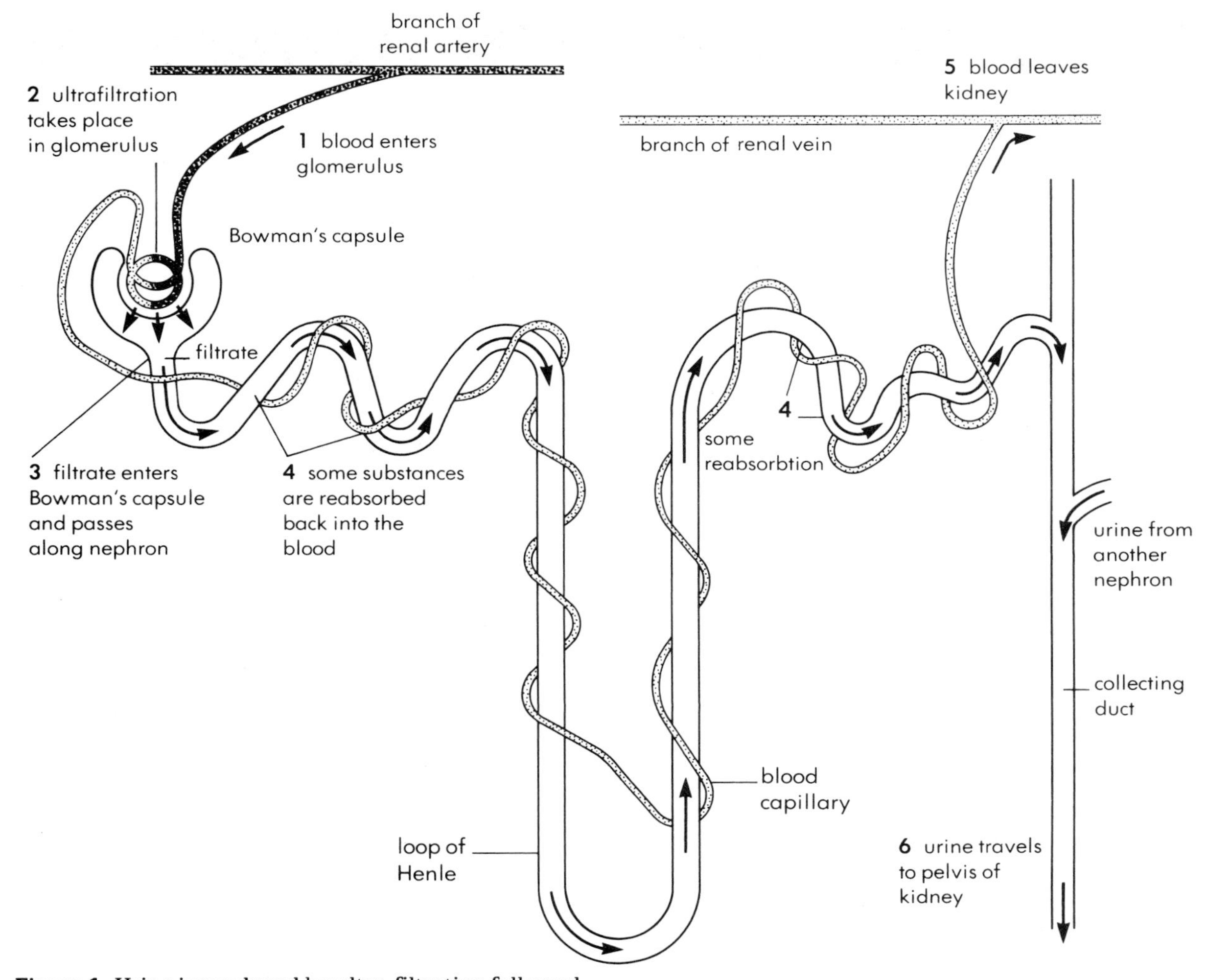

Figure 1 Urine is produced by ultra-filtration followed by selective reabsorption

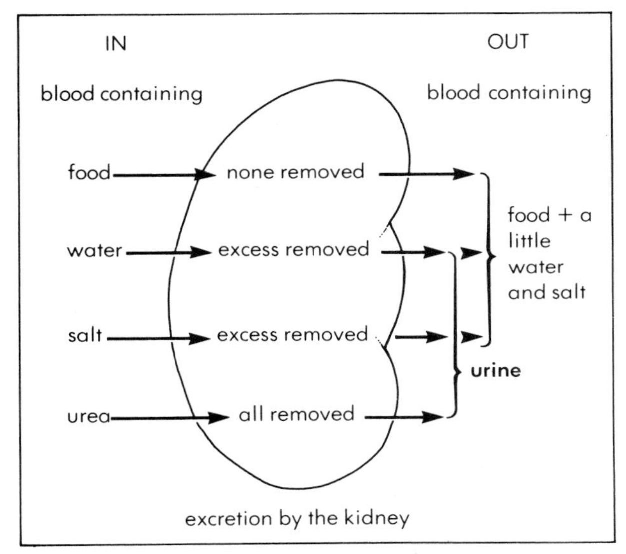

Figure 2 The composition of blood entering the kidneys is different to that of the blood leaving it. What is removed?

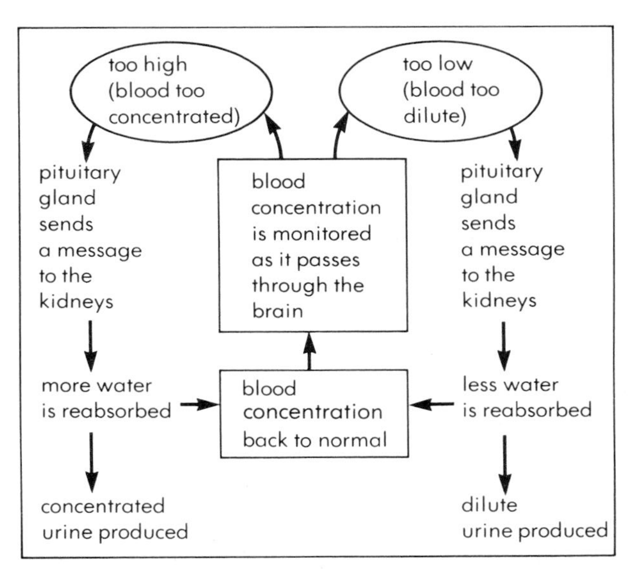

Figure 3 The kidney maintains the correct blood water content

7.5 Blood sugar regulation

What is the blood sugar level?

Glucose is your cells' main source of energy and therefore it must always be available to them. So your body keeps a regular amount of glucose in your blood. This is called the **blood sugar level**.

The blood sugar level is usually maintained at about 1 mg of glucose per cm^3 of blood. The main organs responsible for making sure this level is kept up are the liver and the pancreas, although hormones produced by several other endocrine glands (see page 96) can also alter the blood sugar level.

The work of the liver and pancreas

The blood sugar level, i.e. the amount of glucose in your blood, is checked as your blood passes through the pancreas. If it rises, as happens every time you eat a meal, the pancreas secretes the hormone **insulin** into the blood. Insulin increases the permeability of cell membranes to glucose and activates several enzyme systems inside the cells which can deal with the glucose. One system is in the liver cells where glucose is converted into the molecule glycogen and stored there. Together the liver and pancreas therefore reduce the blood sugar level (figure 1).

If the blood sugar level falls too low, the pancreas stops producing insulin and produces another hormone called glucagon. **Glucagon** stimulates the reconversion of the stores of glycogen in the liver back into glucose. This enters the blood bringing its sugar level back to normal. If this is not enough then other hormones convert fatty acids, glycerol and amino acids into glucose. The extra demands for glucose during exercise are met by the production of **adrenalin** by the adrenal glands (see page 96).

What is diabetes?

Diabetes is the condition where there is a very high level of sugar (glucose) in the blood. This usually results from a failure of the pancreas to produce insulin. Without insulin the body's cells cannot get at enough of the glucose in the blood and therefore die from lack of energy. The brain cells are particularly badly affected because glucose is their only source of energy. Other cells can at least use fat.

The symptoms of diabetes are itchiness of the skin, weight loss, tiredness, irritability and extra visits to the toilet to empty the bladder. When the blood sugar level reaches such a level that the kidneys cannot reabsorb it all, glucose starts to appear in the urine. This is how diabetes is usually detected.

A high blood sugar level is often called **hyperglyceamia**. If it is not treated immediately you will lose consciousness and die.

How is hyperglyceamia treated?

The blood sugar level can be lowered by injecting insulin directly into a vein (an **intravenous** injection). Insulin cannot be taken orally because it is a protein and will be digested. Diabetics usually have to learn how to give themselves regular injections of insulin.

Occasionally, after an insulin injection the blood sugar level falls too much, a condition known as **hypoglyceamia**. Diabetics quickly learn to recognise the symptoms of this; shakiness, excessive sweating, faintness and mental confusion. The solution is to eat some sugar. If this is not done you may lose consciousness and die. Diabetics have to maintain a careful balance between what and when they eat and the insulin they inject.

Questions

1. Why must there always be some glucose in the blood? What is the acceptable level?
2. The blood sugar level rises immediately after a meal. Explain how it is brought back to an acceptable level.
3. Diabetics often have a very high blood sugar level, yet their cells still die from lack of energy. Explain why.
4. A diabetic suddenly felt faint, sweaty and confused. Outline the probable sequence of events just prior to this and suggest what he should do about it.
5. Describe the other ways in which the liver is involved in homeostasis. (See figure 2.)

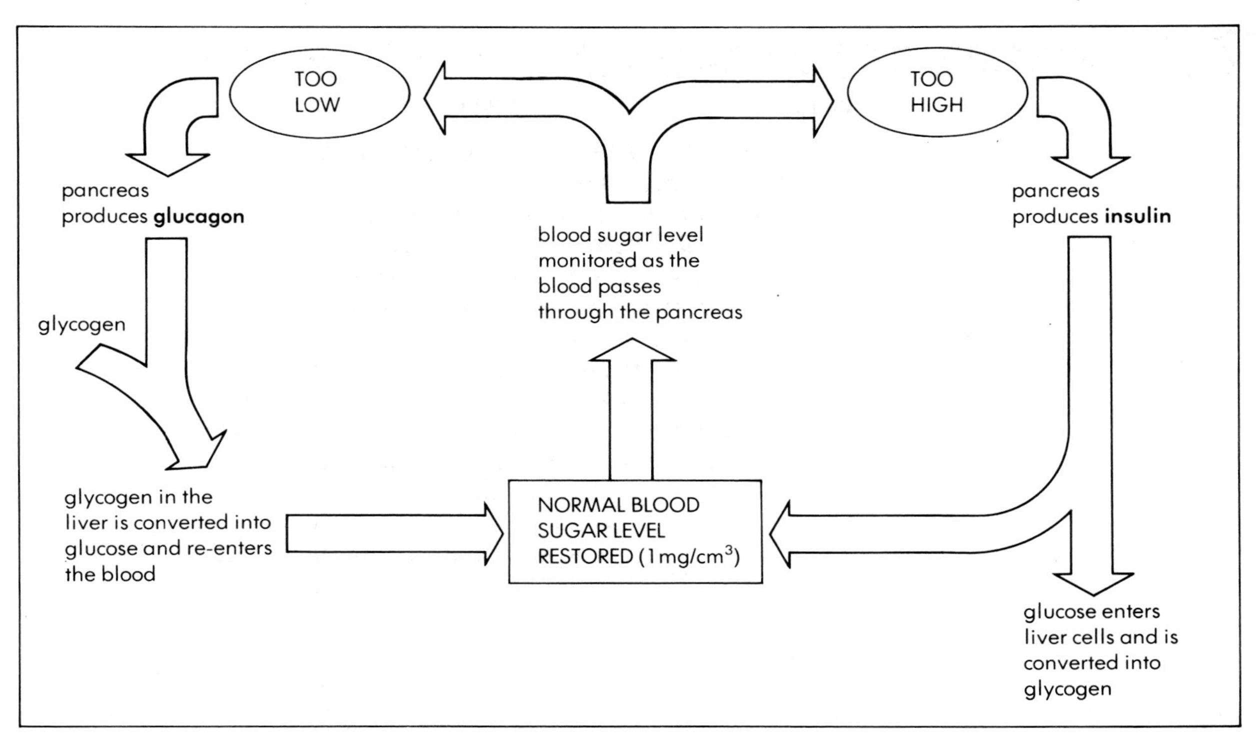

Figure 1 Blood sugar regulation is an example of homeostasis

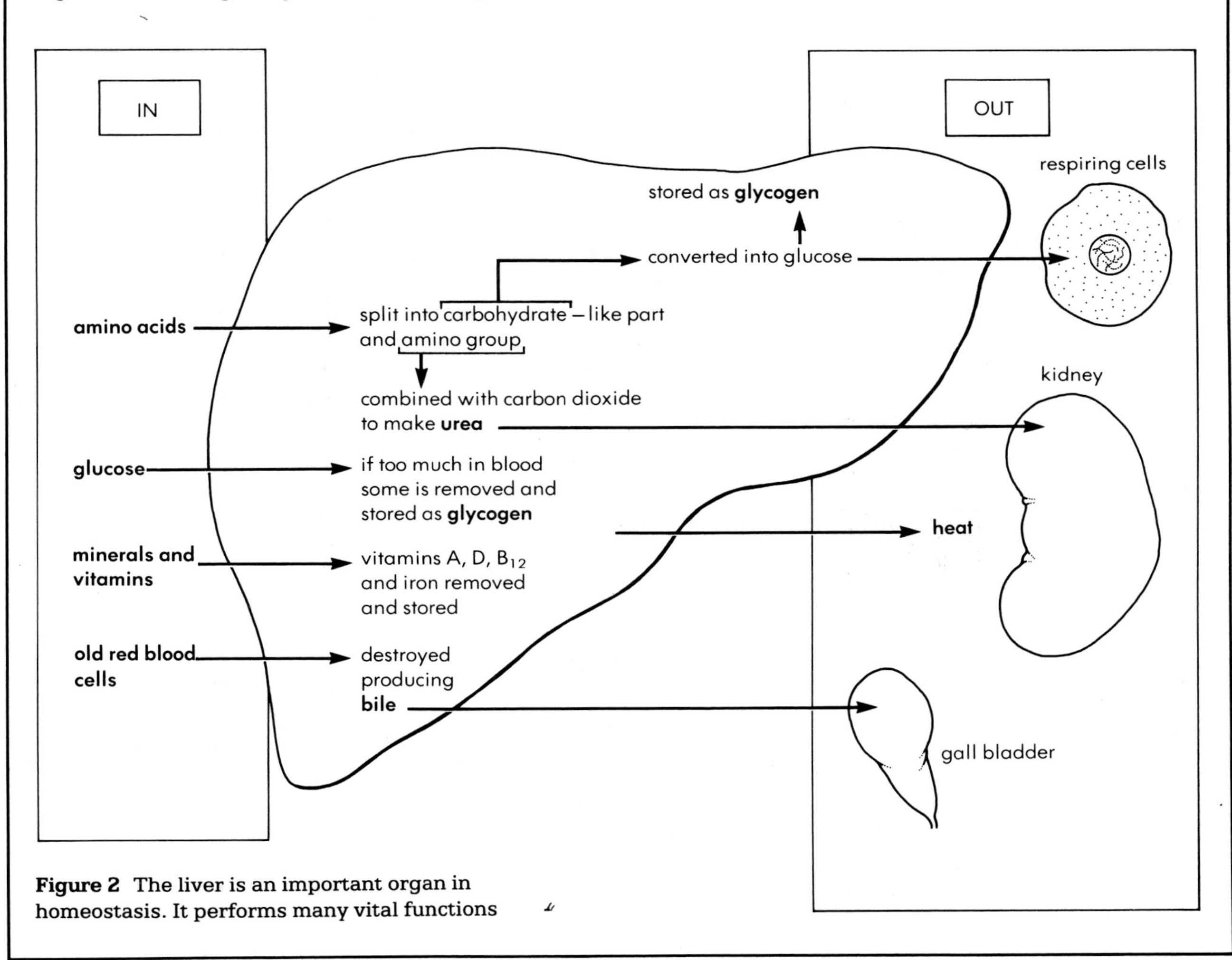

Figure 2 The liver is an important organ in homeostasis. It performs many vital functions

Questions

Recall and understanding

1 The immediate environment around most of your cells is:
A blood C tissue fluid
B water D plasma

2 Which of the following organs is not involved in homeostasis?
A brain C liver
B kidneys D stomach

3 The optimum temperature for most of your body enzymes is:
A 40 °C C 20 °C
B 37 °C D 100 °C

4 Hyperthermia is the condition when your body:
A overheats C dehydrates
B overcools D has too much sugar

5 Deamination of amino acids results in the production of:
A heat C ammonia
B sugar D urea

6 Osmoregulation is the maintenance of a suitable blood:
A pH C osmotic potential
B temperature D sugar level

7 The antagonistic hormone to glucagon is:
A adrenalin C cortisone
B insulin D testosterone

8 Which of the following is not an excretion?
A urea C carbon dioxide
B faecal matter D water

9 A kidney machine works on the principle of:
A dialysis C chromatography
B electrophoresis D centrifugation

10 The presence of protein in your urine is an indication of:
A pregnancy C kidney failure
B PKU D diabetes

Application and interpretation of data

11 A class of 24 pupils recorded their body temperatures. The results are below:

36.8	37.0	37.3	36.6
37.2	36.9	37.1	36.4
37.0	36.9	36.6	37.0
36.5	36.5	37.0	37.2
36.8	36.8	36.7	36.9
37.1	37.2	37.1	36.8

a) Use the results to fill in the table below:

Temperature range/°C	36 – 36.2	36.3 – 36.5	36.6 – 36.8	36.9 – 37.1	37.2– 37.4
Numbers					

b) Which is the most frequent range of temperatures?
c) Calculate the average temperature.
d) Plot the results on a graph.
e) The table below shows the external temperature and energy requirements of some adult men.

External temp/°C	Energy required/kJ
−20	16000
0	13500
20	12000
40	11000

i) What is the relationship?
ii) Explain why.

12 The graph below shows the blood glucose and insulin levels before and after a meal.

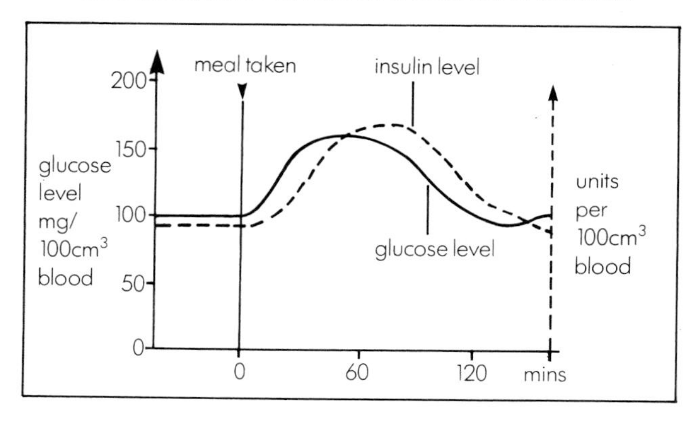

a) What is the normal blood sugar level?
b) What happens to the blood glucose level after the meal?
c) What is the maximum blood sugar level reached?
d) Why does the blood sugar level start to fall after 60 minutes?
e) How does insulin work?
f) During exercise the blood sugar level rises. Which hormone is responsible for this?

13 The table below compares the composition of blood and urine.

	Blood (%)	Urine (%)
Water	90	96
Protein	9	0
Glucose	0.1	0
Urea	0.03	2
Uric acid	0.003	0.05
Creatinine	0.001	0.1
Chloride	0.37	0.6
Sodium	0.35	0.35 → 0.6
Potassium	0.02	0.15

a) From the table write a list of the substances that are excreted by the kidneys.
b) Which substances are more concentrated in the urine than the blood?
c) If glucose appears in the urine what does it indicate?
d) The amount of urine produced before and after exercise was measured.

Time (min)	Urine produced (cm^3)
0	60
30	50
exercise for 30 mins → 60	60
90	5
120	8
150	29
180	45

i) How does exercise affect urine production?
ii) Explain why.

14 The graph shows skin temperature at various distances along a leg.

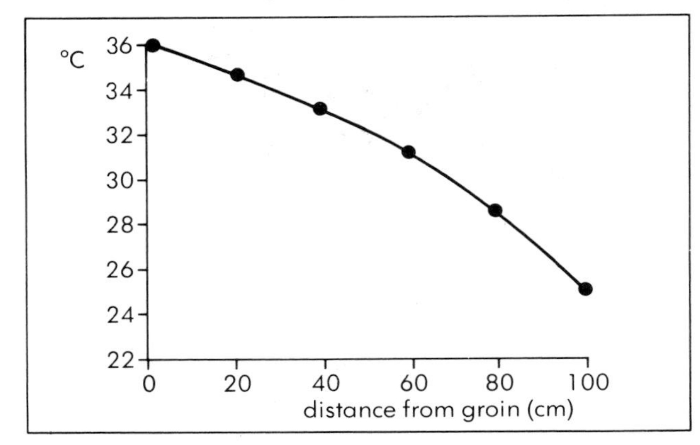

a) What does the graph show?
b) Explain why the skin temperature falls steadily down your leg.
c) What is the overall drop in temperature?
d) Where is your body temperature usually measured?
e) What is the instrument used to measure it called? Describe how you would use this.

15 The table below shows the amount of sweat lost by various people in different climates.

	Sweat lost per day (cm^3)
Office worker – cool day	400
Office worker – hot dry day	1200
Coal miner	4000

a) How does the climate affect the amount of sweat produced and lost? Explain this.
b) If the climatic conditions were hot and humid instead of hot and dry, how would this affect sweat loss?
c) Coalminers have to drink lots of liquid and take salt tablets. Explain why and what will happen if they do not.
d) It takes 2.5kJ of energy to evaporate $1 cm^3$ of sweat. How much energy does the coal miner lose through sweating every day?

CHAPTER 8: *REPRODUCTION*

One of the most wonderful moments in a couple's life is the arrival of a new baby. In this chapter you will learn about the events leading up to this, from the production of the sex cells through to the birth itself. In addition, the changes which take place in the mother's body and how her pregnancy is monitored are covered.

It has taken about 40 weeks for this new-born baby to reach this stage.

Infertile couples helped to have babies

Unfortunately not all couples of child bearing age are able to produce a baby. For many this is the cause of great emotional stress.

What are the causes of infertility?

The causes usually fall into four categories:

- anatomical problems,
- hormonal problems,
- infection,
- psychological reasons.

The table on the opposite page shows some of the problems that can occur.

Problem	Reason for problem
Female	
Blocked fallopian tubes	Development failure Infection Fibroids
No ovum produced	Hormone imbalance Ovarian cyst Emotional stress
Uterus not receptive	Hormone inadequacy Infection
Cervical mucus hostile to sperms	Hormone imbalance Infection
Male	
No sperm or low sperm count	Hormone imbalance Anatomical problem Infection
Sperm abnormal	Hormone inadequacy Anatomical problem
Blocked sperm ducts	Development failure Infection
Failure to deliver sperms	Emotional stress Anatomical problem

How can we help these couples?

Before they can be helped, the doctors need to know the nature of the problem. The couple are referred to a clinic where tests are carried out.

The first test is always on the man as 40% of all problems are due to sperm deficiencies. The doctors are looking for a sperm count of at least 20 million per cm^3 of semen of which 60% of the sperms are normal. If this proves acceptable then tests are carried out on the woman.

Diagnosing the problem may take some time but once it is known, treatment can begin. The following treatments may be recommended.

- **Surgery** to clear a blocked fallopian tube or remove an ovarian cyst.
- **Drugs** to clear up an infection.
- **Hormone treatment** to correct an inadequacy/imbalance.
- **Artificial insemination** to get past the cervix.
- ***In vitro* fertilisation** followed by implantation.

What is artificial insemination?

Artificial insemination involves placing healthy sperm into the woman's uterus at the time of ovulation. The chances are that one of these sperm will fertilise the ovum. Most women become pregnant within three months of starting the procedure.

Sperm from sperm banks

The sperm used in artificial insemination can either be donated by the woman's sexual partner or by any other male. Sperm banks have now been set up where sperm can be stored for up to 3 years. This of course opens up the possibility of selective breeding (page 171) as women can choose to have a baby by any man, provided he has donated a sperm sample to the bank.

What is in vitro fertilisation?

When all other possible solutions have failed both the ovum and sperm can be removed and fertilisation performed in a suitable receptacle (a test tube is usually quoted but in fact these are not used). The resulting embryo can be kept alive and in a very healthy state for a few days giving the doctors a chance to reimplant it into the woman's receptive uterus.

This has now been done successfully on many occasions and produced many normal healthy babies, the so called test tube babies.

Substituting genes

Many diseases are known to be caused by a single defective gene which is passed on from parents to their offspring. In vitro fertilisation opens up the possibility of removing such a gene and replacing it with a new one before the ovum is fertilised. This could solve the problem of inherited disorders such as phenylketonuria and cystic fibrosis. It may even eventually help get rid of these diseases altogether.

Gene therapy like this could also help couples who until now have chosen not to have children because of the possibilities of bad genes coming together. It must be said, however, that there are very strict regulations governing the practice.

When are fertility drugs used?

Fertility drugs are used when the woman is not releasing ova from her ovaries due to a hormone problem. These drugs, many of which are hormones themselves, correct the problem, resulting in ovulation.

Occasionally the drug treatment may result in a multiple birth, but this is rare.

8.1 Kinds of reproduction

Do all animals reproduce in the same way?

There are two forms that reproduction can take (figure 1).

- **Asexual reproduction**. This is reproduction involving only one **parent** organism. The exact **characteristics** of this parent are passed onto its offspring through the **genes**. Even harmful genes are passed on. The offspring is identical in every respect to the parent. It is no better adapted to cope with new situations. For example, a virulent disease could wipe out the whole population because they all have the same low resistance to it.

  Many organisms reproduce asexually when conditions are favourable (lots of food) and build up their numbers rapidly. The precise details of this type of reproduction vary from organism to organism, but all variations involve simple cell division called **mitosis** (see pages 7 and 173).
- **Sexual reproduction**. This requires two organisms of the same species, one male and one female. Each produces a special sex cell called a **gamete** which takes part in the process. Production of these gametes involves a different kind of cell division called **meiosis** (see page 173). It is in these gametes that the parents' characteristics are passed on to the offspring. The details about these characteristics are stored in genes on the **chromosomes** which are in the nucleus of the gamete.

  During the process of sexual reproduction, the nuclei in the respective gametes fuse together and the genes become mixed. This results in the offspring having some characteristics from the male parent and some from the female parent. They will therefore, be slightly different from each parent. For example, some might have a higher resistance to disease so when a virulent disease affects the population, these will survive and go on to produce offspring of their own. Humans only reproduce sexually as do all mammals, fish, birds and many other organisms.

The human reproductive system

The parts of the human body concerned with reproduction form the **reproductive system**. This system is different in the male and female, because they have different functions.

What forms the male reproductive system?

The male reproductive system has three main functions and so its structure is organised to do these (figure 2):

- It needs to produce the male gametes (**sperm**). These are produced in the **testes**.
- It needs to deposit the sperm inside the woman's reproductive system. This is the role of the **penis**.
- It needs to produce the male hormone **testosterone**. This again is produced by the testes.

Additional structures supply nutrients for the sperm to use after it has been released inside the woman.

What forms the female reproductive system?

The female reproductive system has four functions and so its structure is organised to carry these out (figure 3):

- It needs to produce the female gamete (**ovum**). The **ovaries** do this.
- It needs to provide a safe home for the developing baby and source of food. This is provided by the **uterus**.
- It needs to produce the hormones **oestrogen** and **progesterone**. This again is the role of the ovaries.
- It needs to provide food for the newly-born baby. This is produced by the **mammary glands** (breasts).

Questions

1. Draw up a table of differences and similarities between asexual and sexual reproduction.
2. Explain why asexual reproduction has little survival value.
3. Draw a diagram of the male reproductive system. Annotate the parts with their functions.
4. Draw a diagram of the female reproductive system. Annotate the parts with their functions.

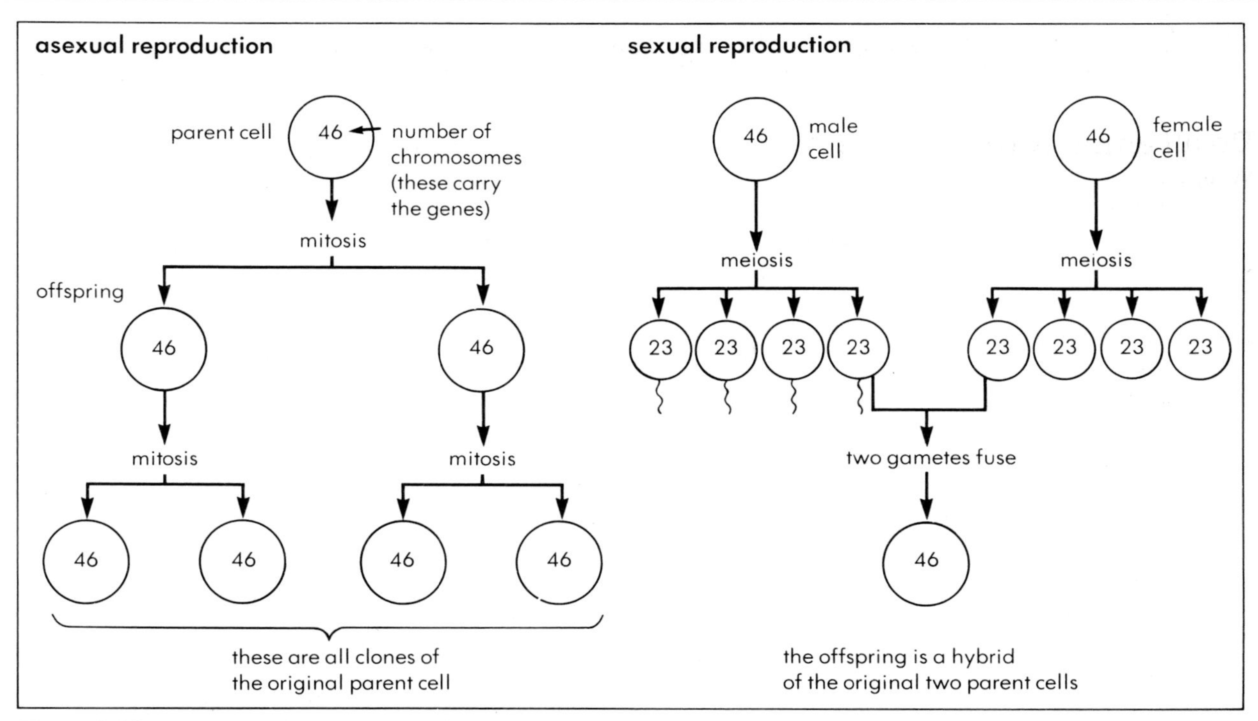

Figure 1 There are two ways a cell can divide

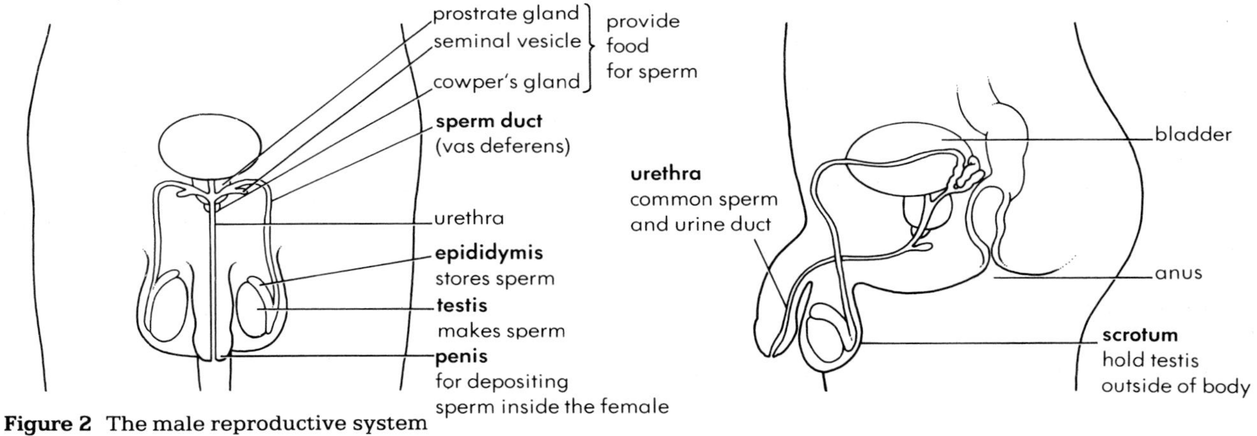

Figure 2 The male reproductive system

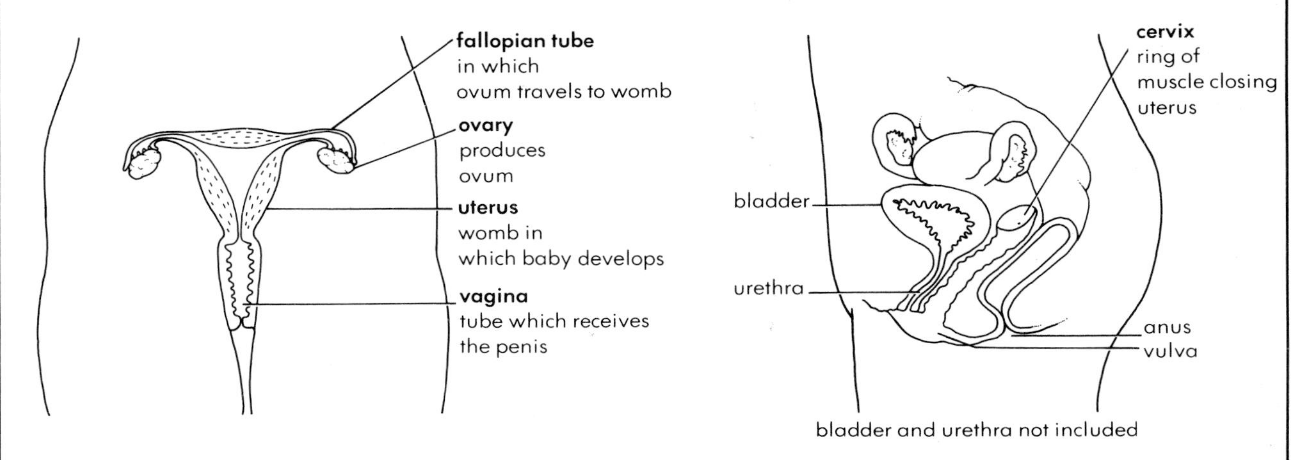

Figure 3 The female reproductive system

8.2 Getting the gametes together

What produces the gametes?

The male gamete is the **sperm cell**. In humans this is produced inside tiny tubes within the testes (figure 1). The production of sperm starts in boys at puberty, usually around the age of twelve to fourteen. Men are able to continue producing sperms until their death. To start with, the freshly produced sperms are stored in the **epididymis** in preparation for their release. If after a short period they have not been released, they are destroyed to make room for fresh sperms.

The female gamete is the **ovum**. It is produced by the ovary (figure 2). When a girl is born, each ovary contains all (and more) of the ova it will ever produce, but these are in an immature state. When a girl starts her periods (**menstruation**), around the age of eleven to fourteen, the ova start to mature. Under the direction of hormones produced by the pituitary gland, one of these ova ripens about every four weeks and this continues from then until she reaches menopause at the age of about 45 to 50 (see page 152). So in between the beginning of menustration and the end of the menopause, a woman is able to become pregnant.

How do the gametes get together?

When an ovum is mature enough, it is released from the ovary – a process called **ovulation**. It is then pushed by cilia along the oviduct towards the uterus, a journey which may take seven days. If the ovum meets a sperm within the first day or so, it can be **fertilised** – that is the nuclei of each gamete fuse together. Otherwise it disintegrates.

What happens during sexual intercourse?

Sperm will only be present in the oviduct if the female has recently had **sexual intercourse**. It is during this act that the man deposits sperm inside the vagina of a woman. This deposition occurs at the climax of intercourse and is called **ejaculation**. At the start of intercourse the man's penis becomes filled with blood, making it hard and erect and thereby easier to insert into the woman's vagina.

During ejaculation the sperm also take with them some liquids made by the **prostate gland**, **Cowper's gland** and **seminal vesicles**. These liquids supply the sperms with food from which they can produce the energy they need in order to swim. The mixture of sperms and nutritive liquids is often referred to as **semen**. About 3 cm^3 of semen is released during ejaculation of which only 10 per cent is sperm. Even so, this contains several hundred million individual sperm cells!

Once inside the woman's vagina the sperms have to make their way to the oviduct – an enormous journey for them. Only a few of the very healthiest make it.

What happens when the sperm and ovum meet?

If they are 'young' enough, when a sperm and ovum meet, fertilisation will take place (figure 3). This is sometimes called **conception**. This can only happen during a certain time in the woman's **menstrual cycle** (see page 150). The fertilised ovum is called a **zygote** and its nucleus contains all the instructions needed to form a new life.

As the zygote continues its passage along the oviduct, it divides several times and by the time it reaches the uterus, it consists of about one hundred cells. This ball of cells is called an **embryo**. The embryo now sinks into the soft uterine lining – a process called **implantation** (figure 4). We say that the woman is now pregnant. From this moment on the embryo will grow by receiving food from its mother's blood. To aid this transfer of food, a special organ called the **placenta** develops. This takes about three weeks.

Questions

1 a) Draw and label the male and female gametes.
b) List five differences and three similarities between the male and female gametes.
c) Where are the gametes produced?
d) What kind of cell division is involved in the production of gametes?

2 What is semen? Explain how and where it is deposited in the female.

3 Explain why only one sperm cell can fertilise an ovum.

4 Why is implantation so critical to the continued development of the embryo?

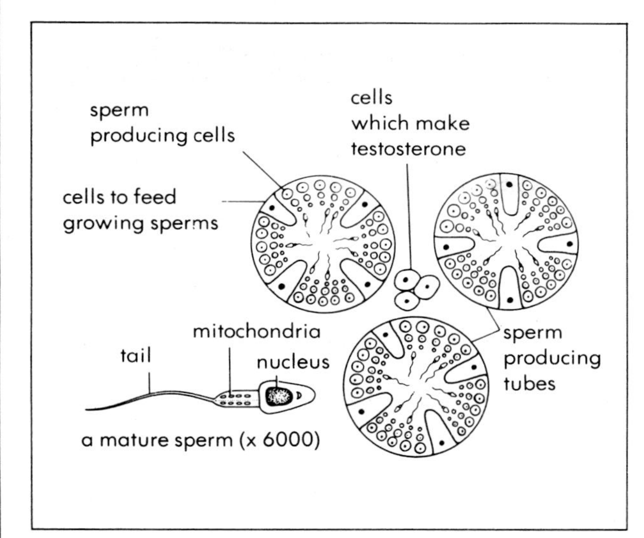

Figure 1 Sperms are produced in the seminiferous tubules within the testes

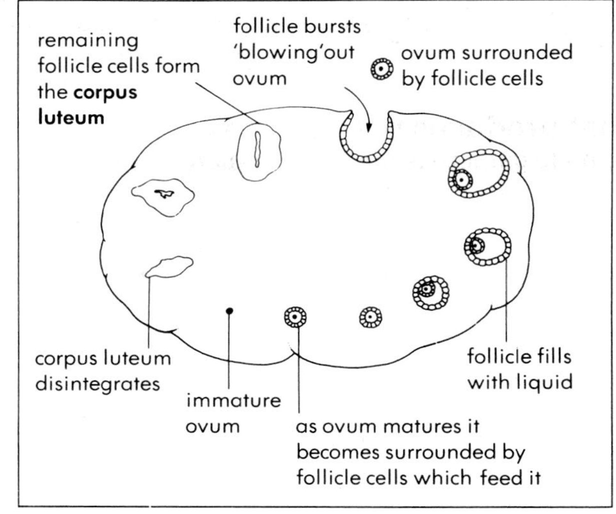

Figure 2 An ovum matures in the ovary

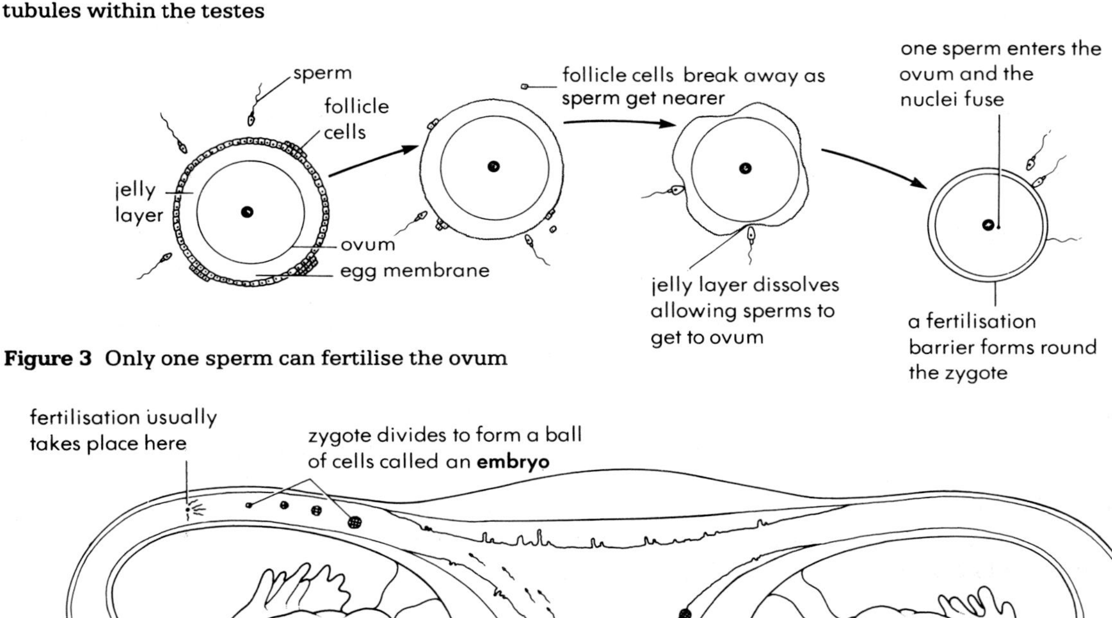

Figure 3 Only one sperm can fertilise the ovum

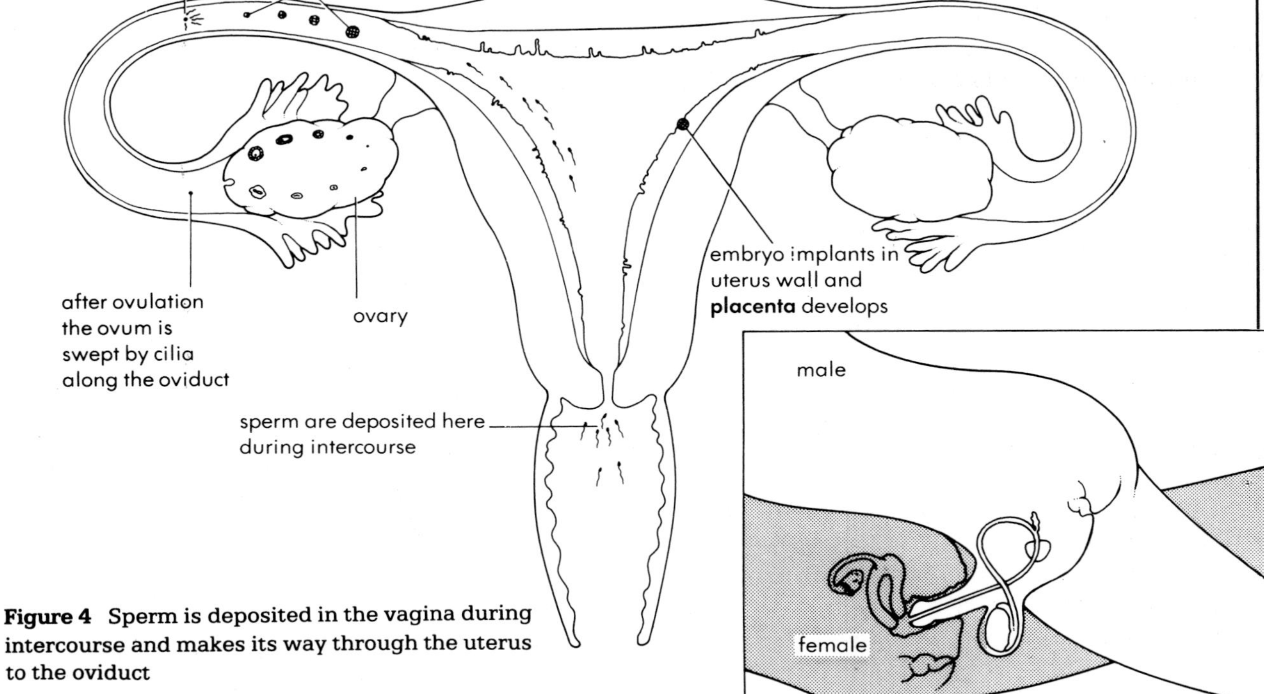

Figure 4 Sperm is deposited in the vagina during intercourse and makes its way through the uterus to the oviduct

8.3 Development of the foetus

What is the placenta?

The placenta is a very important structure. It is necessary for the growth of a healthy baby while it is in the mother's uterus. The placenta develops from some of the cells of the embryo. It starts as a few finger-like projections which penetrate the uterus lining and develops in three weeks into a complex organ. The functions of the placenta are:

- To allow the passage of food substances and oxygen from the mother's blood into the embryo's blood.
- To allow the passage of waste substances including carbon dioxide from the embryo's blood into the mother's blood so that she can excrete them.
- To take over the production of **oestrogen** and **progesterone**, both important hormones during pregnancy (see page 150).

The placenta does the first two of these without the mother's and embryo's blood actually mixing. This prevents many potentially harmful substances in the mother's blood from entering the embryo. A simplified version of the placenta along with the additional structures that develop to house and protect the embryo are shown in figure 1.

What does the placenta prevent from entering the embryo?

The placenta allows food substances, oxygen, waste substances, some alcohol, some drugs (e.g. thalidomide), some germs and some of the mother's antibodies to cross it. It prevents many more drugs and germs which may be in the mother's blood from crossing. The section on antenatal care on page 138 contains more information about this.

Growth and development of the embryo

It takes 38 weeks (from conception) for a zygote to become a baby. This is known as the **gestation** period. Considering the zygote is only one cell about the size of a full stop when first formed, and ends up as a fully formed baby consisting of about 30 million million cells, its growth and development has to proceed very rapidly.

After only four weeks in the uterus it is already 4 mm long and has a beating heart. After five weeks it has grown to 12 mm and has the beginnings of its arms and legs. By the eighth week it has a recognisable human form and is now known as a **foetus** (figure 1). It is now 46 mm long from its head to its bottom. One week later, it is possible to tell whether it is going to be a boy or a girl.

Growth and development of the foetus

After twelve weeks, the foetus is 92 mm long and has all its internal organs. Two weeks later its kidneys actually start working. The foetus is now about 120 mm long (figure 2).

The first kicking movements are usually felt by the mother after sixteen weeks. After twenty weeks the foetus is fully formed, even having eyebrows, fingernails and fingerprints. It is now about 185 mm long and weighs approximately 700 g (figure 3).

By 26 weeks the foetus has grown to about 250 mm (1500 g) and has a reasonable chance of surviving if born. From the 35th week of pregnancy the foetus takes on 14 g of fat a day. Birth usually takes place soon after the 36th week from conception by which time the foetus is 340 mm long and weighs 3200 g. The average full term baby is 360 mm from its head to its bottom and weighs 3400 g.

During pregnancy the womb enlarges from about 5 cm^3 to 7000 cm^3. To make room for this, all the other organs of the abdomen are squashed upwards. So you can understand that the later stages of pregnancy can be quite uncomfortable for women (see page 137).

Questions

1. With the help of a diagram, explain what the placenta is and why it is so important to the continued development of the embryo.
2. How is the embryo connected to the placenta?
3. Copy and complete the following table using the information in the text above.

Age of embryo/foetus (weeks)	Length (mm)	State of development

4. Draw a line graph of the age of embryo/foetus against its size.
5. Calculate the average rate of growth over the 38 weeks for both length and weight.

Figure 1 The structures which help feed and protect the developing foetus

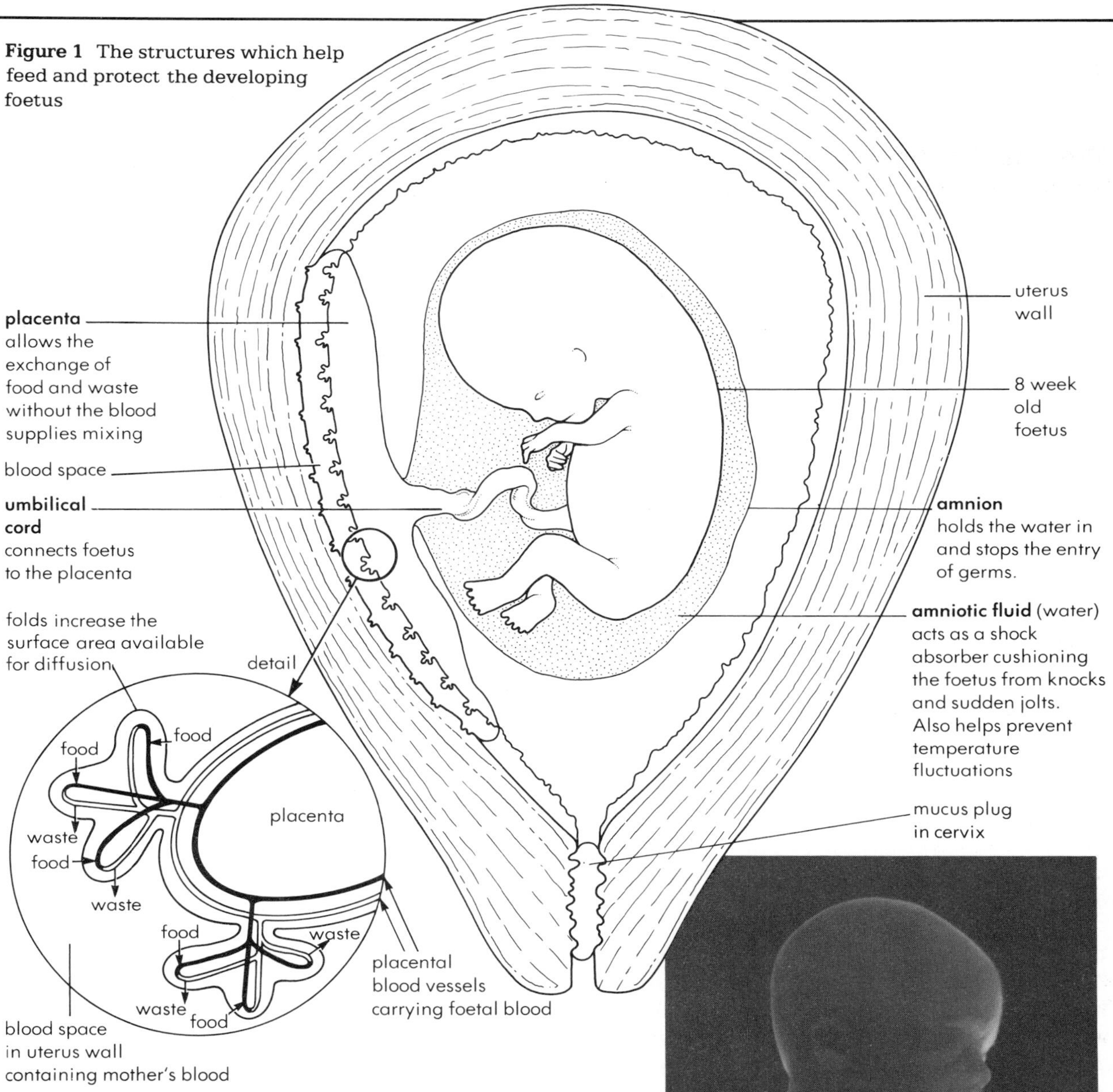

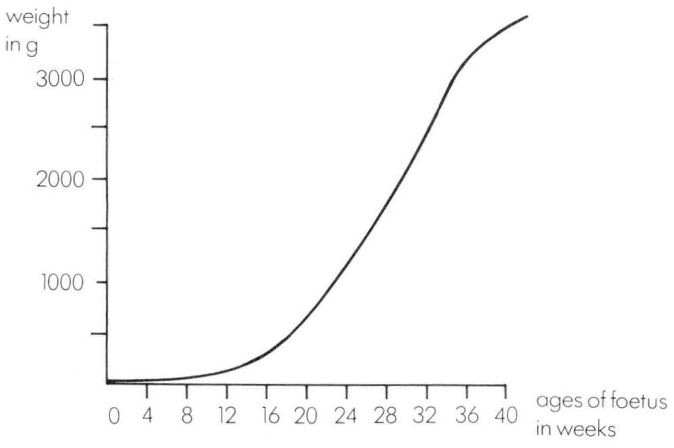

Figure 2 The foetus grows more rapidly after the twelfth week

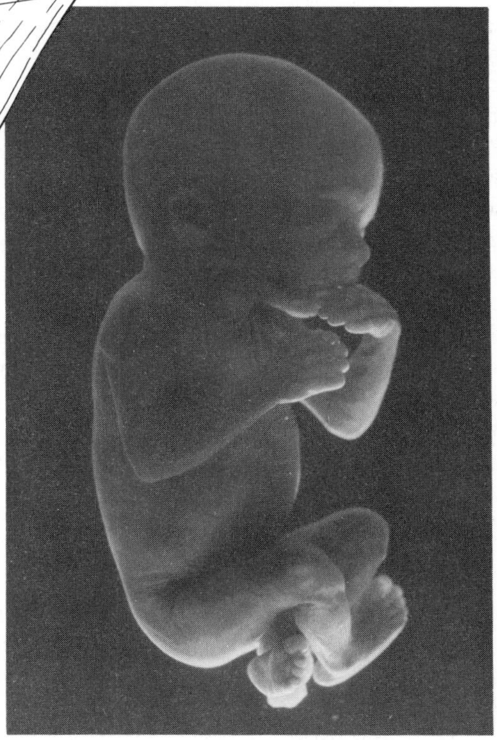

Figure 3 At 5 months old, the foetus is fully formed

8.4 Birth

When does birth take place?

Birth usually takes place in the thirty-eighth week of pregnancy. Women usually get a good idea of the expected date of the birth by working out 40 weeks on from the first day of their last period. The foetus starts to get ready for birth from about the 32nd week by moving into the birth position, that is, turning so its head is nearest the cervix (figure 1).

What happens during birth?

Birth begins with labour. This is the term used to describe the regular contractions of the muscles in the uterus wall brought on by hormonal changes in the blood (see page 150). Labour goes through three stages:

- **Stage 1** This usually lasts between six and twelve hours. It starts with weak labour contractions and the dislodging of the mucus plug from the cervix. The passing of this plug is often called a 'show' indicating that the birth is about to start. The contractions gradually get stronger, eventually opening up the cervix to about 10 cm. The head of the baby now breaks the **amnion** and the fluid is released. The woman will suddenly pass a lot of watery fluid and this is a sign to her that the birth of her baby is not too far off! We call this the 'waters breaking'.
- **Stage 2** This usually lasts between one and two hours. Strong contractions of the uterus wall force the baby through the cervix and vagina and out into the world! As soon as the baby has started breathing, the **umbilical cord** is tied and cut, separating the baby from the placenta, and thus from the mother.
- **Stage 3** The labour contractions may stop for a short period, but eventually start up again, forcing the placenta out. This is called the **afterbirth**.

The average time for a woman's first birth is about fourteen hours. Further births are sometimes quicker.

What is an episiotomy?

During birth the vagina is stretched to its limit, but sometimes this is not enough to allow the baby's head out and therefore the skin between the vagina and anus has to be cut a little. This is called an **episiotomy**. Afterwards, the cut is stitched. This can leave a woman feeling very sore and uncomfortable for several days after the birth.

What if the baby is not in position?

If the foetus fails to get into the birth position, it may have to be delivered bottom-first. This is called a **breech birth**. Breech births need more care and therefore often take longer. Sometimes **forceps** need to be used so that the doctor can guide and pull the baby's head safely through the pelvis. This use of forceps is quite common, especially during awkward births or when the labour contractions are not sufficiently strong. Occasionally, rather than perform a very difficult breech birth or for other reasons which concern the mother and/or baby's health, the doctor may do a **Caesarian birth**. This is the delivery of the baby through a cut in the abdomen wall. It can be done under general anaesthetic or with an **epidural**, that is a local anaesthetic which just numbs the nerves which carry the feelings of pain from the lower abdomen to the brain. Many women prefer an epidural Caesarian so that they are conscious and involved when their baby is born.

Are babies often born too early?

Babies are sometimes born before the end of pregnancy (before the 38th week). The seriousness of this depends on (a) how early it is born and (b) the baby's state of development. If the baby is below 2.5 kg, it is said to be **premature** and treated with special care until it is up to its predicted birth weight. About 8 per cent of babies born in the UK are premature. These babies often experience respiratory and feeding problems, and are kept in a controlled environment such as an **incubator** for some weeks.

Questions

1. What is the birth position? Why is it better for a baby to be born from this position?
2. An older friend of yours is coming to the end of her pregnancy and is very worried about the birth. She has not been to the antenatal classes and does not know what to expect. Explain to her what happens.
3. The same friend says she has heard that the birth is less painful with an epidural. Explain to her what this is and why it is given.
4. Conduct a survey among your friends to find out their birth weights. What percentage of them were premature?

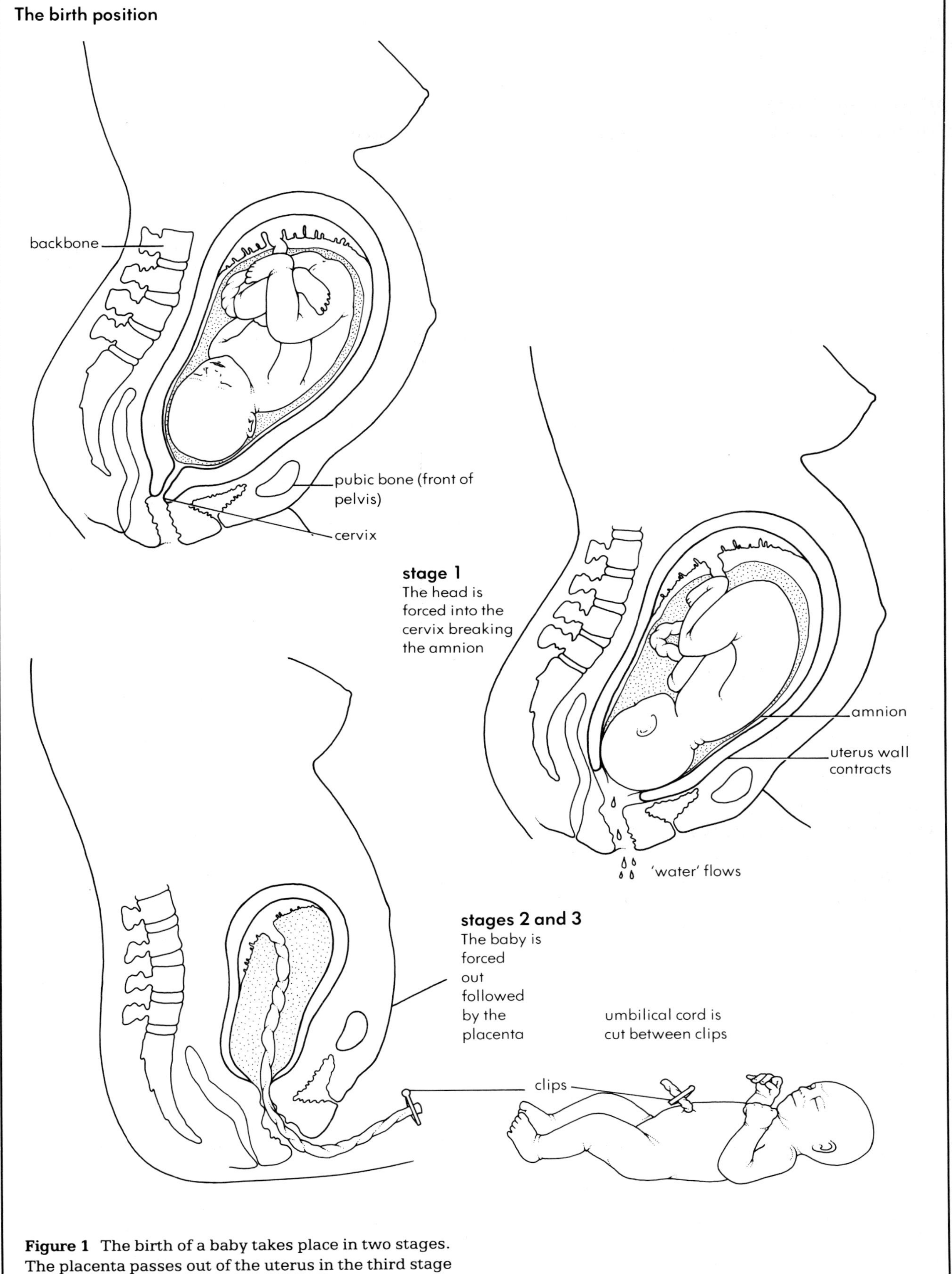

Figure 1 The birth of a baby takes place in two stages. The placenta passes out of the uterus in the third stage

8.5 Antenatal care

Taking care

Having a baby puts an enormous strain on a woman emotionally and physically (figure 1). During pregnancy a woman needs to take good care of herself by staying as healthy as possible and receiving care and support from her partner, friends and family. Usually, as soon as pregnancy is confirmed, the woman is referred by her doctor to an **antenatal clinic** where her health and that of the growing foetus will be checked throughout the pregnancy.

What happens at the antenatal clinic?

The first visit to the clinic is by far the longest. This is because the doctors have to find out a lot about the woman. Later visits are shorter. At first the woman is usually:

- asked a lot of questions about her health, past and present, in order to build up a full medical history. Questions are also asked about her personal circumstances so that any other problems can be foreseen.
- given a full physical examination including measurements of height, weight and blood pressure. These are checked at every visit as they can be good indicators of future problems. For example, a sudden weight gain, especially after the 20th week could be due to **toxaemia**. This is due to the woman's body not coping adequately with the extra work it has to do. Waste substances start to build up in her blood and if these are not dealt with, she may well die.
- given an internal examination to check on the size of the womb. A **cervical smear** may be taken so that changes can be monitored.
- given urine and blood tests. The urine sample is checked for glucose, protein and infections. The presence of glucose is usually an indication of **diabetes**. The presence of protein usually indicates that the kidneys are not coping with the extra load. The blood sample is used to check her blood group and to test for anaemia. It can also show up infections like **rubella**, **syphilis**, **hepatitis B** and **AIDS**. Any of these will certainly complicate the pregnancy and may even give rise to the very difficult question of terminating the pregnancy (an abortion). Blood and urine samples are taken during every visit.

At the end of the first visit, the woman is given a record card called the **co-operation card**. She is told to carry this with her at all times, so if any problems arise, the information is immediately available.

During later visits a routine **ultrasound scan** is done (figure 2). This will confirm the age of the foetus and indicate the position of the placenta. It can also show up any problems in growth.

What is an amniocentesis?

Women over 35 years old, or those with a history of Downs' syndrome, spina bifida, **muscular dystrophy** or **haemophilia** in their family will also be given the chance of an **amniocentesis** (figure 3). This is a special test which can indicate whether her child is likely to be handicapped. This and the ultrasound scan are usually carried out between the 16th and 18th weeks of pregnancy.

After the birth

It is a good idea for a woman to have a **post-natal** check four to six weeks after the birth. This may be done at the clinic or by the family doctor. All the routine tests will be done to make sure the woman's body is returning to normal. For example, checks on her urine will indicate if her kidneys are working properly. Her weight and the size of her abdomen are checked. Both should almost be back to normal.

Is antenatal care just about checking health?

Antenatal care is also about preventing problems and preparing the woman and her partner for birth and parenthood. Advice is given about smoking, drinking alcohol, taking drugs and occupational hazards. Towards the end of her pregnancy, the woman and her partner can go to parentcraft classes, where they will be told what to expect during the birth and taught how to cope with the newborn child (figure 4).

Questions

1. Write a letter to a pregnant friend in America telling her what kind of care she could expect if she lived in Britain.
2. An older friend has just had her pregnancy confirmed. She has been offered an amniocentesis. Explain to her, in writing, how and why these are carried out.

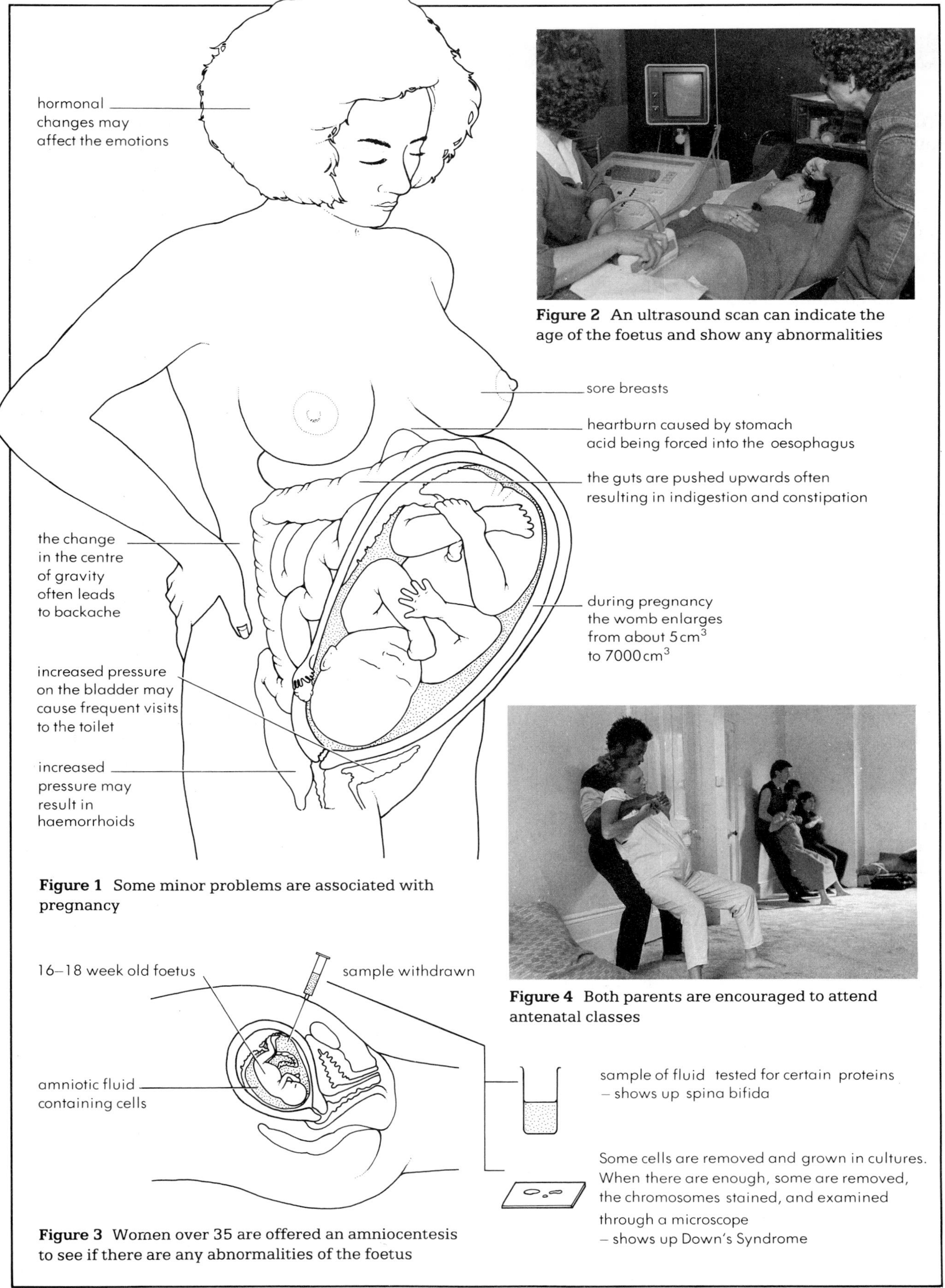

Figure 2 An ultrasound scan can indicate the age of the foetus and show any abnormalities

Figure 1 Some minor problems are associated with pregnancy

Figure 4 Both parents are encouraged to attend antenatal classes

Figure 3 Women over 35 are offered an amniocentesis to see if there are any abnormalities of the foetus

8.6 Drugs and diet

Why is it unwise to smoke during pregnancy?

There is a lot of evidence to suggest that a woman who smokes during pregnancy will produce a smaller, less well developed baby (table 1). This is because chemicals in the cigarettes cause the placental blood vessels to constrict, reducing the flow of blood from which the baby obtains its food. The carbon monoxide taken in also reduces the amount of oxygen in the mother's blood and nicotine actually crosses the placenta and can directly affect the baby. Nicotine is a stimulant and makes the baby's heart beat too quickly, putting great stress on its circulation.

What effect can drinking alcohol have?

There is evidence available which suggests that if there is alcohol in the mother's blood it does get into the baby and this can have a great effect on it. For this reason women who drink regularly and continuously throughout pregnancy run the risk of producing a mentally handicapped and physically retarded child (figure 1). Some researchers have also linked drinking alcohol with a higher incidence of miscarriage.

What else should a pregnant woman avoid?

Many drugs if taken in pregnancy can be dangerous. You probably know about the use of the drug **thalidomide** and the effect this had on babies. Many babies of women who took the drug thalidomide were born with stunted and deformed limbs. The drug was thought to be fairly harmless and was given to pregnant women to help them relax. We know now that thalidomide was able to cross the placenta with horrendous results on the development of the foetus. Unfortunately this knowledge came too late for many.

Before a pregnant woman takes any drugs, she must first consult her doctor. She will assess the situation and advise accordingly. She may even prescribe some drugs but will first take into account the stage of the pregnancy. One thing to come from the tragic thalidomide affair was that it showed that the effects of a drug on the foetus are very dependent on the time at which it is taken during the pregnancy (table 2). Many **antibiotic** drugs can produce devastating effects early on during pregnancy, yet if taken later they are relatively harmless (see page 178).

Another major fear during early pregnancy is that the woman will contract **german measles** (rubella), **syphilis** or AIDS. If she gets german measles during the first twelve weeks of pregnancy, her chances of having a stillbirth, miscarriage or of producing a child with a heart or nervous system defect are increased. Syphilis has similar effects. The effects of AIDS are still being discovered, but it is possible that the new baby could be born with the AIDS virus.

Is a woman's diet important during pregnancy?

It is important that a pregnant woman has a well balanced diet and that she is taking in adequate quantities of nutrients to replace those taken by the developing baby. She will need to eat more **protein** (meat, fish), more **vitamins** (fresh fruit and vegetables), more **minerals** (dairy produce, fruit, vegetables), especially calcium and iron, and also, more **energy foods** (table 3). The antenatal clinic will supply a diet sheet to help. If this is followed the woman should gain about 12.5 kg during the pregnancy (table 4).

What about money?

All pregnant women receive free dental care, free prescriptions and some dietary supplements, such as iron and calcium tablets. All working women are also entitled to **maternity pay** from the 30th week of pregnancy onwards and time off work for visits to the antenatal clinic. Employers must give women the legal minimum maternity leave and some employers make their own better arrangements with their staff. There are other benefits available for those who need them. The DHSS should advise women about these.

Questions

1. A good friend of yours is pregnant yet still smokes twenty cigarettes a day and goes out drinking three times a week. She believes that pregnancy should not alter her lifestyle. How would you persuade her that what she is doing is not sensible?
2. Many women suffer from malnutrition during pregnancy. Explain why and what a woman can do to avoid it.
3. Design a poster encouraging women to attend antenatal clinics.

Number of cigarettes smoked each day	Average birth weight when	
	mother smoked during pregnancy	mother stopped smoking before pregnancy
0	3.39 kg	3.39 kg
5 to 9	3.20 kg	3.39 kg
20 to 30	3.18 kg	3.36 kg

Figure 1 Does smoking during pregnancy affect the size of a baby? (BMA)

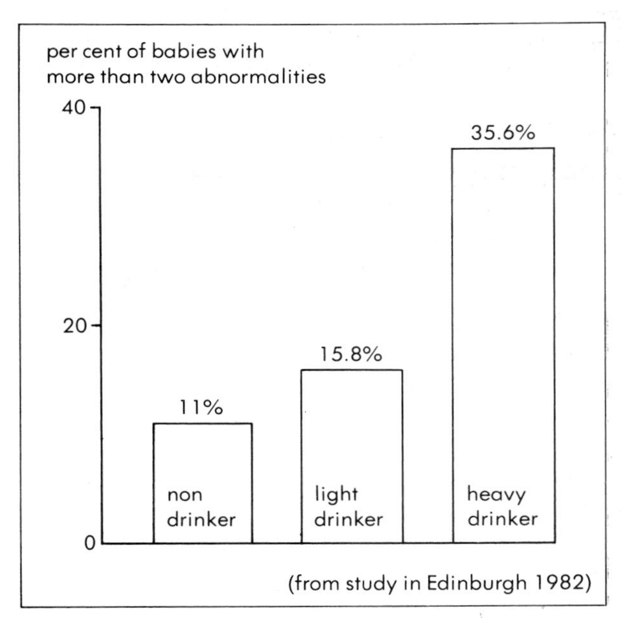

Figure 2 Does drinking alcohol during pregnancy affect a baby?

Days into pregnancy when drug was taken	Organ in which malformation appeared
20 to 24	thumb, ear
25 to 30	arms
26 to 31	heart, gall bladder, and duodenum
28 to 34	legs
36+	rectum

Table 1 The effects of thalidomide depend on when it was taken

Place weight is gained	Average gain (kg)
Foetus	3.4
Uterus	0.9
Placenta	0.6
Amniotic fluid	0.8
Other fluid	1.4
Extra blood	1.5
Breasts	0.4
Body fat	3.5
TOTAL	12.5kg

Table 3 Weight gain during pregnancy

Nutrient	Amounts for average	
	non-pregnant	pregnant woman
Energy (kJ)	9000	10 000
Protein (g)	54	60
Calcium (mg)	500	1200
Iron (mg)	12	13
Vitamin A (μg)	750	750
Vitamin C (mg)	30	60
Vitamin D (μg)	—	10

Table 2 How a woman's diet should alter during pregnancy

Figure 3 Abnormalities caused by thalidomide

Questions

Recall and understanding

1 The production of gametes involves:
A mitosis C fission
B conjugation D meiosis

2 The male sex hormone is:
A oestrogen C progesterone
B testosterone D adrenalin

3 The placenta develops from the:
A embryo C ovum
B uterus D sperm

4 The correct definition of a premature baby is:
A it measures less than 32 cm
B born too early
C not fully formed
D it weighs less than 2.5 kg

5 An amniocentesis is normally performed during which week of pregnancy?
A 10 – 12 C 16 – 18
B 12 – 14 D 22 – 24

6 Clones are produced by:
A mitosis C fusion
B meiosis D conjugation

7 A sperm contains a lot of mitochondria so that it can:
A fuse with the ovum
B live longer
C produce energy to move
D grow

8 A corpus luteum develops from the:
A follicle C ovary
B ovum D sperm

9 The average weight gain during pregnancy should be:
A 20 kg C 10 kg
B 12.5 kg D twice your own weight

10 A parent cell has 46 chromosomes. How many will its offspring produced by meiosis contain?
A 23 C 92
B 46 D none

Interpretation and application of data

11 The data below refers to the birth weight of 24 babies. (Figures are in kg.)

3.2	3.4	2.6	3.5
3.4	3.5	2.6	3.0
2.7	2.8	3.4	3.1
3.0	2.9	3.5	3.4
3.5	3.0	3.0	3.8
2.2	3.1	2.9	3.5

a) Calculate the average birth weight.
b) Copy the table headings and use the data to complete it.

Birth weight (kg)	2 – 2.4	2.5 – 2.9	3 – 3.4	3.5 – 3.9
Number of babies				

c) What is the modal birth weight?
d) How many babies were premature?

12 The figures below are taken from a survey done by the BMA on smoking during pregnancy.

	Numbers of cigarettes smoked per day	**Average birth weight of baby**
Non smoker	0	3.39 kg
Light smoker	5 – 9	3.20 kg
Heavy smoker	20 – 30	3.18 kg

a) Do the figures suggest there is a relationship between smoking and birth weight? If so what is this?
b) The study also persuaded some women to give up smoking during their pregnancy.

	Number of cigarettes that used to be smoked	**Average birth weight of baby (kg)**
Non smoker	0	3.39
Light smoker	5 – 9	3.39
Heavy smoker	20 – 30	3.36

Was it worth giving up? Explain

c) In the study, the non-smokers were a control group. What does this mean?

d) The data below refers to birth weight and social class.

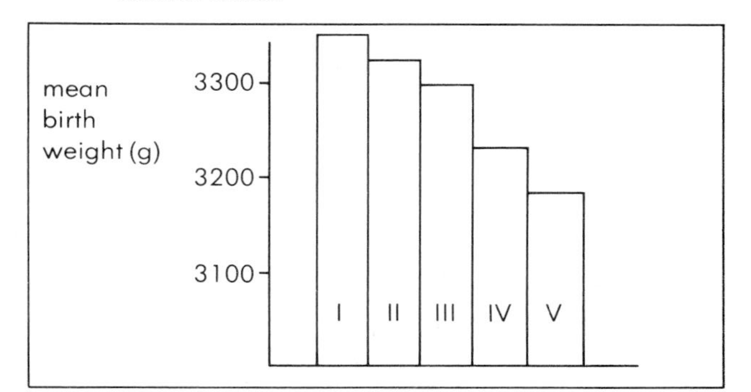

i) Which social class produces babies with the lowest birth weight?

ii) Which social class contains most women who smoke?

13 The figures below refer to the number of legal abortions per 1000 women in the UK.

Year / Age (yrs)	1969	71	73	76	80	81	82	83	84
Under 16	1	2	3	4	4	4	4	4	4
16 – 19	9	19	25	26	34	34	34	33	36
20 – 34	32	60	68	61	79	80	79	79	86
35 – 44	10	17	19	16	19	19	19	18	19
45 +	-	1	1	-	1	1	1	-	-
Unknown	1	2	2	2	-	1	- -	-	-
TOTAL	53	101	118	109	137	138	137	136	146

a) Which age group has the most abortions?

b) Which age group shows the biggest increase in abortions over the nine year period?

c) Are the overall number of abortions increasing or decreasing?

d) Plot the totals on a graph. Use the graph to predict the number of abortions in 1994.

e) For what reasons can you have a legal abortion?

14 The figures below refer to sperm samples taken from nine men.

Man	A	B	C	D	E	F	G	H	I
Number of sperm (millions) in 1 cm^3 of semen	20	40	7½	3	96	19	26	45	0
Percentage of sperm that are active	40	80	30	10	75	60	45	5	0
Percentage of sperm that are normal	80	20	60	60	90	70	10	75	0

A man is considered fertile if there are over 20 million sperm to every 1 cm^3 of semen, 40 per cent of these are active and at least 60 per cent are normal.

a) Which men are fertile?

b) Which men are infertile?

c) Which man is the most fertile?

d) Which man has probably had a vasectomy?

e) State the reasons for doing such a count.

15 The table below shows the amounts of calcium and iron a pregnant women requires.

	For first 28 weeks	**For last 12 weeks**
Calcium	1.2 g per day	1.5 g per day
Iron	13 mg per day	15 mg per day

a) Why do pregnant women need extra calcium and iron in their diets?

b) Name two good sources of calcium.

c) Name two good sources of iron.

d) Milk contains 120 mg of calcium per 100 cm^3. How much milk would the woman have to drink every day to get her calcium needs:

i) during the first 28 weeks of pregnancy,

ii) during the last 12 weeks of pregnancy?

CHAPTER 9: GROWTH AND DEVELO

Although every child contains a 'blueprint' for its future growth and development, this can be influenced by many other external factors. For example, the amount and quantity of available food.

This chapter is all about how the blueprint is followed and how it can be modified by external factors. It also looks at ageing and some of the things that can go wrong with your body as you get older.

New parts for old

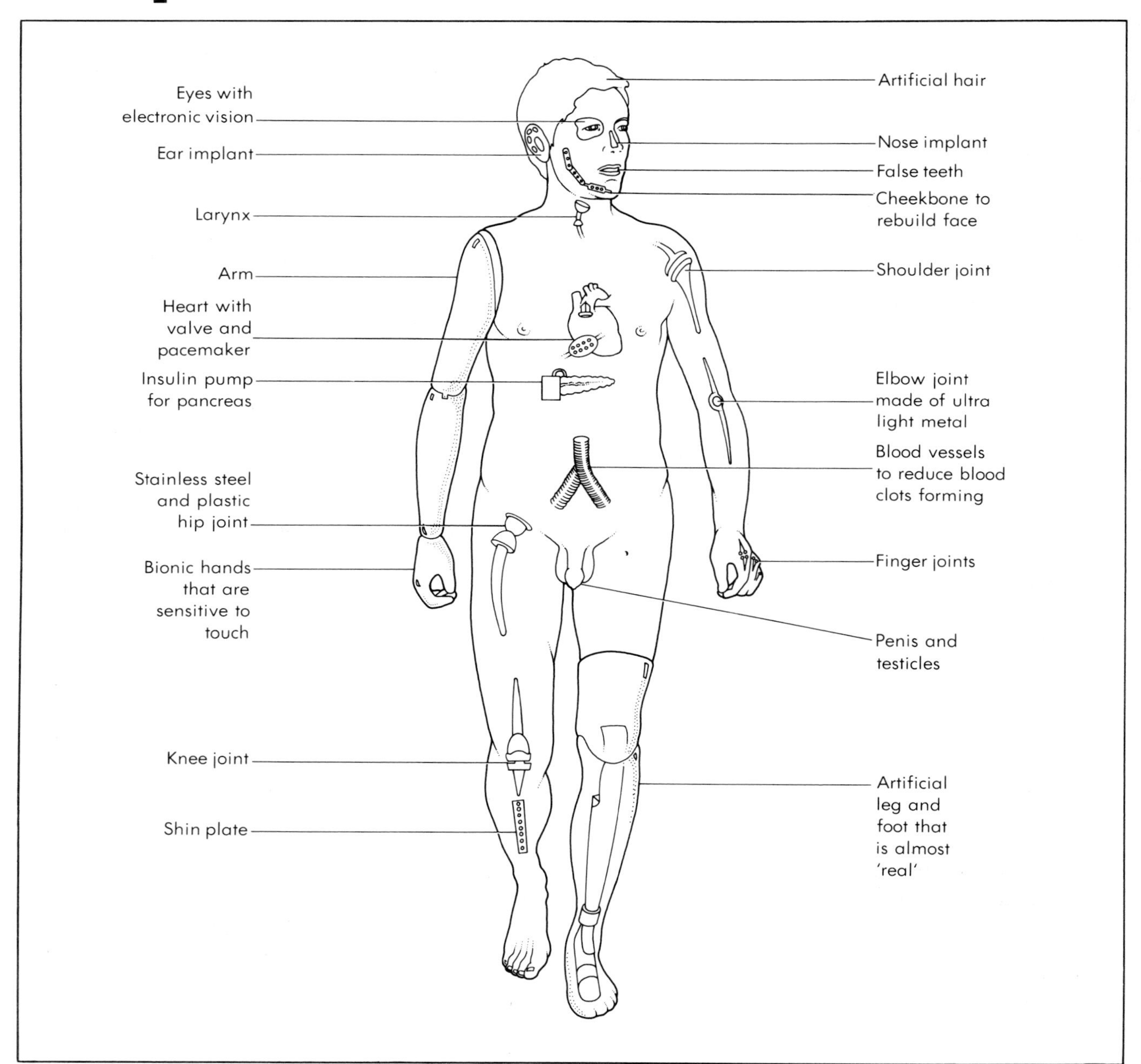

Parts of the body that can, or will in the future, be replaced by artificial equivalents.

Frankenstein – a terrible reality?

For years the story of Frankenstein has both fascinated and frightened many people of all ages. The reason – he was made from bits of others, some dead and some alive. At the time when Frankenstein was supposed to have been made, even minor surgery was difficult and to put someone together from bits was a total fantasy. But now we can almost do it. What is more, if we cannot transplant parts there are now many artificial ones available. This section looks at what exactly we can do nowadays.

Which parts of the body can we transplant?

There is an important difference here between tissues and whole organs. Tissue **grafting** is a fairly common practice whereas an organ is only **transplanted** if the person would die without it.

- Some tissues which are grafted:
  1. **Blood** Blood transfusions are routine now. There is even a service (The National Blood Transfusion Service) which collects blood from donors and stores it for future use (page 48).
  2. **Skin** The main reason skin grafts are performed is to cover burns. The skin is usually taken from the thigh of the same person, but skin from others, even people who have recently died, can also be used.
  3. **Bone** Slivers of bone are often put into areas where new bone is needed. This seems to stimulate the growth of new bone.
  4. **Connective tissue** One particular kind of connective tissue taken from muscles is particularly useful for strengthening weak areas such as hernia sites.

- Some organs which are transplanted:
  1. **Heart** The first human heart transplant was performed in 1967 by Dr Christian Barnard. By 1984, 264 heart transplants had been carried out in the UK and over a thousand worldwide. The heart is now considered one of the easiest organs to transplant.
  2. **Kidneys** More kidney transplants are performed than any other organ. The great advantage with kidneys is that it is possible to take one from a living person. This increases the chances of the transplant being successful.
  3. **Liver** The liver is the hardest of all transplants to perform. One of the problems is that the liver needs to start working immediately as there is no substitute available which can take over for a while.
  4. **Lungs** Lungs are often transplanted with a heart as to separate them would put too much of a shock on the heart resulting in failure.
  5. **Pancreas and other endocrine glands.**

Tissue grafts are more successful than organ transplants. Success with organs is still measured in months rather than years except for kidneys.

Why is the success rate with organs so low?

Organs for transplant must come from other people. They will therefore be treated as foreign by the new body and rejected. Doctors tackle this problem in two ways.

1. By getting as good a tissue match as possible. All cells have antigens stuck to their surfaces. If your body recognises these, it will not produce antibodies to destroy them. If however your body does not recognise the antigens, it will try and destroy them by producing antibodies.

   A good tissue match is one which has the same, or very nearly the same, antigens as the recipient's tissues. For example, a person who needs a new heart has the antigens ADRXZ on the surface of his cells. There are two possible donors.

   Tissue typing shows one of these to have the antigens AFTVZ and the other to have the antigens ADSXZ. The surgeon will choose the one with the ADSXZ because this is the closest match. It is not perfect but it does stand less chance of rejection and less drugs have to be taken to prevent this.
2. By using drugs to prevent the production of antibodies. The disadvantage of this is that you leave yourself open to attack by many other antigens in the form of germs.

What artificial parts are available?

There are now well over 30 artificial parts available to the transplant surgeon. These range from whole limbs to delicate pieces of electronic machinery such as heart pacemakers.

9.1 The newborn baby

The birth of a baby

A baby's birth is probably quite a shock for it! Its whole environment is now suddenly different. There are often lots of people around, lights, noises – all very different from the stable, watery rhythm of the womb. Not only this but the baby now has to start surviving on its own. When the umbilical cord is cut the baby's only source of oxygen is the air so it must use its own lungs to breathe. This involves some major adjustments to the baby's circulatory system. Until birth the lungs only needed to receive enough blood to supply them with food for growth and development. After birth, they must receive all the blood (figure 1).

What can a newborn baby do?

Compared to many other animals, a human baby is quite helpless. It struggles to communicate, cannot walk, has few, if any, of its homeostatic mechanisms working at maximum efficiency and has not yet got fully developed senses or nervous system. For example, the eyes have only a small field of vision and cannot focus beyond 20 cm or so. The hearing is even worse, but quickly improves. The best developed sense is smell and it is through this that the baby first learns to recognise its mother and others.

A newborn baby has little control over its muscles, but does show several survival reflexes (see figure 2). These are soon replaced by deliberate movements as the baby gains control.

The importance of the parents

The parents have to do a lot for their baby. They have to feed it, protect it and care for it. They must provide a suitable environment in which their baby can grow and develop. They need to help it learn to talk, walk and eventually how to look after itself. No other animal takes care of its young for as long as the human animal does.

What does a baby eat?

Within 24 hours of its birth, the baby needs feeding. For the first three to four months of its life, its food consists entirely of milk. This contains all the nutrients the baby needs with the exception of iron. The baby should already have plenty of this, which it took from its mother during her pregnancy. Most women can breast feed their baby, but if this is not possible there are several good commercially produced substitute milks available (table 1). A baby fed properly on any of these will grow just as well as a breast fed baby.

When the baby is drinking about two pints of milk a day, it is gradually transferred to solid food. This is usually at about four months but as with everything, it varies from baby to baby. This process is called **weaning**. It is very important that the parent gets a baby's diet right as an inadequate, unbalanced diet can interfere with growth and development and overfeeding can lead to obesity later on in life. As a general guide, a baby's diet must be healthy and at the same time it should:

- supply **enough energy**.
- supply **enough protein for growth** (both animal and plant proteins).
- supply **adequate minerals for growth** (e.g. calcium for bones).
- supply a **selection of vitamins and minerals** to maintain health.

How fast should a baby put on weight?

To start with, a newborn baby will lose a little weight, but then it should increase at about 900 g per month for the first six months. The baby's weight gain will be monitored very carefully by the doctor or health clinic and any irregularities sorted out quickly.

Questions

1 Why do you think some babies scream when they are first born?
2 Describe with the help of a diagram, the changes that take place to the foetal blood circulation at the time of birth. Explain why these are necessary.
3 Draw up a table of advantages and disadvantages of breast and bottle feeding.
4 Imagine you had to plan a diet for a 12 month old baby. What guidelines would you use?
5 Explain why a baby's life is only as good as its parents make it.

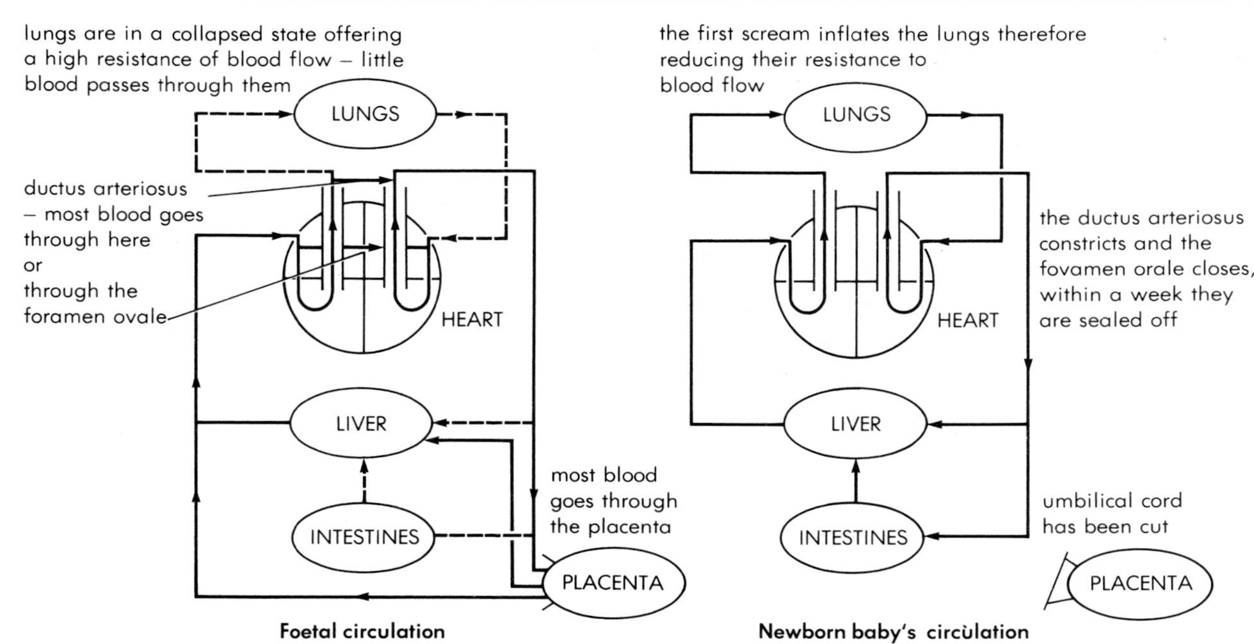

Figure 1 The circulatory system of a newborn baby has to make some major adjustments in the first few moments after birth

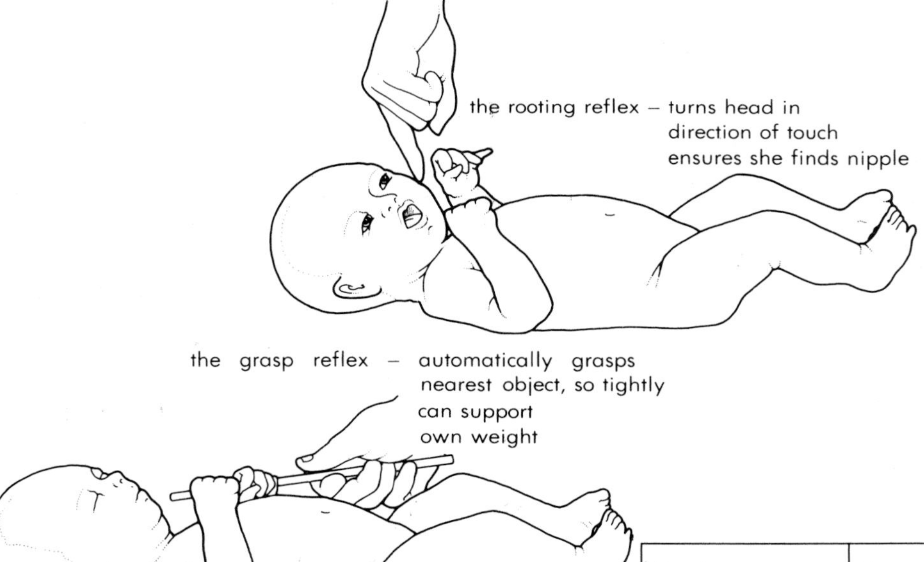

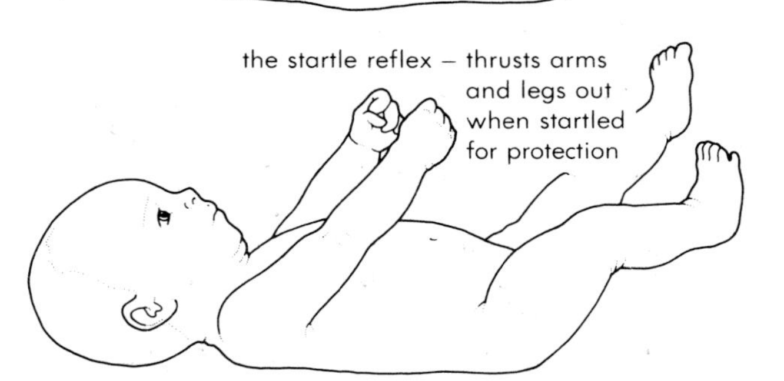

Figure 2 Newborn babies have several survival reflexes

Nutrients per 100 cm^3	Human milk	Typical prepared milk	DHSS recommend-ations
Carbohydrate (g)	7	7.2	4.8–10
Protein (g)	1.3	1.5	1.2–2
Fat (g)	3.6	3.6	2.3–5
Vitamin A (μg)	60	69	40–150
Vitamin C (mg)	3.8	6	3
Vitamin D (μg)	0.8	1	0.7–1.3
Calcium (mg)	35	59	30–120
Phosphorus (mg)	15	35	15–60
Energy (kJ)	274	281	270–315

Table 1 Prepared milk is very close in composition to human milk

9.2 The child

The newborn baby

The major organs of a healthy newborn baby are all fully formed and most are functioning, although some are not fully efficient. After birth the child continues to grow and its organs continue to develop until they are working at maximum efficiency (figure 1). As the organs develop, so do its **sensory functions**, **motor skills**, and **intellectual abilities** (see page 148). When the child reaches eighteen years, it is a fully formed adult and growth stops.

What makes each child different?

Every newborn baby contains an inbuilt programme for its future growth and development; that is, a pre-determined set of instructions. These instructions are contained within its **genes**. Some of the instructions will be the same in every baby in that they are a basic recipe for a typical human (as opposed to any other organism). Others are very specific instructions which make the child a unique individual. These will depend upon who the parents are. Every child displays characteristics from both parents. For example, it may grow tall like its father, but have curly hair like its mother. The exact features it displays from each parent will depend upon which genes were passed on in the respective gametes (see page 176).

What else influences a child's development?

Although a baby's genes are very important, they are not the only things that influence its growth and development. Some of the others are:

- **Food intake**. Both quantity and quality of food can affect growth and development. Overeating can lead to obesity and its related problems, whereas an inadequate intake of food may produce both stunted growth and development. For example, if a baby does not get enough energy foods, its nervous system often fails to develop fully.

  By far the commonest influence is that of malnutrition. Eating the wrong foods leads to deficiency diseases, many of which are growth-related (see page 20).
- **Hormones**. There are four hormones which directly affect growth and development (and one which indirectly affects it).

  1. **Growth hormone** produced by the pituitary gland. This influences the rate at which tissues grow, especially bone tissue. A child who overproduces growth hormone during childhood will grow too much, too quickly. This may result in **giantism**. Underproduction results in **dwarfism**.
  2. **Thyroxine** which is produced by the thyroid gland. This influences tissue development and cellular metabolism. A child who overproduces thyroxine during childhood may become mentally unstable. Underproduction results in poor physical and mental growth.
  3. The sex hormones **oestrogen** and **testosterone**. These two hormones are produced by the ovaries and testes respectively and are responsible for most of the changes which take place at puberty (see page 151).
  4. **Insulin** which is produced by the pancreas. This influences the amount of glucose in the blood and therefore indirectly affects growth.
- **The environment**. Light, pollution and disease are just some of the environmental factors that can affect growth and development. For example, light is needed for your skin to make vitamin D. Vitamin D is essential for the absorption of calcium which itself is needed for proper growth of bones. Environmental factors are dealt with more fully in other sections.

Questions

1. Many people claim that they can recognise which child belongs to which parents simply by comparing their features. State your reasons for agreeing or disagreeing with them.
2. When identical twins are separated at birth and reared apart from one another, they often turn out different. What are the possible reasons for this?
3. Study the graphs opposite.
   a) At what ages do boys and girls grow most rapidly?
   b) Explain how the body proportions alter during childhood.
   c) Organs grow at different rates throughout childhood. Give an example of this.

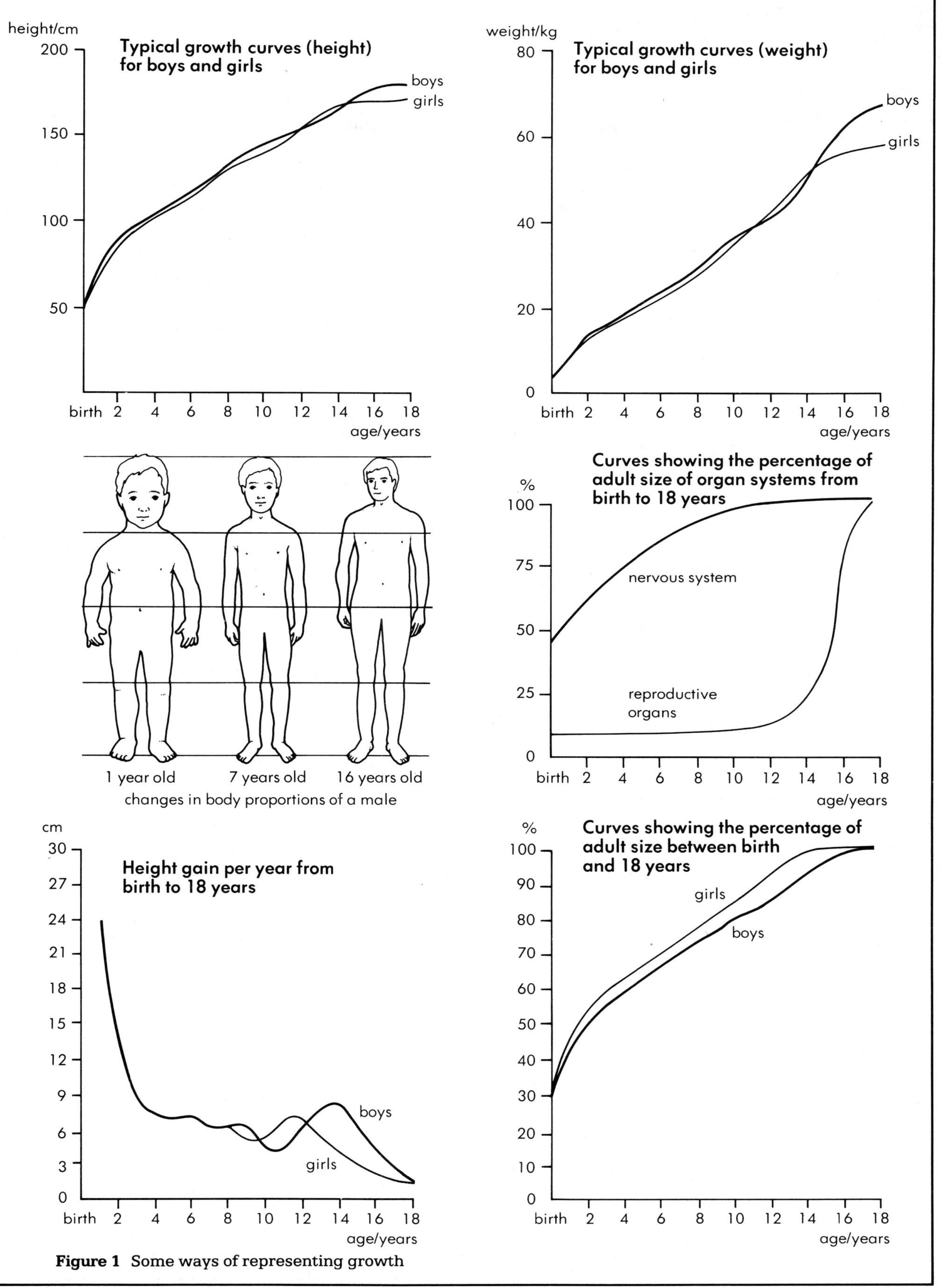

Figure 1 Some ways of representing growth

9.3 The growing child

When does a child start to walk?

As a child grows in size and the parts of its body mature, it begins to lose its helplessness and to do things like feeding itself. It learns skills like walking and talking.

Walking requires very strong muscles and the ability to co-ordinate all the different ones involved. Much practice is needed to do this and the child who experiments and gets lots of encouragement will usually end up walking quite early. However, few children start walking before the end of their first year, simply because their muscles are not strong enough. For some children, learning to walk can take as long as 20 months.

The development of co-ordinated movement is called **motor development** (figure 1).

When does a child start talking?

You may have noticed that very young babies make quite a lot of different sounds. To start with, when they are a few months old, they gurgle, laugh and cry to show pleasure or displeasure or other emotions. This is a way of **communicating** without words. Later between ten months and eighteen months most children will start to copy sounds and say and understand simple words. A two year old can usually join words together and by three years of age a child will usually have a vocabulary of more than 500 words. But all of this varies from child to child.

Why do some children throw tantrums?

Tantrums are a normal part of growing up. They are a part of every child's **emotional development**. Probably the first emotions to be displayed are love and affection, closely followed by anger and frustration. Children often start to have tantrums at about eighteen months. They are the child's way of getting attention. The normal sequence of mental, emotional and social development is outlined in figure 2. It is important to realise that this can be seriously affected by early 'bad' experiences and greatly helped by good parental attitudes.

Playing

All children like to play and this should be encouraged as much as possible. Play helps the child's physical, mental, emotional and social development.

- **Physical development**. The activity of play e.g. building bricks, pulling a toy, helps the child develop strong muscles and better co-ordination.
- **Mental development**. Children learn through their senses. Play can stimulate all senses and give children new chances to learn. It also helps develop intellect and encourages creativity. Having to talk to others during play helps with communication skills.
- **Emotional development**. Play can provide a necessary outlet for emotions such as anger and aggression.
- **Social development**. Play with parents and other children encourages co-operation, communication and friendship. It helps children understand that they have responsibilities, as well as helping the child to understand the roles of others and themselves.

When does a child start school?

In the UK formal schooling begins when the child is five years old, but parents can if they wish send their child to school well before this. **Nursery schools** are available to children from the age of two and in some areas the Social Services provide playgroups and nursery classes for even younger children.

Questions

1. What is motor development? Draw or describe two pieces of equipment which will help a child with its motor development.
2. Draw up a table which shows when the main changes in mental, emotional and social development take place in a pre-school child.
3. Suggest some ways in which parents can help with a child's mental, emotional and social development.
4. Briefly describe why play is so important to young children. Draw or describe suitable toys for (a) a two year old, and (b) a four year old child.
5. It is now possible for children to get formal schooling from two years old. Plan a day's nursery schooling for a three year old child. State your reasons for including each activity.

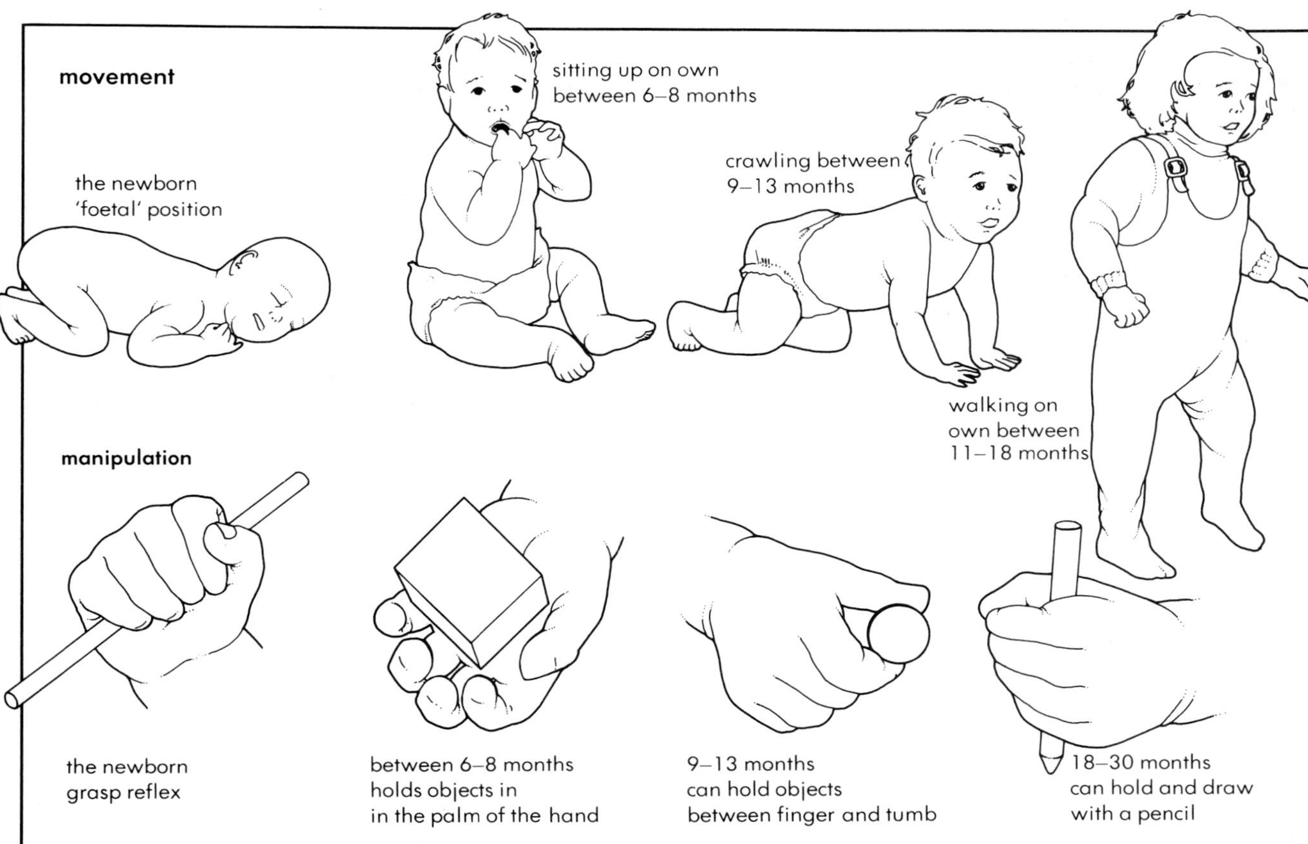

Figure 1 There are well recognised stages in the motor development of a child

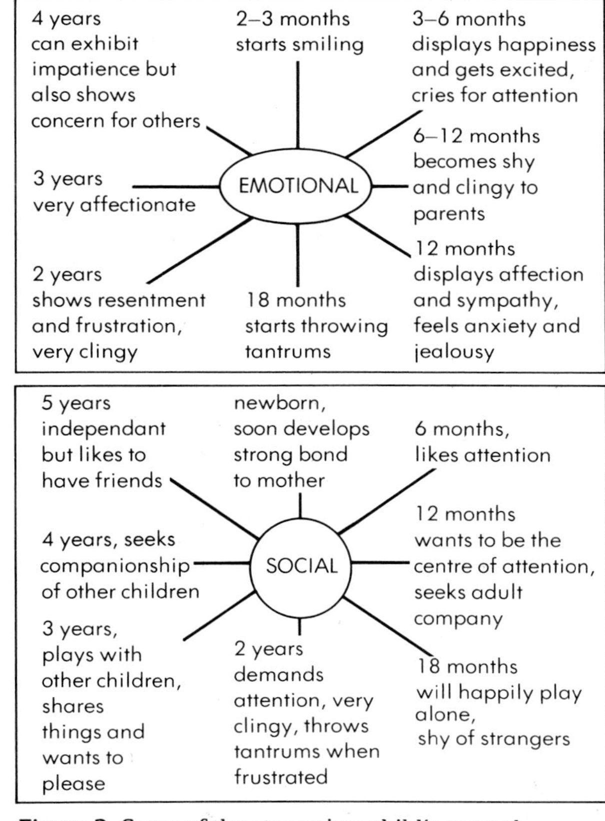

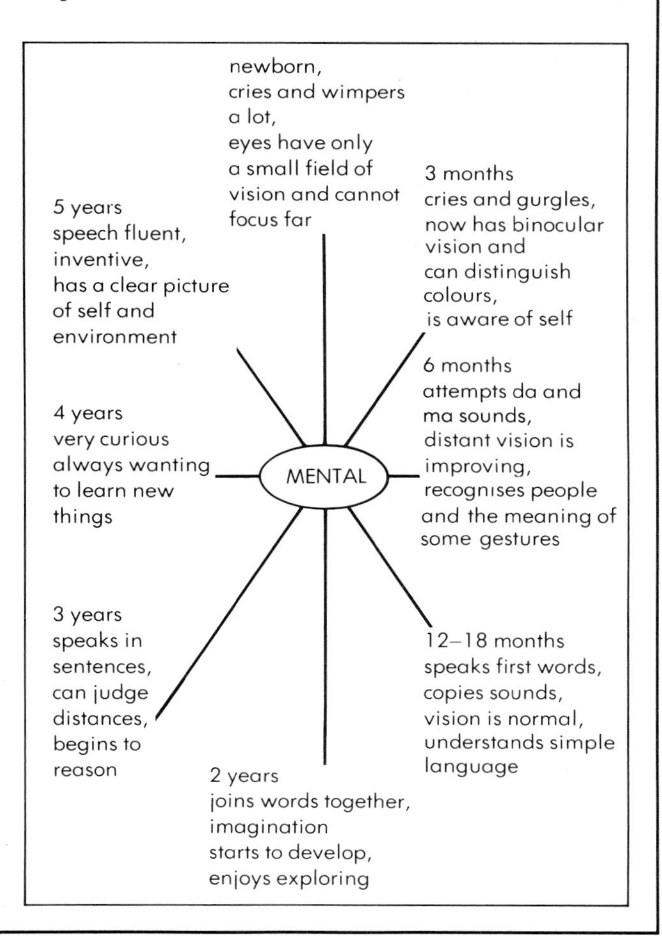

Figure 2 Some of the stages in a child's mental, emotional and social development

9.4 The adolescent

What is adolescence?

The early teenage years can be a very difficult time for a person. Many changes take place in your body and you may feel emotionally different. This stage, called adolescence, is the time during which a person passes from childhood to a sexually mature adult. It is a period of rapid physical growth and development.

The period over which the sex organs mature is often called puberty. Some of the other changes that take place during puberty are shown in figure 1.

The female cycle

One of the changes which takes place in the female body as she reaches sexual maturity is the starting of the **monthly period**. From puberty up until menopause, a woman will discharge blood through her vagina approximately every 28 days. This discharge usually lasts for about five days.

Where does this blood come from?

To understand where this blood comes from you need to understand how the female body prepares itself for pregnancy. These preparations are controlled by the hormones **oestrogen** and **progesterone** and they are repeated every time an ovum is produced. Here is a summary of what takes place (see also figure 2):

- The cycle starts at the beginning of the period (also called the **menstrual flow**). This is day one. For the next five days, the unused uterus lining, together with some blood, is discharged through the vagina.
- While this is going on, a new ovum begins to ripen in the ovary. At the end of the period, the **follicle cells** surrounding the ripening ovum start producing oestrogen. This oestrogen tells the uterus to start preparing a new lining for the ovum to implant into, if it is fertilised.
- On the 14th day, ovulation takes place and the ovum starts its journey along the fallopian tube. The remains of the follicle (the **corpus luteum**) reduces its production of oestrogen and starts to produce a second hormone, called progesterone. This progesterone tells the uterus that an ovum is on its way and therefore, it must complete its preparations.
- If by day 24 a fertilised ovum has not arrived at the uterus, the corpus luteum degenerates and therefore stops producing the two hormones. In the absence of these, the new uterus lining starts to break down and is discharged on day 29 – this becomes day 1 of the next cycle.

The timings given here are for the average female but the length of a cycle can vary considerably. Not every woman has her period, regularly, every 28 days.

Can life go on as normal during a period?

This monthly discharge of blood should not interfere with any normal activity. However, it does require more attention to personal hygiene. An absorbent pad should be worn to catch the flow of blood. This pad can be worn externally as a sanitary towel or internally in the form of a tampon. It must be changed regularly and disposed of hygienically.

What happens if the ovum is fertilised?

If the ovum is fertilised the woman is pregnant. The corpus luteum persists and continues to produce oestrogen and progesterone for a further three months. By this time, the placenta has formed and this takes over the job of producing the two hormones for the rest of the pregnancy. The presence of these hormones in the blood ensures that the uterus lining remains intact and continues to develop, and that no more ova are produced. The high levels present in the blood also stimulate breast development. Just before the birth the woman's progesterone level falls and this signals the **pituitary gland** to produce two more hormones: **prolactin** and **oxytocin**. Prolactin tells the breasts to start producing milk whereas oxytocin initiates labour.

Questions

1. What is puberty? What initiates the changes which accompany puberty?
2. What are the similarities and differences between puberty in boys and girls?
3. List the events leading up to the female monthly period.
4. What happens if the ovum is fertilised and implants?

Changes in a boy at puberty

pituitary gland
starts producing a hormone between the ages of 10 and 16

hormone tells testes to

1 start making sperms

2 start making testosterone

testerone causes the following changes:
a sex organs enlarge
b muscles enlarge
c body hair develops
d more sweat produced
e voice breaks (larynx enlarges)

Changes in a girl at puberty

pituitary gland
starts producing a hormone between the ages of 9 and 15

hormone tells ovaries to

1 start ripening eggs

2 start making oestrogen

oestrogen causes the following changes:
a sex organs enlarge
b hips and breasts enlarge
c body hair develops
d more sweat produced
e menstrual cycle starts

f Both sexes undergo a growth spurt and increase in size

Figure 1 Many changes take place in boys and girls during puberty

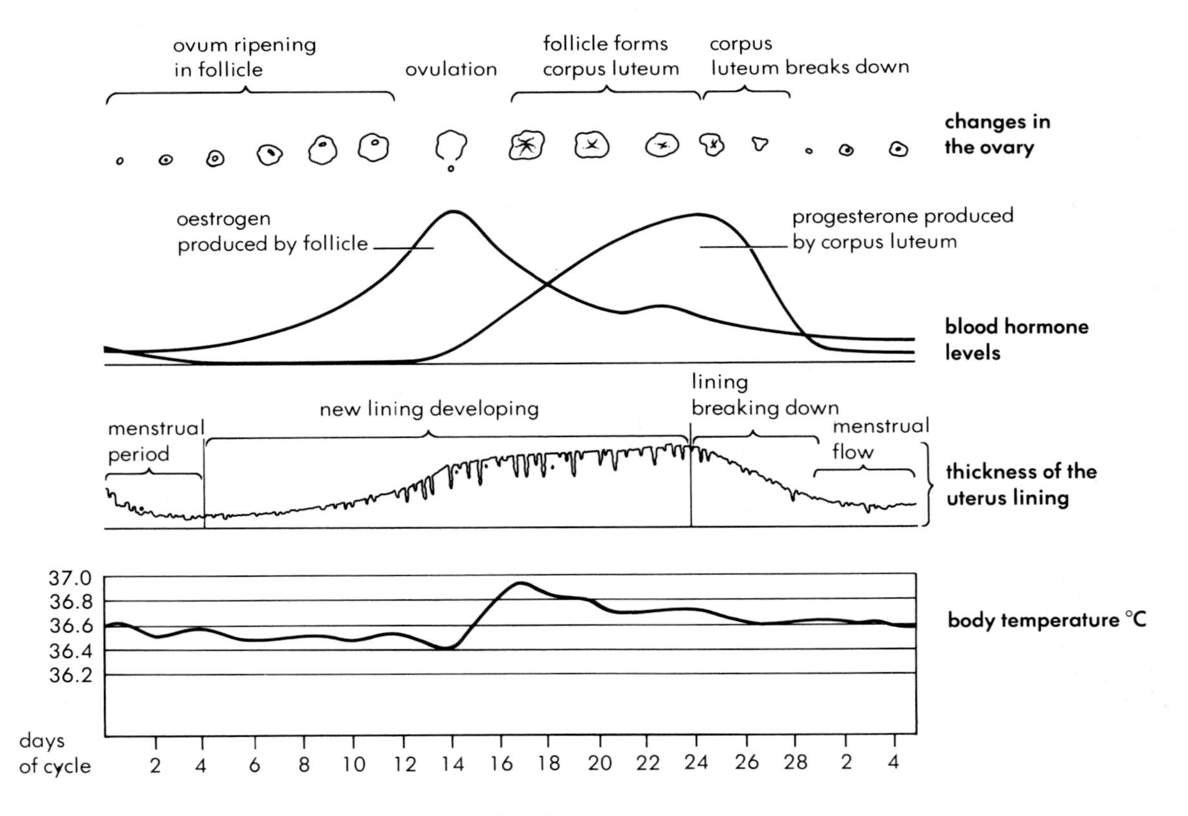

Figure 2 The changes caused by hormones during the menstrual cycle

9.5 Ageing

What happens as you grow older?

You begin to age as soon as you have finished growing. You stop growing at about sixteen to eighteen years of age. Ageing is a natural process in which the cells of your body's organs gradually become less efficient and eventually die. This process will happen to everybody but the rate at which it happens varies from person to person and depends on how you live your life. Some of the signs of ageing are:

- **Wrinkles** Your skin loses its elasticity and suppleness and stops being able to return to its original shape when stretched (figure 1).
- **Grey hair** The skin cells which produce the hair pigment stop working, resulting in non-pigmented hair.
- **Thinner, weaker muscles** As muscle cells die, they are not replaced.
- **Brittle bones** Calcium is removed from bone, but not replaced, leaving large spaces. Figure 2 shows how your body's composition changes.
- **Poor eyesight** The lens loses its elasticity, affecting your ability to focus. Old people develop long sight.
- **Stiff joints** Joints are used less and lose their suppleness
- **Always feeling cold** Some of your body's homeostatic mechanisms start to fail.
- **Senility** Your brain cells die at the rate of 10 000 per day, resulting in brain shrinkage and loss of mental functions.
- **Women lose their ability to have children.** Ova are no longer produced and the monthly period ceases.

What causes the changes during ageing?

Many of the changes during ageing are part of an inbuilt plan; they are in the genes. Menopause is one such change. Other changes are the result of the person's lifestyle and/or environment. Stiff joints, for example, can result from lack of exercise. Dry skin could be due to too much sun.

What is menopause?

Menopause is the stage of a woman's life when menstruation ceases. It represents the end of a woman's fertile period and is often called the 'change of life', for obvious reasons. It may be a gradual process, lasting several years or may just last a few months. It can be quite a difficult time for women. By the age of 51, most women have been through it.

Menopause is the result of hormonal changes within the body. The production of these hormones is genetically controlled.

Can the rate of ageing be slowed down?

Many of the changes associated with normal ageing can be delayed to some extent by:

- **Remaining physically active** Activity helps your circulation and muscles, increases metabolism, joint suppleness and stimulates your body's natural repair systems (see page 154).
- **Remaining mentally active** The more you use your brain, the longer it will remain alert.
- **Remaining healthy** Disease causes premature ageing and early death. Bad habits can do the same (see page 156).

Questions

1. Think ahead to your retirement age. Describe some of the changes you would have expected to have happened to your body.
2. What do you think could have caused these changes?
3. One change all women have is the menopause. Explain what this is and why it might be a difficult time for a woman.
4. For the last 100 years or so, the percentage of the UK population over 65 years old has been steadily increasing (figure 3). Suggest reasons for this.
5. The average life expectancy is different in different areas of the world (figure 4). Suggest reasons for this.
6. It is important for older people to remain physically and mentally active. Explain why and suggest how this can be achieved.
7. The main causes of premature death in old people are coronary heart disease, respiratory diseases and cancer (figure 5). Suggest how a person's lifestyle can contribute to these.

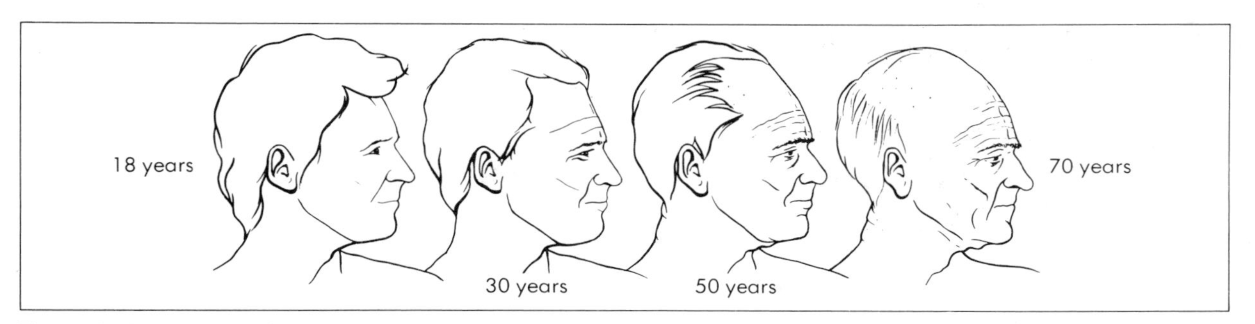

Figure 1 What are the facial changes to come about with age?

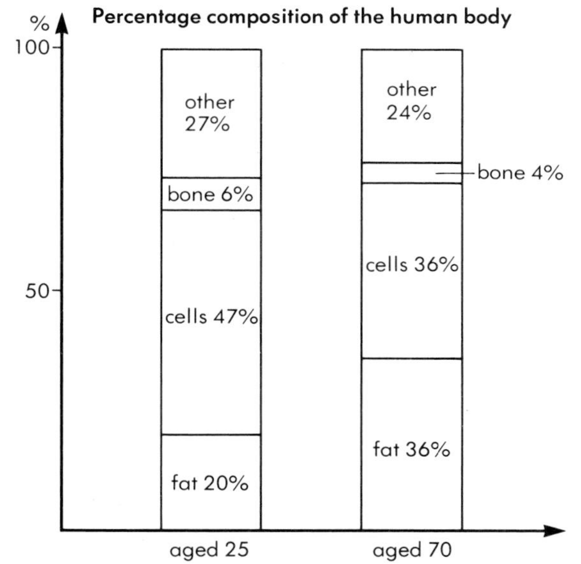

Figure 2 Your body composition changes with age

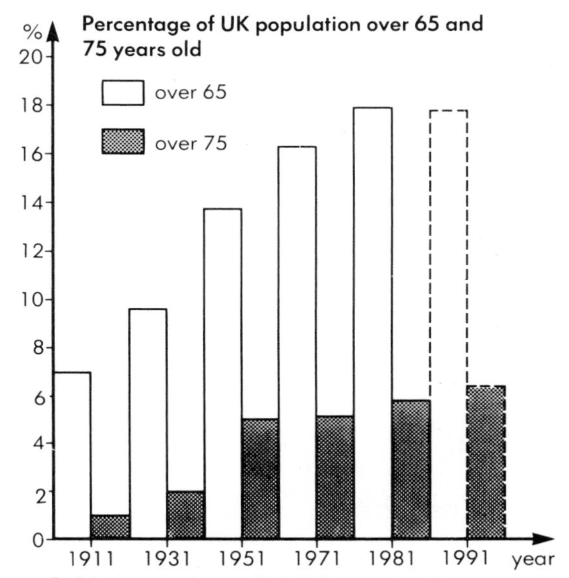

Figure 3 More people are living longer in this country

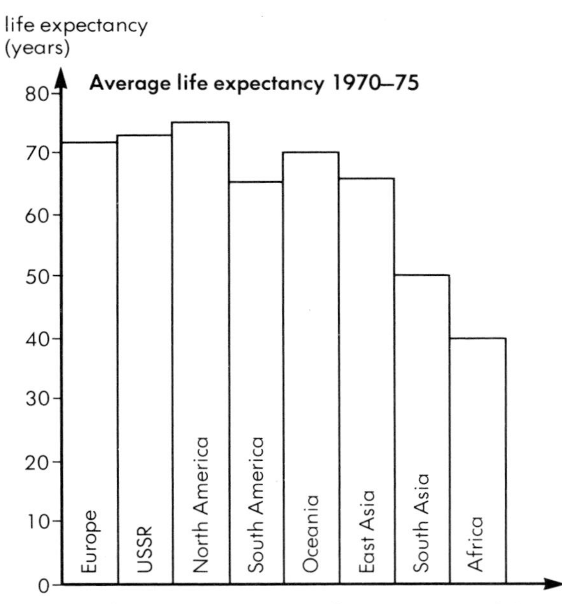

Figure 4 Life expectancy varies from one part of the world to another

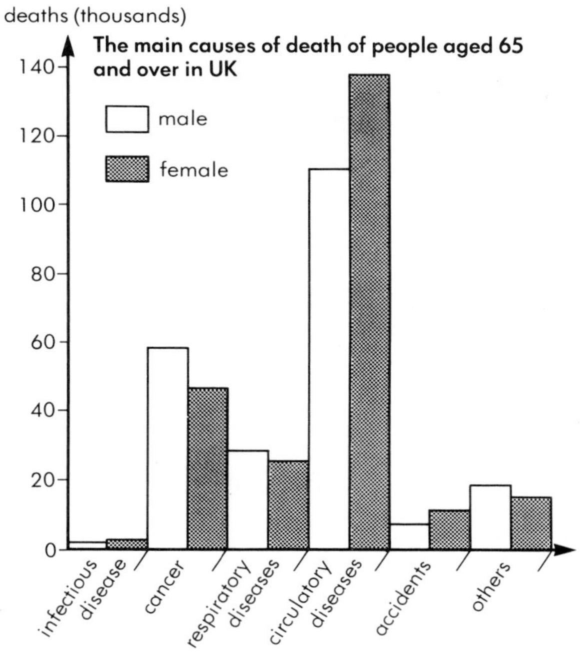

Figure 5 Circulatory diseases are the main cause of death in old people

9.6 Exercise and rest

Why should I exercise regularly?

Regular exercise will keep your body working at its maximum efficiency. It does this by:

- Improving its general fitness.
- Stimulating its natural maintenance and repair systems so that, for example, wounds heal more quickly.
- Raising its **metabolic rate** (see page 10), so that food is used up more quickly. In general, fat people usually have a low metabolic rate, whereas thin people have a high one.
  Exercise can be used to reduce weight, but you need to do a lot of exercise to really make a difference. For instance, it takes nearly an hour's jogging to work off the energy contained in a 60 g portion of cheese.
- Improving the circulation of the blood by lowering your pulse rate and blood pressure. It also reduces the amount of fatty substances in the blood and strengthens the heart muscle. These help prevent heart disease.
- Improving muscle tone and posture (see page 108).

What is fitness?

Physical fitness is a measure of your ability to cope with the physical demands of everyday life. It can be divided into three sections.

- **Stamina** This is your ability to keep going without running short of breath or collapsing from exhaustion.
- **Suppleness** Every joint in your body is capable of a range of movements brought about by the muscles near it. A top gymnast will use the full range of movements, whereas the average person will only use a few of them. The gymnast's joints will be more supple than the joints of the average person. This is because the gymnast uses them more often and for the full range of movements.
- **Strength** Your strength is determined by the size of your muscles relative to your body size. A fit person tends to have more and bigger muscles.

How can fitness be measured?

The easiest way to judge your fitness is to measure your pulse rate. In general, the lower your resting pulse rate and the less your pulse rate quickens after three minutes of step up exercise, the fitter you are (figure 1).

The chart opposite gives an indication of how effective various activities are for developing fitness (table 1).

How important is rest?

No matter how fit you are, there always comes a time when your body is exhausted and needs to rest. The best kind of rest is sleep. This allows your body to remove the waste which has accumulated during the day and to make any necessary repairs.

The amount of sleep a person needs varies (figure 2) and is not as important as the quality of sleep. There are two kinds of sleep: (1) **plain sleep** and (2) **dream sleep**. A good night's sleep must consist of alternate periods of these. For example, 90 minutes plain sleep followed by five minutes dream sleep, followed again by 90 minutes plain sleep, and so on.

Questions

1 Describe how lack of regular exercise will affect your body.

2 What is meant by fitness? Assess your own fitness (see figure 1).

3 a) What form of exercise could you use to improve stamina?
b) What form of exercise could you use to improve strength?
c) What form of exercise could you use to improve suppleness?
d) What is the best kind of exercise to lose weight?

4 Assess the benefits of attending keep fit sessions for:
a) a busy housewife,
b) an overworked executive,
c) an old age pensioner.

5 Use the other sections of the book to help you answer this question.
A 45 year old man smokes heavily, drinks regularly, eats far too much and takes very little exercise. His son is a typical active twelve year old. Compare the state of their bodies.

Exercise	Average energy expenditure (kJ per min)	Stamina	Suppleness	Strength
Golf	10–20	poor	fair	poor
Walking slowly		fair	poor	poor
Tennis	21–30	fair	good	fair
Gymnastics		fair	excellent	good
Jogging		excellent	fair	fair
Cycling slowly		good	fair	good
Walking quickly		good	poor	poor
Football	over 30	good	good	good
Swimming		excellent	excellent	excellent
Squash		excellent	good	fair
Cycling fast		excellent	fair	good
Disco dancing		good	excellent	poor

Table 1 Which exercise is best?

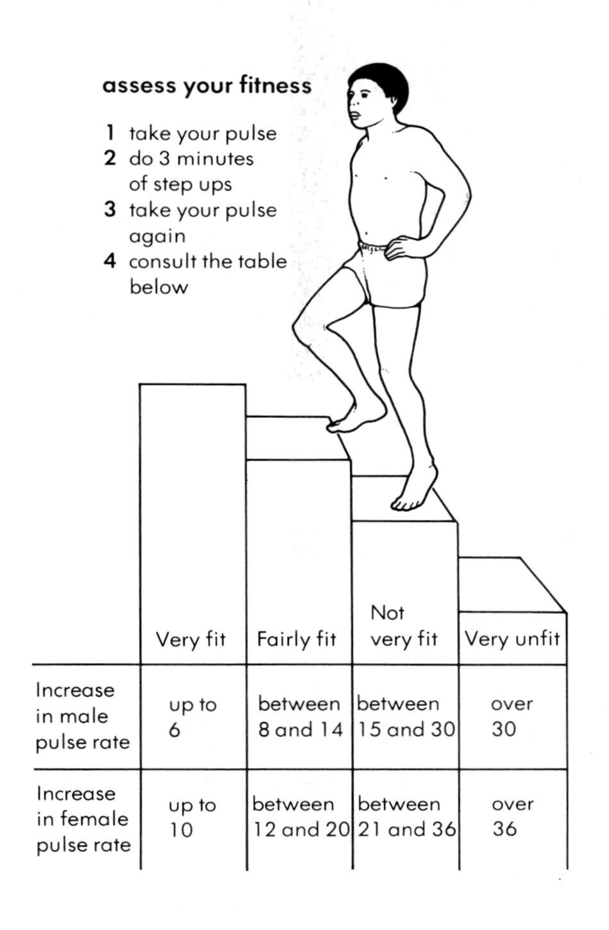

Figure 1 Do this test to assess your fitness

number of hours sleep required

0 2 4 6 8 10 12 14 16 18

1 wk 6 m 1 yr 4 yr 10 yr 20 yr 60 yr age

Figure 2 You need less sleep as you get older

9.7 Social diseases

The effects of disease

One of the major causes of premature ageing is disease. In the UK, it is responsible for about 90 per cent of all early deaths.

Are some diseases caused by the ageing process?

Diseases such as arthritis, coronary heart disease and some cancers are associated with middle to old age, but younger people can also get them. These diseases are all dealt with in the relevant sections.

Most of the other diseases usually associated with older people fall into two main categories: **social diseases** and **occupational diseases**.

What are social diseases?

Social diseases are related to the way we live and in particular, our habits. They include those caused by smoking, drinking, eating, drug taking and sexual activity.

What are sexually transmitted diseases (STDs)?

The sexually transmitted diseases are all infectious diseases which are transferred from person to person during sexual intercourse. Details of some of the more common ones are shown in table 1.

Sexually transmitted diseases are far more widespread than people think. **Gonorrhoea**, for example, has now reached **epidemic** proportions in Europe. One reason for this is that the early symptoms are so slight, especially in women, that they often go unnoticed. The disease may, therefore, be passed on to many more people before it is treated.

AIDS is also rapidly reaching epidemic proportions and is destined to become the most serious of all the STDs (see page 228).

Are these diseases dangerous?

If caught early, most can be cleared up immediately, but if the early signs are ignored or not noticed, then the results can be horrific. For example, **syphilis** may cause blindness, deafness, mental illness, heart disease and even death.

AIDS is the exception in that no matter how early it is diagnosed, it cannot at present be cured, and for the majority of sufferers this means death. The only really effective way of fighting AIDS is to make sure you do not catch it.

How can you avoid catching an STD?

There is really only one answer to this: avoid casual sex. Be very selective about your partner(s) and always maintain high standards of hygiene.

Contrary to many stories you may hear most STDs cannot be caught from toilet seats, towels or dirty clothing, because when exposed to the air, the microbes that cause them die. The only place these microbes can live is inside a body.

Unfortunately, your body does not build up immunity to these diseases, so you can catch them again even if you have been treated and cured before.

What should you do if you catch an STD?

If you are worried that you may have caught any of the sexually transmitted diseases then see your doctor or go to a local clinic at once. The longer you leave it, the worse it gets. The advice and treatment is strictly confidential, although the doctors would like to know from whom you may have caught it, so that they can also be helped, and the spread of the disease can be stopped.

Questions

1. What are sexually transmitted diseases? List as many as you can.
2. Some people think that the spread of STDs can be controlled by making everyone take a blood test every year. What do you think?
3. A close friend of yours has written to you saying she is worried about a small sore spot near her vulva. She asks for your advice. Write a letter back to her.
4. The government is spending a lot of money informing people about AIDS. Suggest why.

Disease	Organism which causes it and method of spread	Time for first symptoms to show	First symptoms	Later symptoms
Gonorrhoea (clap)	bacterium – spread by sexual contact only	2 to 10 days	*Men*: urination becomes painful and more often, may be a yellow discharge from the penis. *Women*: symptoms are rare, often there is only an increase in the vaginal discharge.	the vas deferens and fallopian tubes may be damaged causing sterility, much later the heart, liver and joints may be damaged
Syphilis (pox)	bacterium – spread by sexual contact only	10 days to 12 weeks, average 21 days	painless sore (chancre) appears on the body, usually on the sex organs – this may be inside a woman's sex organs and therefore not noticed – the chancre disappears after a few days and a skin rash may develop.	the bacteria may attack the brain, nervous system, ears, eyes, liver and heart – this may result in blindness, deafness, mental illness or even death
Urethritis (NSU)	unknown– spread by sexual contact	usually 2 to 3 days but can be a week or more	clear white or greyish discharge from the urethra, painful urination	symptoms return every few months, the joints and eyes may become sore,
Candidiasis (thrush)	fungus – spread by sexual contact	usually 2 to 3 days but can be a week or more	*Men*: rare in men, tip of penis may become red and sore *Women*: irritating white vaginal discharge, soreness and itching	symptoms continue for as long as the fungus is present
Trichomoniasis (TV)	protozoan – can be caught from toilet seats and so on	up to 1 month	*Men*: rare *Women*: yellowish green discharge from the vagina, soreness and itching	symptoms continue for as long as the protozoan is present
Genital herpes	virus – spread by sexual contact	a few days	stinging sensation on the sex organs, small vesicles appear eventually forming an ulcer – ulcers may be on penis, scrotum, vulva or cervix	ulcers keep reappearing every few weeks
Acquired immune deficiency syndrome (AIDS)	virus – spread by contact with body fluids such as blood	6 months to 6 years (average 28 months)	always tired (for weeks), swollen lymph glands, excessive weight loss, fever which lasts for weeks, pink/purple blotches on skin, diarrhoea which lasts for weeks, depression.	malignant cancers of the skin and connective tissues secondary infections which usually result in death

Questions

Recall and understanding

1 Which essential nutrient for babies is not found in milk?

A iron C protein
B calcium D sugar

2 Boys and girls both go through a growth spurt between the ages of:

A 14 – 16 years C 0 – 2 years
B 16 – 18 years D 0 – 18 months

3 Motor development is the development of:

A emotions
B reasoned thought
C co-ordinated movement
D sociability

4 The period over which the sex organs mature is called:

A menopause C fertile period
B menstruation D puberty

5 Which of the following is not a secondary sexual characteristic?

A sex (gender) C enlarged sex organs
B body hair D menstrual cycle starts

6 The hormone which initiates milk production is:

A oxytocin C oestrogen
B prolactin D progesterone

7 The end of a woman's fertile period is called:

A menstruation C adolescence
B puberty D menopause

8 Which of the following blood vessels is not found in a one year old child?

A aorta C coronary artery
B vena cava D ductus arteriosus

9 Which of the following diseases can you not develop immunity to?

A syphilis C German measles
B measles D polio

10 A small increase in pulse rate after exercise indicates:

A fitness C stamina
B illness D weak heart

Interpretation and application of data

11 The table below shows the approximate composition of cow's milk and human milk (per 100 cm^3)

	Cow	Human
Water (g)	87	87
Carbohydrate (g)	4.8	7
Protein (g)	3.4	1.3
Fat (g)	4.2	3.6
Vit A (μg)	15	60
C (mg)	1	3.8
D (μg)	0.7	0.8
Calcium (mg)	120	35
Iron (mg)	0.1	0.076
Energy (kJ)	283	274

a) State three ways in which cow's milk differs from human milk.
b) Which of the substances shows the biggest percentage difference?
c) How can cow's milk be made more suitable for feeding to a baby?
d) State three reasons apart from the nutritional ones for breast feeding a baby.
e) Look at the following table:

Age	Average daily protein required (g/kg body weight)	Average body weight (kg)
Newborn	3.6	3.2
Child	1.6	30
Young adult	0.7	70
Old man	0.5	65

i) How much human milk does a baby need to drink to get all the protein it needs?
ii) Why would it be unrealistic for a child to get all the protein it needs from milk?
iii) Why does the amount of protein required fall as you get older?

12 The figures below refer to per cent of adult size reached.

Age (yrs)	6	8	10	12	14	16	18
General growth	45	50	55	60	75	90	100
Lymphoid tissue	100	120	170	185	160	125	100
Nervous system	85	90	95	98	99	100	100
Sex organs	12	12	12	12	20	75	100

a) Plot the data on a graph.
b) From what age does the amount of lymphoid tissue start to decline?
c) At what age is the growth of the sex organs most rapid?
d) What is this growth period called?
e) At what age is the growth of the nervous system most rapid?
f) Why does the nervous system grow so rapidly early in life?

13 The graph shows the changes in hormone levels following conception.

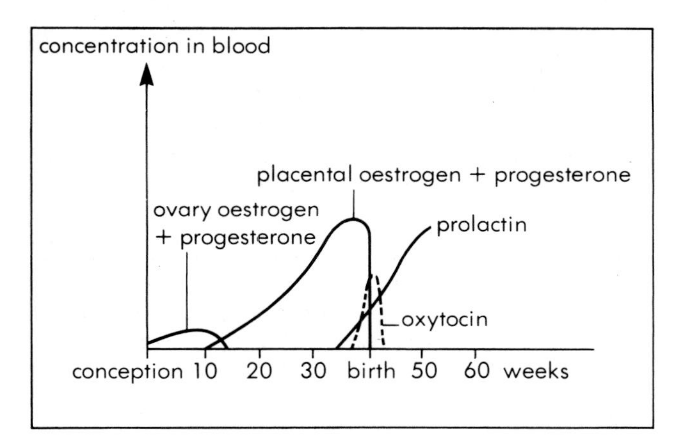

a) Approximately how long does the ovary go on producing oestrogen and progesterone?
b) What takes over the production of those two hormones?
c) What happens when the production of these hormones stops?
d) What do you think oxytocin does?
e) Which hormone do you think causes the production of milk?

14 The table shows the number of new cases of STDs in the UK. (Figures are in thousands).

		1971	1981	1984
Syphillis	male	2	3	3
	female	1	1	1
Gonorrhoea	male	43	37	34
	female	20	21	20
Herpes	male	3	7	11
	female	1	5	9
NSU	male	65	99	112
	female	14	33	44

a) Work out the totals for each year.
b) Plot a graph of these totals.
c) Do the figures suggest that STDs are on the increase or decrease?
d) Predict the numbers of STD cases expected in 1990.
e) Which STDs in particular are increasing?
f) There appears to be more cases of STDs in men than women. Suggest why.

15 The graph below shows how smoking can affect fitness.

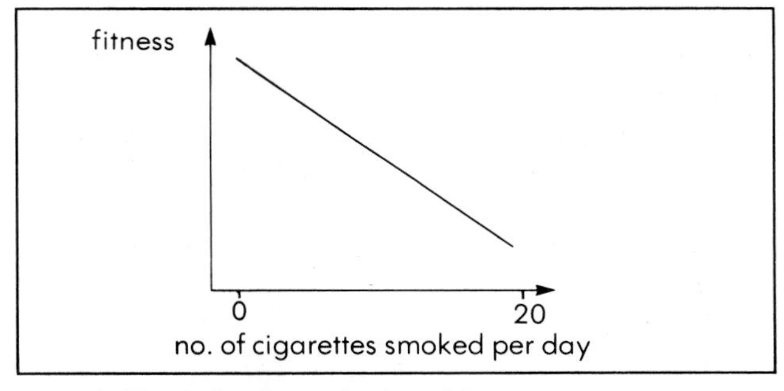

a) Explain the relationship.
b) Why do you think a person who smokes a lot can never be fully fit?
c) The graph below shows the relationship between rest and fitness.

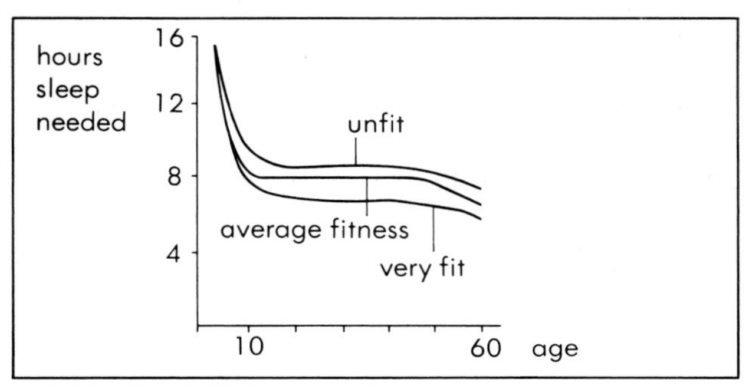

i) Explain the relationship.
ii) During which period of your life do you need most sleep?
iii) During which period of your life do you need least sleep?

CHAPTER 10: GROWTH OF POPULA

As long ago as 1798, the Reverend Thomas Malthus made a prediction that eventually the population of the world would stop increasing because of the shortage of food.

He may well be proved right, but probably not in the way he meant. Most people in the world today do not get enough to eat, but this is not due to a world food shortage. It is more due to political difficulties, poverty and ignorance.

In the following chapter, human population growth is discussed more fully. Predictions are made about the future and possible solutions outlined.

People, people everywhere!

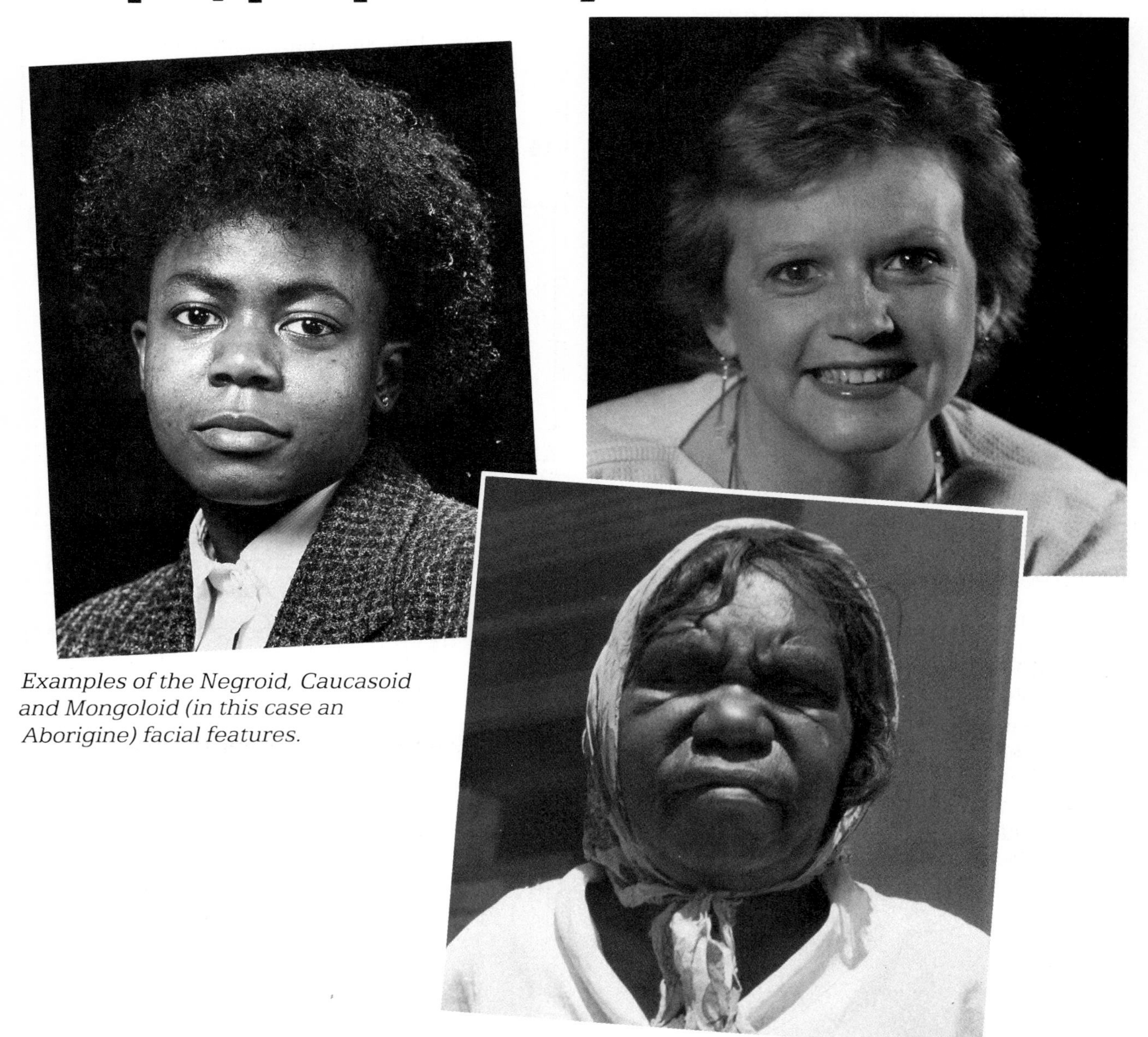

Examples of the Negroid, Caucasoid and Mongoloid (in this case an Aborigine) facial features.

This section looks at how human population growth has contributed to the development of the many different human races.

The next time you go for a walk count the number of different kinds of dog that you see and try and work out how they are different. The time after, do the same with people. All the different kinds you see are called **breeds**.

What is a breed?

The simplest way to describe a breed is as a group of organisms that have the same observable characteristics. The number of different breeds of people that you counted will therefore depend upon the characteristics you chose. If for example you chose hair colour, you will have probably counted at least five. If however you chose hair type, you will only have counted four.

With humans, the term **race** is usually used in preference to breed.

How many human races are there?

Anthropologists (people who study humans) can never agree on this. Some say thirty-four, whereas others say just three, four or seven.

If physical characteristics such as skin colour, hair texture, position and shape of facial features etc are used, three main groups shown in the photos can be seen: the Caucasoids, Negroids and Mongoloids. Their characteristics are described in table 1.

If blood groups are used to divide up humans, five main groups appear. These are the Caucasoids, Negroids, Mongoloids, Amerindians and Australoids.

How did the different races come about?

No one really knows the answer to this but it is thought that they all developed from one breed due to the selective pressure of the environment and social selection.

To illustrate this imagine a group of people (one breed) living together quite happily in a fertile but relatively sunless place. As the population gets bigger food starts to become scarce and so some of them decide to leave the main group in search of pastures new. They eventually settle in another fertile but this time very sunny place. They now have plenty of food once more but they also all suffer from sunburn.

Table 1 The 3 main human races based on physical features

Race	Physical features	Examples
Caucasoids	fair skin, fine wavy hair, hair on body and face, straight eyes, thin to medium lips, thin nose	English Germans Egyptians Iranians Indians Basques
Negroids	dark skin, woolly hair, little body or facial hair, straight eyes, thick lips, broad flat nose	American negroes Nigerians Ethiopians Congo pygmies
Mongoloids	yellowy brown skin, straight hair, little or no body hair, slant eyes, medium lips, medium to broad nose	Chinese Japanese Indonesians American indians Aborigines Melanesians

The mutation of genes

After a few generations some of the children start developing more pigment in their skin and they no longer have to worry about the sunburn. They are also more attractive, not having great blisters all the time. So because of this one change these few people are (a) better suited to survive in their new environment and (b) are chosen as mates more often than the others.

The change was caused by a **mutated gene** (see page 180) and therefore is passed onto their children. Eventually more and more of the breakaway group are born with brown skin. If other characteristics change in the same way they will soon become a completely different breed.

Is one race more intelligent than another?

Hitler certainly believed this, but there is in fact no evidence to support his theories. All races compete very successfully in their own environment. There are however still people like Hitler who believe that a super race should be produced by encouraging clever people to breed more than less clever people. They are called **eugenicists**.

10.1 Overpopulation

What is the world population now?

The world is fast becoming over-populated. At present there are nearly 5000 million people living on the earth and this is being added to by approximately 200 000 a day. At this rate of increase, the population will have doubled in about 68 years time and in 500 years time, there will be standing room only! Figure 4 shows the annual growth rate for different parts of the world.

How do we know how many people are on the earth?

These days most countries in the world hold a **census** every ten years or so. This usually takes the form of a questionnaire which each household must fill in. It asks how many people live there, their ages and their status. In addition, all births and deaths must be reported to the **Registrar General's Office**. All this information is used to work out the population size and is also used to calculate the population's **vital rates**.

What are vital rates?

Vital rates are values such as **birth rate** and **death rate**. They are usually expressed as numbers per 1000 people in the population and are extremely useful to organisations which need to do forward planning. For example, a local authority may need to predict how many schools it will need in five years time. Vital rates can also be used to indicate whether a population is growing, static or shrinking. The infant mortality rate is often used as an indicator of the state of development of a country.

Why is the world population increasing so rapidly?

Calculations have shown that the human population is still growing at an ever increasing rate (figures 1 and 2). There are many possible reasons for this. Some of these are:

- Better medical care has meant that far fewer babies die at birth or when they are very young. In the UK the **infant mortality rate** (the rate at which babies die) has fallen by just over 30 per cent in the last ten years. The average fall for the world as a whole is 35 per cent.
- Better medical care has increased the life expectancy of a person. In the UK the average life expectancy of a man has risen from 48 to 70 years and a woman from 50 to 76 years over the last one hundred years.
- Better medicine has almost wiped out some diseases and dramatically reduced the death rate of others.
- More health education has made people aware of what is good and what is bad for them.
- Better agriculture means that there is more food to go round and this has been shared out.
- There has been more sharing of knowledge and resources.
- Fewer people are being killed by wars.

Questions

1 A typical population growth curve shows four phases (figure 1).
a) Explain these four phases.
b) What is meant by environmental resistance?
c) Draw the growth curve for a population which meets no environmental resistance.
d) Which phase are we in?

2 The present average growth rate of the developed countries is about 1.25 per cent per year. How long will it take for its population to double?

3 The present average growth rate of the developing countries is about 2 per cent per year. How long will it take for its population to double?

4 Towards the end of 1970s the government started closing schools down and stopped training so many teachers. Explain why and how they got the information they based their decision on.

5 Draw a pie chart to show the distribution of the world population in 1985 (use table 1).

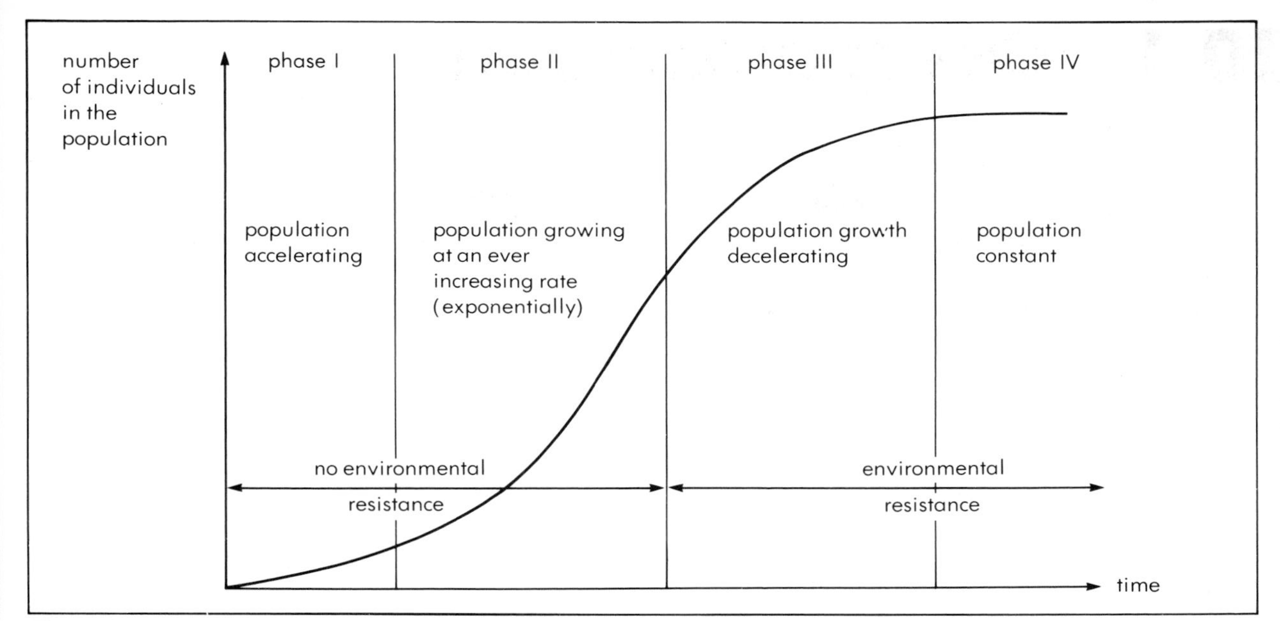

Figure 1 A typical population growth pattern

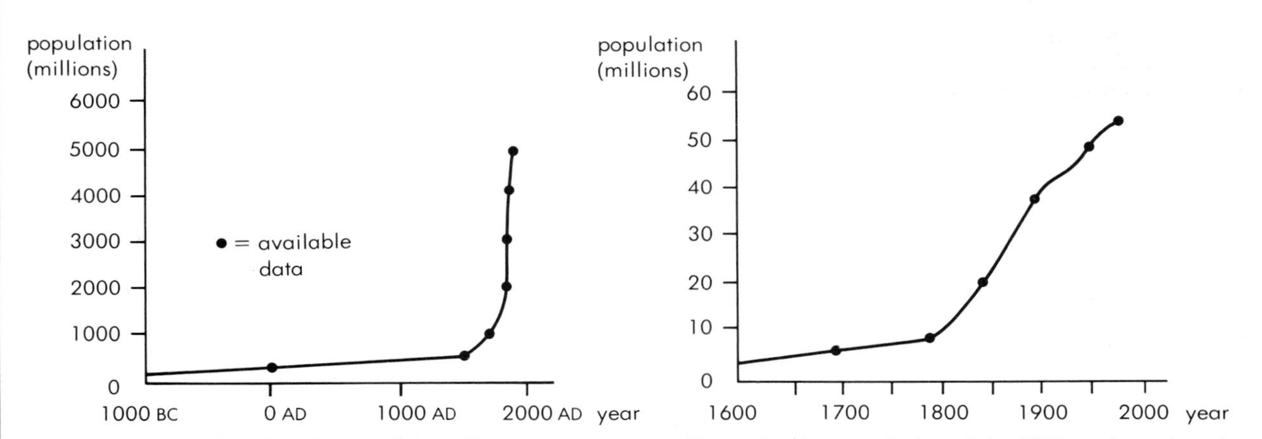

Figure 2 World population growth is still in phase II

Figure 3 The population of the UK has risen sharply since the 18th century

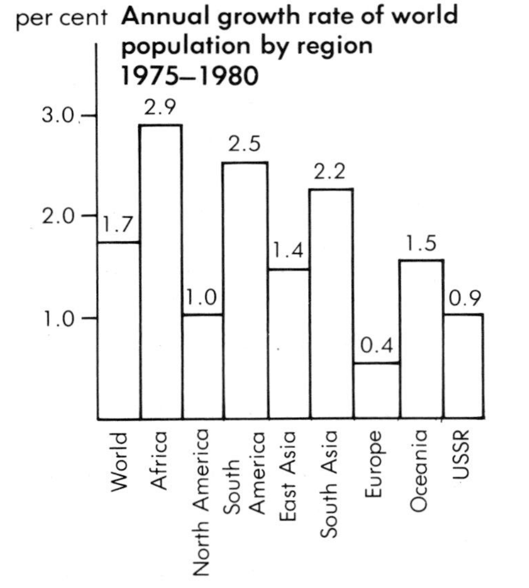

Figure 4 Africa has the highest population growth rate in the world

Region	1950 (millions)	Percent of world total	1985 (millions)	Percent of world total
Africa	222	8.9	553	11.4
Latin America	165	6.6	406	8.4
North America	166	6.6	263	5.4
East Asia	571	22.8	1252	25.9
South Asia	695	27.8	1572	32.5
Europe	392	15.7	492	10.2
Oceania	113	4.5	25	0.5
USSR	180	7.1	278	5.7

Source: Department of International Economic and Social Affairs 1985.

Table 1 A comparison of world population by region, 1950 and 1985

10.2 Population control

What might happen in the world?

The world as yet has never been seriously over-populated and so we can only predict what might happen if the world did get too overcrowded. These might be:

- **Not enough food** to go round. Although agriculture is getting better and new sources of food are being discovered (chapter 2), eventually there will come a time when the rate of population increase overtakes the rate of food production. Figures presently available show that in many countries although the rate of food production is keeping up with the rate of population increase, the amount of food available per person is actually falling.
- **More pollution** More people create more waste, either directly, as in the case of sewage, or indirectly, as with sulphur dioxide. These pollutants could eventually destroy the planet (see chapter 14).
- **War** In certain animal populations, an increase in numbers results in a change in behaviour. This is almost certainly due to the increase in stress. Overcrowding in the human population may lead to irrational behaviour and more war.
- **Shortage of resources** If timber goes on being used at the present rate, it will run out in 50 years. Coal will run out in 150 years and natural gas even sooner. These figures are based on the present population size. If it continues to increase, there will be more demand and these resources and others will run out even sooner.

What is the Club of Rome?

Figure 2 shows what could happen if present trends continue. This model was put forward by a group of scientists, economists, industrialists and politicians who meet regularly to discuss such issues. They call themselves the **Club of Rome**.

It is clear that unless we can control the growth of the human population, a ridiculous and frightening situation will eventually be reached. What is needed is a **worldwide population policy**.

What sort of things would such a policy contain?

There are only two ways of preventing overpopulation; either increase the death rate and/or decrease the birth rate. Many governments have been trying to reduce the birth rate for years now. Some of the ways this is done are rather worrying. Here are some techniques countries have employed:

- To educate people so that they become more aware of the situation and their choices.
- To educate people on methods of contraception; sex education is rapidly becoming part of the curriculum in schools.
- To provide, free of charge, many of the forms of contraception. In the UK condoms are free from family planning clinics.
- To legislate against having more than two children. China adopted a policy in 1968 whereby people are penalised for having more than *one* child.
- To pay men to be sterilised. This has been operating in India for many years now.

Even without these measures the United Nations has predicted that the world population will eventually stabilise. However, by this time it will be seriously overcrowded.

Questions

1. In 1000 years time it has been predicted that there will be 1500 people per square metre of the earth's surface. Explain with reasons whether or not you think it will ever get to this.
2. Imagine you have been put in charge of making sure that it does never get to this state. Write down all the things you would consider doing to stop it happening.
3. India has for many years been paying men to be sterilised yet its population is still rising very rapidly. Why do you think this policy has not worked?
4. Some people have suggested that people who are fed up with living or are very ill should be allowed to die if they want to. What do you think?
5. Nature has always operated on the principle of survival of the fittest. With this in mind do you think it is right to help people who cannot help themselves, such as the starving millions in the Sudan? State your reasons.

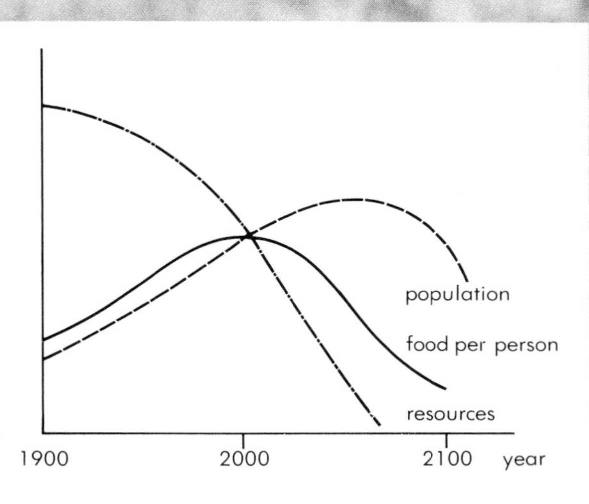

Figure 2 A model of what could happen if present trends continue (Club of Rome)

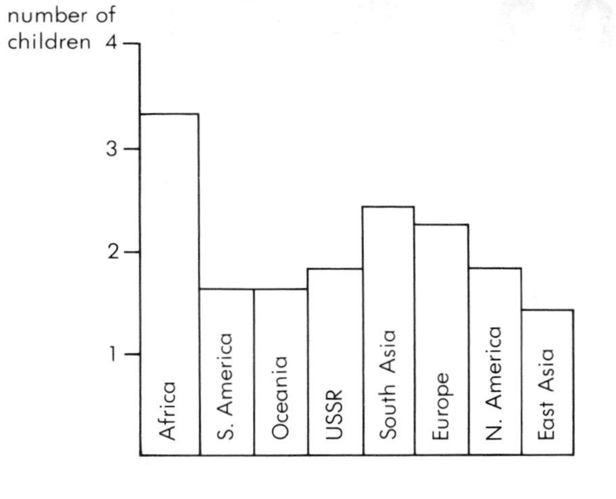

number of children
4
3
2
1
Africa
S. America
Oceania
USSR
South Asia
Europe
N. America
East Asia

Figure 3 The number of children needed per family to maintain present population levels (1987 figures)

Area of world	Year
East Asia	1990
North America	1995
Europe	2020
South Asia	2020
USSR	2030
Oceania	2035
South America	2040
Arica	2045

Table 1 United Nations predictions of when the net reproductive rate will equal 1

Figure 1 Many areas of the world are already overcrowded

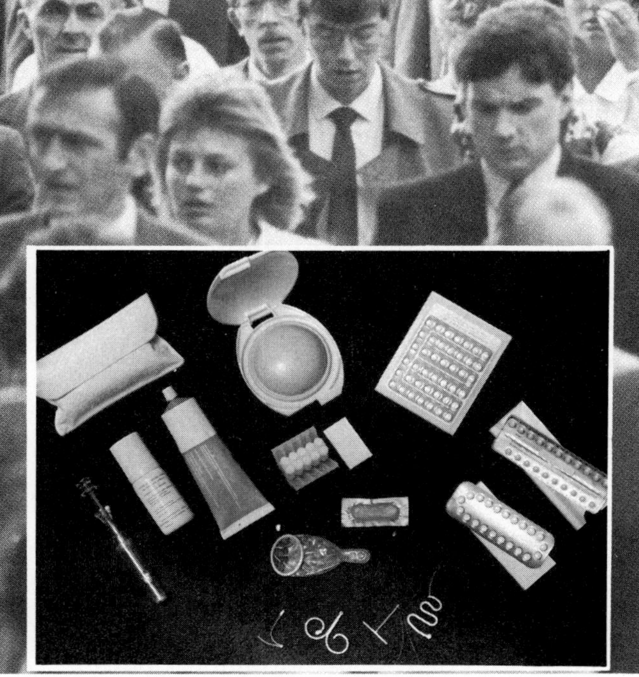

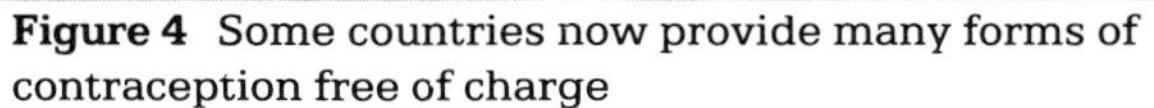

Figure 4 Some countries now provide many forms of contraception free of charge

10.3 Birth control

What is contraception?

Contraception means avoiding conception - *contra/ception*. There are many methods of doing this, some more reliable than others. Also some are safer than others (see table 1).

- **Withdrawal** The man withdraws his penis immediately *before* ejaculation so this takes place outside the vagina.
- **The safe period** The woman predicts when ovulation will take place each month by taking her body temperature. Once this has been decided, she must not have intercourse for the five days before it and the five days after it. This is known as the **fertile period**. The remaining days of the cycle are known as the **safe period** and it is only during this time that intercourse should take place (figure 1).
- **The condom** The condom is a very thin sheath of rubber which is rolled over the penis before intercourse. On ejaculation, the sperms are caught in the sheath (figure 2). A **spermicidal cream** should always be used with a condom. This is the only form of contraception which gives protection against sexually transmitted diseases.
- **The cap or diaphragm** The cap is a dome-shaped piece of rubber which fits over the entrance to the uterus (the cervix). It should be fitted before intercourse and left in position for six hours afterwards (figure 3). A spermicidal cream should be used with it.
- **The intra-uterine device (IUD)** This is also called the coil. It is a piece of carefully shaped plastic which is inserted through the cervix into the uterus where it can be left for many months (figure 4). No-one really knows why it works, but it seems to stop fertilised eggs implanting. A variation of this is the morning after coil. This can provide protection if inserted up to 72 hours after intercourse.
- **The pill** There are two main kinds of contraceptive pill: the **combined pill** and the **progesterone only pill** (mini-pill). The combined pill contains the hormones oestrogen and progesterone. It works by preventing the release of an egg from the ovary.
  The progesterone only pill works by thickening the mucus in the entrance to the uterus. This makes it almost impossible for the sperms to get through.
  Both types of pill must be taken regularly and according to the instructions if contraception is to be successful.
- **The sperm absorbing sponge** A sponge soaked in spermicide is placed over the cervix before intercourse. This absorbs the sperm and can be taken out later to be washed and re-used.
- **Sterilisation** Male sterilisation is called **vasectomy**. It involves cutting and tying the sperm duct so that no sperms can be released during ejaculation (figure 5).
  Females can be sterilised by a similar operation to cut the fallopian tubes (figure 5). Usually these operations are irreversible.

What other methods are being developed?

There are some new methods of contraception being researched at the moment. But whether they are better remains to be seen. These are:

- **The male pill** This prevents the production of sperm.
- **The hormone releasing ring** This is fitted inside the vagina in much the same way as the cap. It releases hormones into the body which fool it into thinking it is pregnant. No more ova are therefore released. A new ring must be fitted every month.
- **The hormone pellet** A small pellet impregnated with hormones is implanted under the skin. As it dissolves, it releases the hormones into the body.
- **The hormone injection** A progesterone injection is given several times a year. Although this method has been in use for several years now it has still not been perfected.

Questions

1. There are many methods of birth control. Before choosing the right one, people must weigh up the advantages and disadvantages. Produce a table listing these.
2. Which is the only form of contraception which gives protection against AIDS?
3. Can abortion be considered as a method of birth control? Give reasons for your answer.
4. Do you think the male pill is a good idea or not? Give reasons for your answer.

Methods	Pregnancies in every hundred users	Some disadvantages
No contraceptives	40 (average)	
Withdrawal	17	may not withdraw in time, spoils the pleasure
Safe peiod	up to 15	needs careful record keeping – cannot have sex at certain times
Condom + spermicidal cream	4	condom may cause loss of sensation, may slip off.
Cap + spermicidal cream	3	cap has to be inserted well before intercourse, may move
IUD	3	needs fitting by a doctor, may slip out unnoticed
Combined pill	almost 0	are various minor side effects, must be taken daily
Progesterone pill	2	are various minor side effects, must be taken at same time every day
Sterilisation	0	not reversible

Table 1 The failure rate of contraceptives

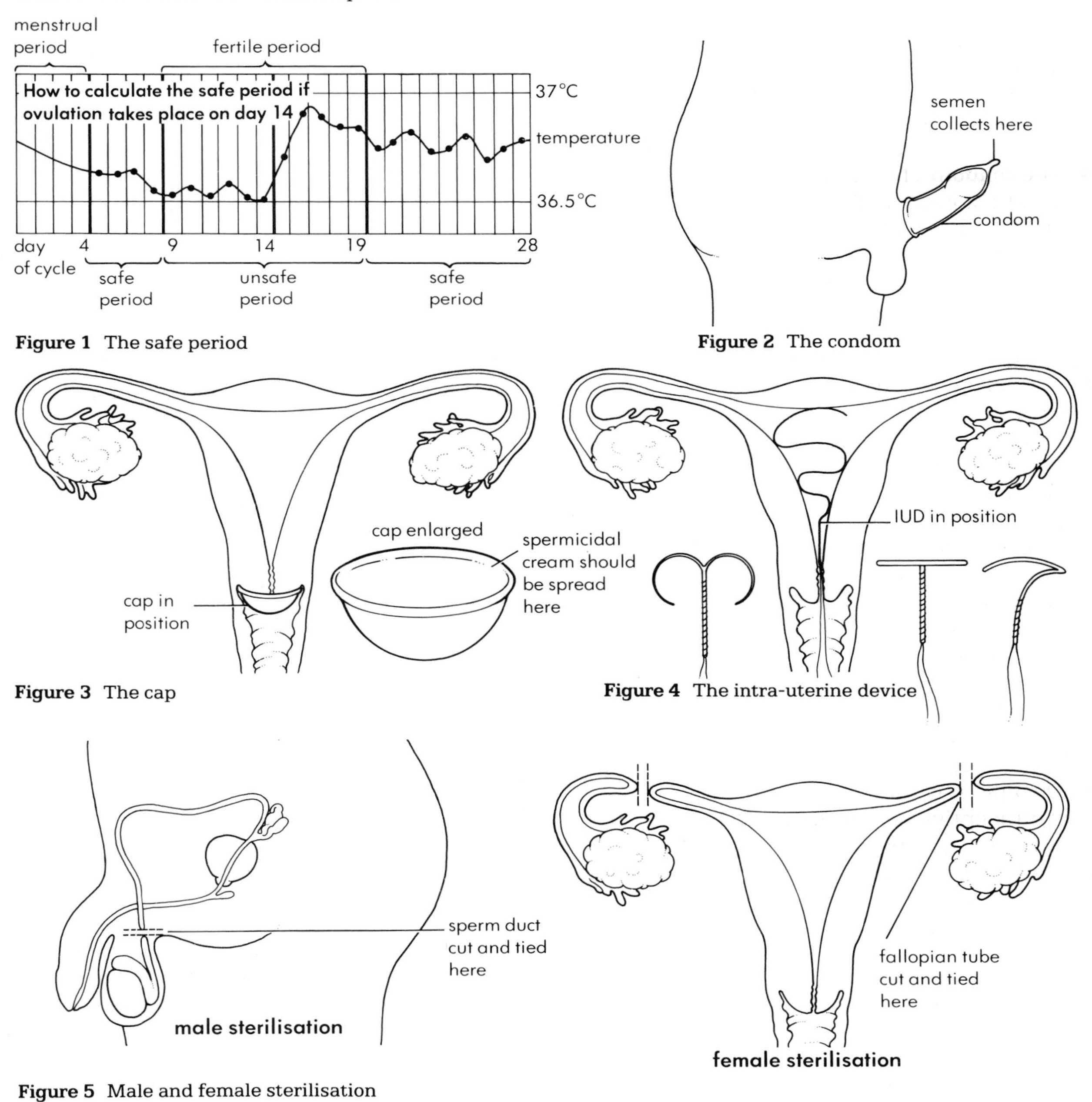

Figure 1 The safe period

Figure 2 The condom

Figure 3 The cap

Figure 4 The intra-uterine device

Figure 5 Male and female sterilisation

Questions

Recall and understanding

1 The office where records of births and deaths are kept is:
A The Office of Population and Censuses
B The Registrar General's Office
C The Tax Office
D The Post Office

2 The infant mortality rate is a measure of:
A the death rate of children
B the infant birth rate
C the provisional death rate
D the death rate of under 1 year olds

3 The human population is increasing:
A slowly C exponentially
B negatively D more and more slowly

4 The only form of contraception which is 100 per cent effective is:
A sterilisation
B the contraceptive pill
C withdrawal
D the safe period method

5 The world population is increasing at the rate of 200 000 per day. How long before there are 1 million more people?
A 25 days C 5 days
B 5 hours D 5 years

6 The annual growth rate of the world is about 1.25 per cent. There are 5000 million people at present. How many will there be in four years?
A 5125 million C 5500 million
B 5050 million D 6000 million

7 As the population increases, which of the following will also increase?
A the chance of a war
B food production
C manufactured goods
D pollution

8 A population starts to decline when it meets:
A predators
B a stronger species
C parasites
D environmental resistance

9 Just prior to ovulation:
A the body temperature rises
B the period starts
C a headache develops
D you need to visit the toilet

10 The difference between the birth rate and death rate is called the:
A increase C net population size
B growth rate D gross population size

Application and interpretation of data

11 A good way of showing age distribution in a population is to draw a population pyramid.

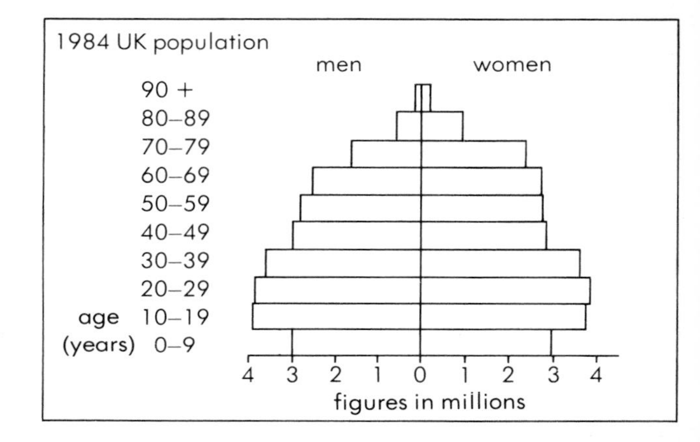

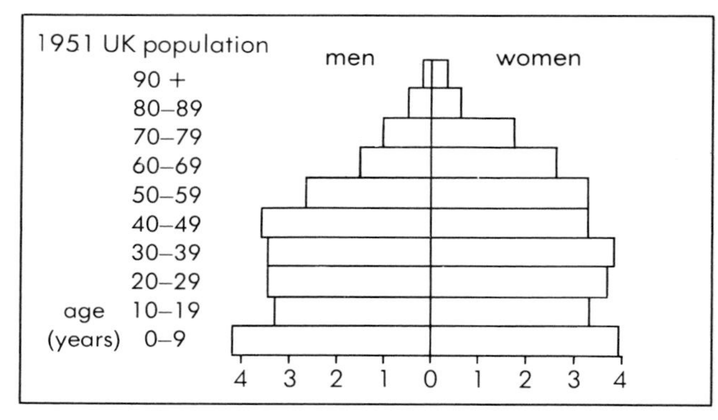

a) Which age range contains the largest number of people in
i) 1984
ii) 1951?

b) In 1951, the local authorities started building a lot of schools. Why?

c) How does the 50s population compare with the 80s with respect to:
i) size
ii) age distribution
iii) ratio of men to women?

12 The infant mortality rate (IMR) is a good indicator of the state of development of a country:

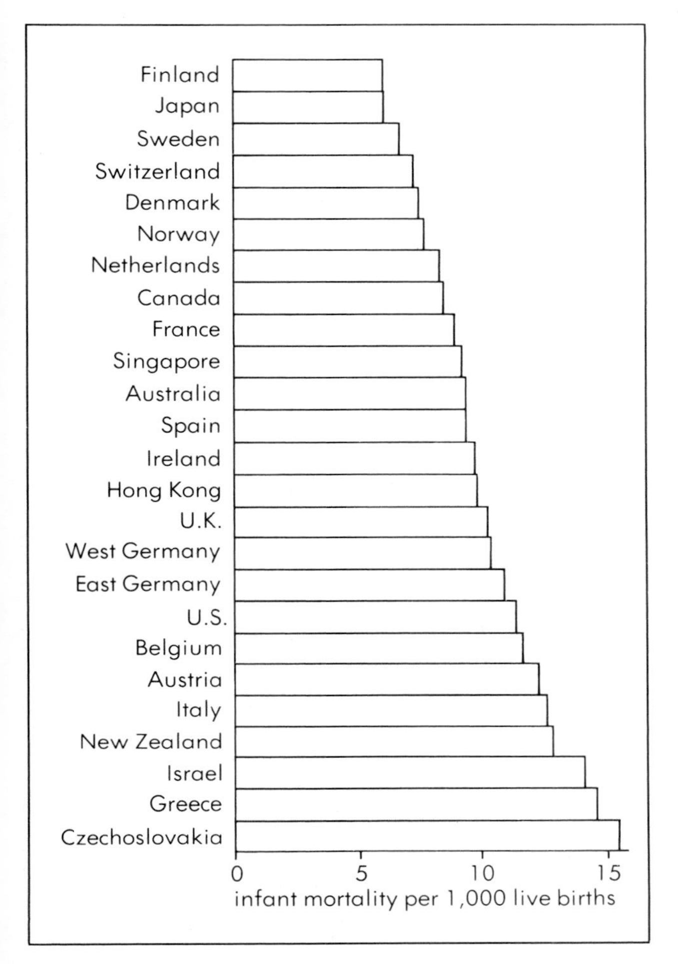

a) What is meant by the IMR?
b) What is the overall range of the IMR for the countries shown?
c) What is the IMR of the UK?
d) Countries like Africa, India and Bangladesh have IMRs over 30. What does this tell you about their state of development compared to the countries above?
e) The IMR for the UK over the last 10 years:

Year	1975	1979	1980	1981	1982	1983	1984
IMR	14.7	12.6	11.9	10.9	10.6	10	10

i) What do these figures tell you about the health care for infants in this country?
ii) The World Health Organisation IMR target is 20. How do we compare with this?

13 The doubling time of a population depends upon the annual percentage increase.

Percentage increase/yr	Doubling time (yrs)
0.5	140
1.0	70
2.0	35
3.0	23
4.0	18
5.0	14
10.0	7

a) Plot these figures on a graph.
b) The doubling time for the UK population between 1850 and 1930 was 80 years, and between 1930 and 1975 was 45 years. What was the annual rate of increase for these two periods?
c) The average growth rate for the developed world is 2 per cent. How long will it take for the population to double?
d) The average GR for the undeveloped countries is nearer 3.0 per cent. How long will it take for this to double?
e) Explain why there is such a difference.

14 The chart below shows the distribution of the world population by regions in 1982.

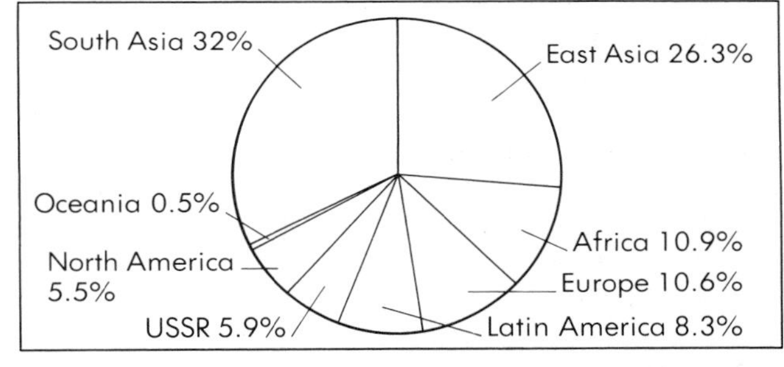

a) What proportions are in the developed and underdeveloped world?
b) The world population at the moment is 5000 million. How many are in the developed and underdeveloped world?
c) Consider the table:

	Percent aged	
	under 15	**60 or over**
UK	21	20
Greece	22	17
Australia	25	14
USSR	36	13
Egypt	40	6
India	39	6
Brazil	37	6

i) Which of the countries have a low life expectancy?
ii) List three reasons for the differences between Australia and India.

CHAPTER 11: *INHERITANCE AND VARI*

Your body is built to a 'blueprint'. This blueprint is stored in the nucleus of each cell. It takes the form of genes.

Each gene contains the information for one particular characteristic of your body. You inherited your genes from your parents, some from your mother and some from your father. If you have children they will inherit some of your genes. In other words they will have some of your characteristics, just as you have some of your parents'.

The following chapter explains how genes are passed on from parents to children and why this always results in everybody being different.

You may also have heard that humans are closely related to the apes. This chapter also covers how all life forms could have developed from one common ancester by a process called **evolution**.

The white streak in the hair of all male members of this family is inherited.

Scientists harness bacteria

This section looks at a topic linked to all this – a scientific technique called **genetic engineering**.

What is genetic engineering?

The genes in a cell tell it how to work and what products to make. Scientists can now transfer genes from one organism to another and thereby add or take away instructions. The technique is called genetic engineering.

What are its main uses?

There are many current applications of genetic engineering, and many more are still in the research stage. The following four areas are of particular interest.

- **To produce high quality products such as hormones, enzymes and antibiotics.**
  Human growth hormone (HGH) used to treat growth disorders such as dwarfism (page 97) is produced in this way. The HGH gene is placed into a fast growing bacterium. This bacterium then grows and multiplies, soon resulting in millions of bacteria, all containing copies of the HGH gene. Given the right conditions they can be made to produce HGH.
  It is hoped that soon the blood clotting factor VIII which is absent in some haemophiliacs can also be produced in this way. At present this is obtained from human blood given by donors (page 49). On rare occasions this has been found to be contaminated with the AIDS or Hepatitis B viruses. Producing it by genetic engineering would completely eliminate this dreadful problem.
- **To produce better and more productive crop plants.** Genes can be inserted into plants to give them more resistance to diseases and to enable them to grow on low nutrient soils.
  Some bacteria contain a gene that enables them to use atmospheric nitrogen to make proteins. Scientists are trying to isolate this gene and transfer it to plants. If successful this could save the farmers a fortune on nitrogen fertilisers.
- **To extend the life of antibody-producing cells (lymphocytes) and enable antibodies to be produced in the laboratory.** Normally antibodies can only be produced by injecting an antigen into an animal (page 226). Any attempt to grow lymphocytes and produce antibodies outside the body has always failed.
  However, a new development in genetic engineering can now combine the lymphocyte with a cancer cell. The new cell formed can be grown in the laboratory and will produce antibodies. These are called **monoclonal antibodies**.
  Many scientists believe this to be the most important development in genetic engineering.
- **To remove unwanted genes from organisms.** Many plants contain genes that actually lower their resistance to disease. In the next decade or so it should be possible to replace these genes with better ones.
  Taken to its extreme, it may soon even be possible to substitute genes in fertilised ova before implanting them in the uterus. This could prevent many hereditary diseases and give new hope to people who choose not to have a family because they might pass on a hereditary disease.

What is genetic breeding?

Genetic engineering is really an extension of **genetic breeding** which itself has been practised for centuries. It involves breeding organisms which have favourable characteristics in the hope that the offspring will inherit these. Some would call it artificial selection (page 186).

Most of our crop plants and domestic animals have been produced by genetic breeding. Some of them now bear no resemblance to the wild stock they originally came from (see page 186).

Genetic engineering has taken the guesswork out of genetic breeding.

Why are some people worried by genetic engineering?

Genetic engineering gives scientists the ultimate power, that is to create new life forms. The worry is that one of these new organisms might escape from the laboratory and turn into a killer. It is possible that it could even kill off the human race!

When the potential of genetic engineering was first realised, many scientists refused to use it until strict safety regulations and codes of conduct were drawn up. That's how worried they were!

11.1 Genes

What is a gene?

Do you ever wonder what makes you different from your friend. Why do you have blue eyes? Why does your friend have black hair? Characteristics like these depend on the **genes** people inherit from their parents during the reproductive process. A gene is a set of instructions which the body can use to produce a particular characteristic. For example, a gene may contain the instructions to produce a certain eye colour. The usual way to describe this is to say that the gene **codes** for eye colour. Another gene might code for hair colour and so on. Genes are often called the **units of inheritance** because it is these that carry the instructions for characteristics from the parents to their children.

How are the instructions in genes carried out?

The instructions that a gene contains are actually the blueprints for the manufacture of enzymes. The presence or absence of an enzyme can be the difference between a characteristic being present or not. So if you have brown eyes, it is because you have a gene which contains the blueprint for the enzyme which helps in the production of the brown pigment (see figure 1).

How many genes do you have?

The more complex an organism is, the more genes it has. A human being is one of the most complex organisms and therefore has a lot of genes (i.e. instructions for its assembly and functioning). These genes actually form part of the **chromosomes** within a cell nucleus.

In humans, there are enough genes to form 46 chromosomes. They are in fact in pairs, the two chromosomes of a pair looking exactly alike (see page 182). We therefore call them **homologous pairs** of chromosomes. Homologous means 'the same'.

Every cell nucleus except the gametes, contains all 46 (23 pairs) of these chromosomes. They are said to be in the **diploid** state. The gametes only contain 23 chromosomes, one from each homologous pair. These are in the **haploid** state.

Why do the gametes only contain 23 chromosomes?

The gametes are produced by a special kind of cell division called **meiosis**. Meiosis separates the chromosomes of each homologous pair and directs them into different gametes (figure 2). Each sperm produced will therefore contain half the man's chromosomes and each egg will contain half the woman's chromosomes, one from each homologous pair. When these gametes fuse at fertilisation, the chromosomes are able to pair off again, but now each homologous pair is made up of one chromosome (and its genes) from the father and one chromosome (and its genes) from the mother.

How is meiosis different to mitosis?

Meiosis is only used to produce gametes and always results in haploid cells. Mitosis is the type of cell division which is used for growth and asexual reproduction. During mitosis, the number of chromosomes remains unchanged. Figure 2 shows how the chromosomes behave during these two kinds of cell division.

Questions

1. Every business has an office. Within this office is a filing cabinet and inside this are files. These files contain information which enables the business to function. Compare this to a cell. In a cell, which parts are equivalent to the office, filing cabinet and files?
2. With the help of a diagram explain how genes produce their effects.
3. Why are genes sometimes called the 'units of inheritance'?
4. Your cells contain 23 homologous pairs of chromosomes. How many chromosomes is this?
5. Here are some statements about meiosis and mitosis. Arrange them in two lists based on which type of cell division they best apply to:
   - produces two identical daughter cells
   - produces four haploid cells
   - homologous chromosomes pair off
   - produces clones of the parent cell
   - takes place in all cells
   - takes place in gamete producing cells

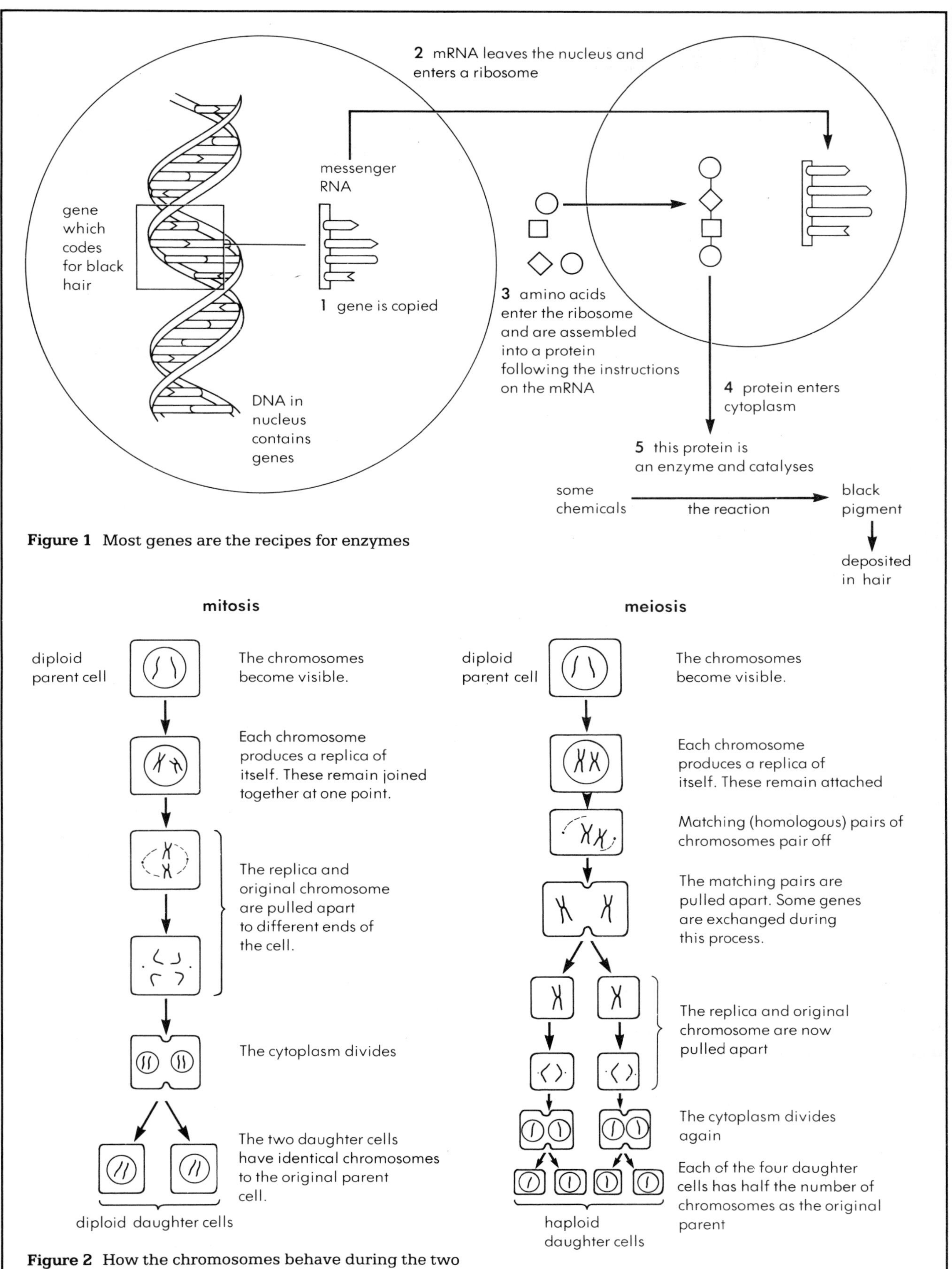

Figure 1 Most genes are the recipes for enzymes

Figure 2 How the chromosomes behave during the two kinds of cell division

11.2 Mendel and inheritance

Is each characteristic controlled by one gene only?

A characteristic is usually controlled by two genes which are on the different chromosomes of a homologous pair. When these chromosomes are separated during meiosis the two genes which control the characteristic are also separated.

Some of the more complex characteristics such as skin colour are controlled by several genes. These are very difficult to study and so the examples given in this book are all characteristics controlled by two genes, one on each chromosome of an homologous pair.

How can a characteristic be controlled by more than one gene?

Much of our knowledge of genetics comes from the early work of an Austrian monk called **Gregor Mendel**. Between 1856 and 1865 he carried out a series of breeding experiments with pea plants. Using **pure bred** tall and dwarf plants, he first investigated the characteristic of height. Pure bred plants are ones which have retained the same characteristics for many generations. From results he concluded that the height of a plant was controlled by two genes. He further reasoned that these two genes were on separate chromosomes and although they both coded for height, the exact information contained within them could be different. For example, one gene could code for tallness and the other for dwarfness.

Mendel repeated his experiments using several other visible characteristics of pea plants and on each occasion he found that the characteristic was controlled by two genes, but the exact information contained within these genes could be different. He called these different forms of the gene **alleles**. Mendel further reasoned that one of the alleles was usually more **dominant** than the other and this allele was always the one which the cell used as its instructions. The 'weaker' allele he referred to as the **recessive** allele. Details of one of Mendel's breeding experiments is shown in figure 1.

Genotype and phenotype

Normally when you describe a person, you describe their appearance; for example you say if they are tall or short, fat or thin. In genetics language this is called the person's **phenotype**.

It is also possible to describe a person, or any organism, in terms of the genes it contains. Geneticists call this its **genotype**. The genotype of a particular characteristic can be described as being **heterozygous** or **homozygous**.

Heterozygous means that the two genes coding for a particular character contain different information, i.e. are alleles. Homozygous means they contain the same information. **Homozygous dominant** means that they are both dominant genes and **homozygous recessive** means that they are both recessive genes (see figure 2).

Questions

1 Gregor Mendel was famous for his work on pea plants. Explain how this work has helped us to understand inheritance more fully.

2 In your opinion what was the most important discovery Mendel made?

3 A *tall* man who was *homozygous dominant* for the characteristic married a *small* woman who was also *homozygous* and produced two children who were both *tall*, but *heterozygous*. The words which are italic either describe the phenotype or the genotype of the people. Which describe which?

4 The table below shows the results of some of Mendel's crosses involving pea plants. From the results predict which of the two alleles in each cross is the dominant one.

Character investigated	Cross	2nd generation offspring
form of seed	smooth × wrinkled	5474 smooth, 1850 wrinkled
colour of seed coat	grey-brown × white	705 grey-brown, 224 white
colour of cotyledons	green × yellow	2001 green, 6022 yellow

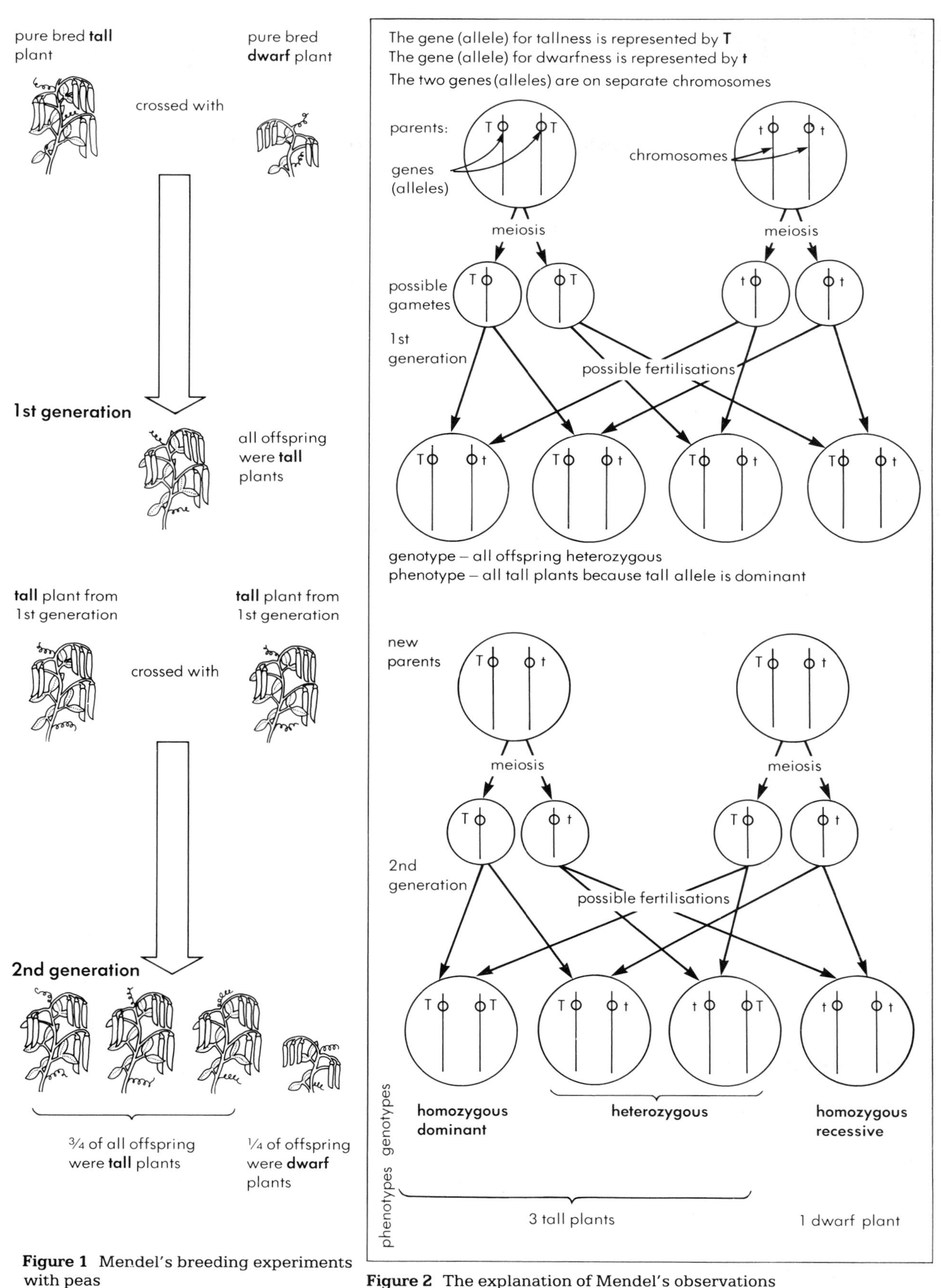

Figure 1 Mendel's breeding experiments with peas

Figure 2 The explanation of Mendel's observations

11.3 Inheritance

What will a baby look like?

If you know exactly what the genotypes (and resulting phenotypes) of the parents are and which is the dominant form of the characteristic, it should be possible to predict some of the features a baby will possess. Taken to its extreme, it is even possible to select genotypes for breeding, so that you can be certain that only good characteristics are passed on. This is done a lot with plants and some animals, but there are laws which stop geneticists trying it with humans. Figure 1 shows how eye colour is passed on from parents to their children.

What decides if a baby will be a boy or girl?

The genes which control the sex of a person are found on one pair of chromosomes called the **sex chromosomes**. In a female these chromosomes are identical and can be represented by the letters **XX**. In a male, however, one of the chromosomes is smaller than the other. This is represented by the letter **Y** and a male's sex chromosomes are therefore **XY**. Figure 2 shows how these chromosomes are passed from parents to offspring.

Do the sex chromosomes carry genes which have nothing to do with sex?

Although the X chromosome carries other genes, the Y chromosome does not appear to. This means that some characteristics in a man are coded for by only one gene. These characteristics are said to be **sex-linked**. Examples are **haemophilia** and at least one form of **muscular dystrophy**.

What is haemophilia?

Haemophilia is a blood disorder caused by the presence of a sex-linked recessive gene. The blood of people who have this disorder will not clot. Females who are homozygous recessive or males who have only the one recessive gene will have haemophilia. A heterozygous female is called a **carrier** because although she does not have the disease, she does have a recessive gene that can be passed on to her children. Figure 3 illustrates how haemophilia can be inherited.

The inheritance of blood groups

A person's blood group is determined by the presence or absence of blood factors A and B and the production of these is controlled by the genes. There are three possible alleles in this case: A, B and O. Alleles A and B are both dominant forms and allele O is the recessive form. Only two of the possible alleles are represented in the genotype. If these two are A and B (both dominant) then the blood group will be AB because both blood factors will be made. This is an example of **co-dominance**. If the two alleles are both O (homozygous recessive) then neither of the factors will be made and the blood group will be O (figure 4). If either A or B is present then the blood group will be A or B, respectively.

Are there such things as 'bad' genes?

There are some genes which everyone could do without. Most of these are **mutated** genes such as those causing haemophilia, **sickle-cell anaemia** and the metabolic disorder **PKU** (see page 180).

Questions

1. A man who is homozygous for brown eyes marries a woman who is homozygous for blue eyes. Predict what colour eyes their children will have.
2. A man and a woman, both heterozygous for brown eyes produce a blue-eyed baby. Explain this.
3. Your sister is pregnant and convinced that it is going to be a boy because boys run in the family. Write down how you would convince her that the baby could equally well be a girl.
4. You know you are a carrier for haemophilia and have therefore decided not to have any children. Your parents are desperate to become grandparents and cannot understand your decision. Explain it to them.
5. Your friend's brother has recently gone through a divorce. He was able to show that his ex-wife's child was not his by having a blood test done. Explain why this procedure sometimes works, but not always.
6. Do you think scientists should be allowed to experiment on humans so that one day we will eventually be able to breed out bad genes?

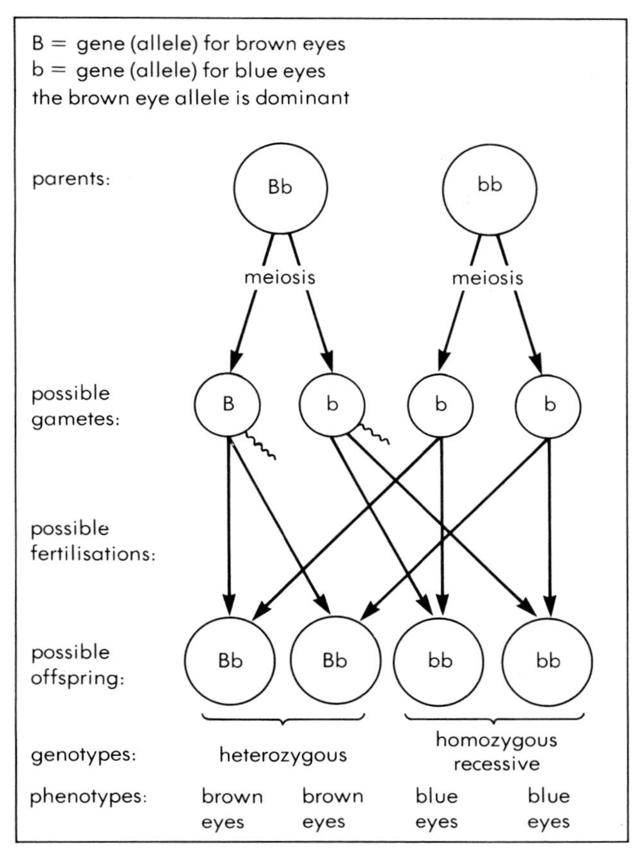

Figure 1 How eye colour is inherited

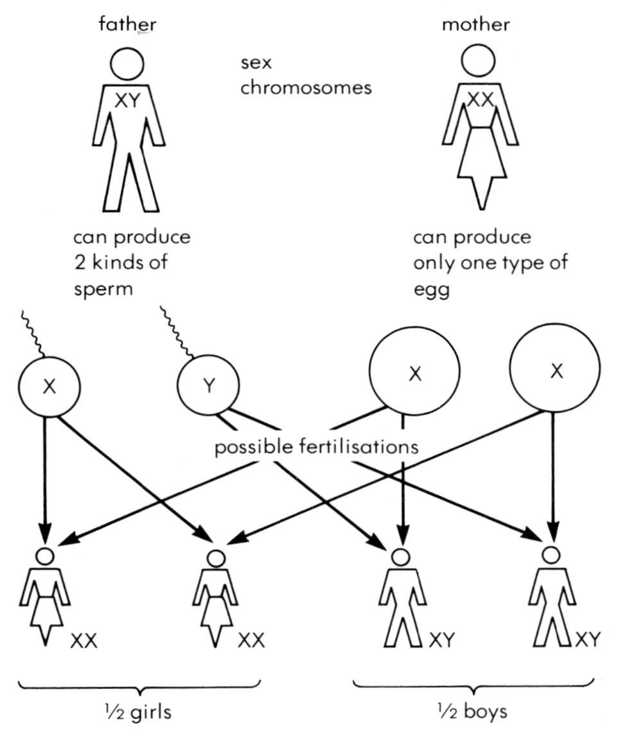

Figure 2 How sex is inherited

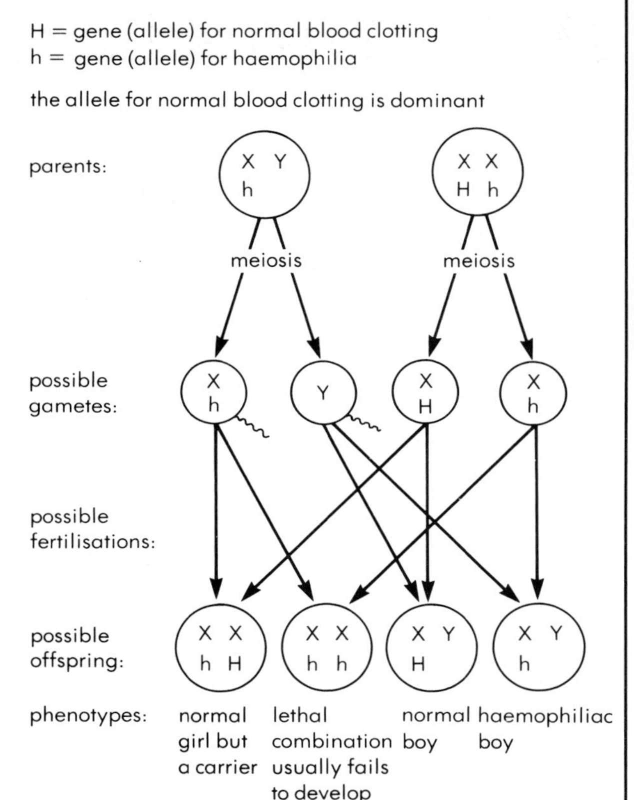

Figure 3 How haemophilia is inherited

Blood groups – some possible combinations

1 Blood group of parents A and B
genotypes = AA BB

possible gametes	A	A
B	AB	AB
B	AB	AB

possible offspring genotypes

Offspring must be blood group AB

2 Blood group of parents A and B
genotypes = AO BO

possible gametes	A	O
B	AB	BO
O	AO	OO

Offspring can be any of the blood groups

3 Blood group of parents AB and O
genotype = AB OO

possible gametes	A	B
O	AB	BO
O	AO	BO

Offspring can be blood group A or B

Figure 4 The inheritance of blood groups

11.4 Variation

Why is there so much variation?

There are about 5000 million people on the earth at present and every one of these is different. Even those produced by the same parents (except identical twins) are different. This is because the chances of any two people, even from the same parents, inheriting exactly the same genes are so small that it can be considered impossible. Everyone will, therefore, have a different set of genes and will be different. This however is not the whole story as it does not take into account environmental influences – what happens to a person once they are born. For example a baby boy may well have the genotype coding for tallness, but unless he gets enough food, he will never grow tall. Variation is therefore the result of two things:

- the genes
- the environment.

Discontinuous and continuous variation

Variation within a population of organisms falls into two categories.

- **Discontinuous variation** is the kind of variation where you get the two 'extremes' and no intermediates. Examples are sex and blood groups. In discontinuous variation the differences are entirely due to the genes present and the environment has no influence (see figure 2).
- **Continuous variation** is the kind of variation which has intermediates, usually because of the environmental influences. Examples are height and intelligence (see figure 3). Continuous variation can also be almost entirely due to the genotype, only in this case several genes are involved, as happens with skin colour. This is determined by the effects of several pairs of genes interacting.

How does the environment affect variation?

Studies involving **twins** have provided most of the information available about how a person's genetic characteristics can be affected by the environment. People with the same blueprint can turn out differently (figure 4). The main two environmental influences after birth seem to be diet and disease. These are dealt with in chapters 9 and 13.

Environmental influence can also occur before birth, when the foetus is developing in the womb. It is known that the phenotype of a child is very dependent upon aspects of the mother's environment and in particular, the drugs that she takes during pregnancy. Most of these drugs can pass through the placental membranes and enter the embryo/foetus and like the thalidomide drug, can cause horrific damage (see page 139). Drugs can be grouped according to when they produce their worst effects:

- In the first few weeks of pregnancy. These drugs usually cause the abortion of the embryo. e.g. **immunosuppressants** and **anti-cancer** drugs.
- Up to sixteen weeks into pregnancy. The organs are still developing up to this time and so any drugs can cause major malformations of these. The part affected depends on the time of administration and the dosage, e.g. **thalidomide, cortisone**.

  Tetracyclines taken during early pregnancy can cause yellow milk teeth and poor bone growth. **Streptomycin** and **neomycin** can cause deafness.
- After sixteen weeks. Because the organs are already formed, the effects of these are not so critical, e.g. **steroids** may cause masculinisation.
- Immediately before birth, usually resulting in respiratory stress, e.g. **anaesthetics** and **analgesics**.
- At any time, e.g. **sulphonamides** can lead to jaundice and **anticoagulants** can cause haemorrhaging.

Questions

1 When identical twins are born, they contain an identical blueprint for their future growth and development. Explain therefore why twins reared apart often end up different (figure 4).

2 Divide the following into two groups; those which are examples of continuous variation and those which are examples of discontinuous variation:
sex, height, weight, blood group, eye colour, hair type.

3 The effects of drugs on the growth and development of an embryo/foetus depend on when they are taken. Explain this with suitable examples.

Figure 1 It is easy to tell these are members of the same family even though they are not identical

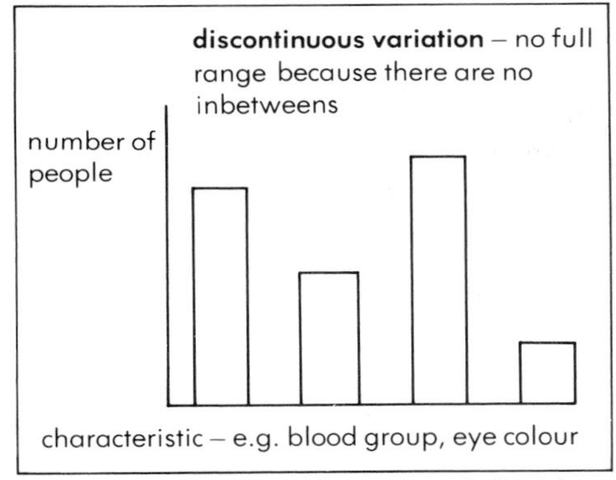

Figure 2 With discontinuous variation the different forms of a character are easily distinguishable within a population

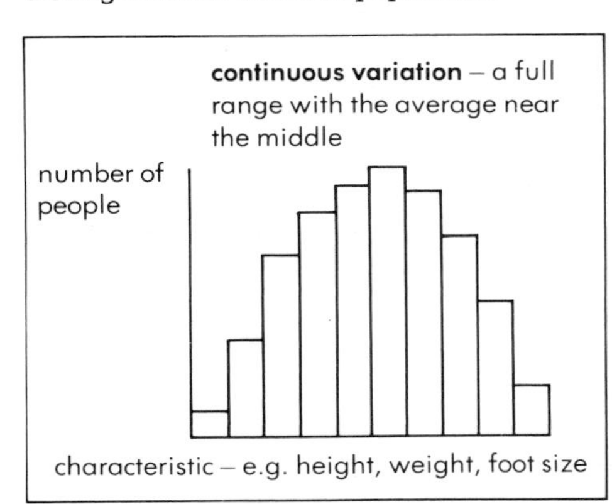

Figure 3 With continuous variation, the different forms of a character are often difficult to tell apart

Figure 4 What differences and similarities can you see in the appearance of these twins?

11.5 Variation and evolution

What are the main causes of genetic variation?

The genes a person ends up with will depend on:

- The genes their parents have to pass on.
- What happens during meiosis. With 23 homologous pairs of chromosomes, you can produce eight million different gametes (figure 1). If you also take into account that during meiosis some of the genes on homologous chromosomes are swopped, then many more than eight million different gametes are possible!
- Which sperm fertilises the ovum. Out of the eight million plus potential sperms and eight million plus potential ova, any two could produce the zygote. The chances of the same two taking part in successive fertilisations is 64 billion to 1 – pretty huge odds against this happening!
- **Mutations** of the parent genes/chromosomes.

What are mutations?

A mutation is a spontaneous change in the genetic material of an organism. The most important changes are those which can be passed on to the offspring. In humans, this means the mutation must occur to the genetic material in the cells which produce the gametes.

Mutations can occur to single genes or to whole chromosomes.

- **Gene mutations** Haemophilia, sickle cell anaemia and phenylketonuria (PKU) are all the result of gene mutations. PKU is a disease of young children which is caused by the absence of an important enzyme. Without this enzyme, an amino acid which is vital to the normal development of the nervous system cannot be made. Instead, the substrate is converted into a toxic substance which also interferes with the development of the nervous system resulting in the child becoming mentally retarded. The reason the enzyme is not present is because the gene which normally codes for it has mutated. When two of these mutated genes come together, the enzyme is not made.

  Figures 2 and 3 show more unusual results of gene mutations.
- **Chromosome mutations** One of the most well known chromosome mutations is that of **Down's Syndrome**. It is really a fault which occurs during meiosis and which results in the child having a full extra chromosome.

  The possession of this extra chromosome has enormous effects on the phenotype. A Down's Syndrome child will have a flat face with slanting eyes and prominent nostrils, a large tongue, small hands with short fingers and some degree of mental retardation (figure 4). In severe cases there may also be heart and respiratory problems. Down's Syndrome is more common in children born to older women, or possibly older men.

  If a similar thing happens with the sex chromosomes it is possible to get combinations like XXY. This would result in a sterile male who may develop female features – a very distressing condition.

How can we prevent these mutations happening?

It is not possible to prevent mutations occurring. In many ways this would not be a wise thing to do anyway because mutations are thought to be the basis of **evolutionary** change (see page 182). It is possible however to prevent bad genes from being passed through a family. Parents who may be at risk of producing a defective child can attend **Genetic Counselling Units** where an in-depth investigation will be done and the appropriate advice given. If high risk parents do decide to have a child, the counsellors will then spend time on preparing the parents for raising a handicapped child and help with schooling arrangements.

Questions

1. The gamete-producing cells of an organism contain four pairs of homologous chromosomes. Calculate how many different zygotes are possible from the fertilisation of the different male and female gametes.
2. During antenatal care many women have an amniocentesis done. The main reason for this is to identify a Down's Syndrome child. Explain exactly how the doctor can tell if the child has Down's Syndrome.
3. A good friend of yours really wants a family, but has just been told by a genetic counselling unit that her chances of producing a child with muscular dystrophy are very high. Write down the pros and cons of her having a baby. Do you personally think she should start a family?

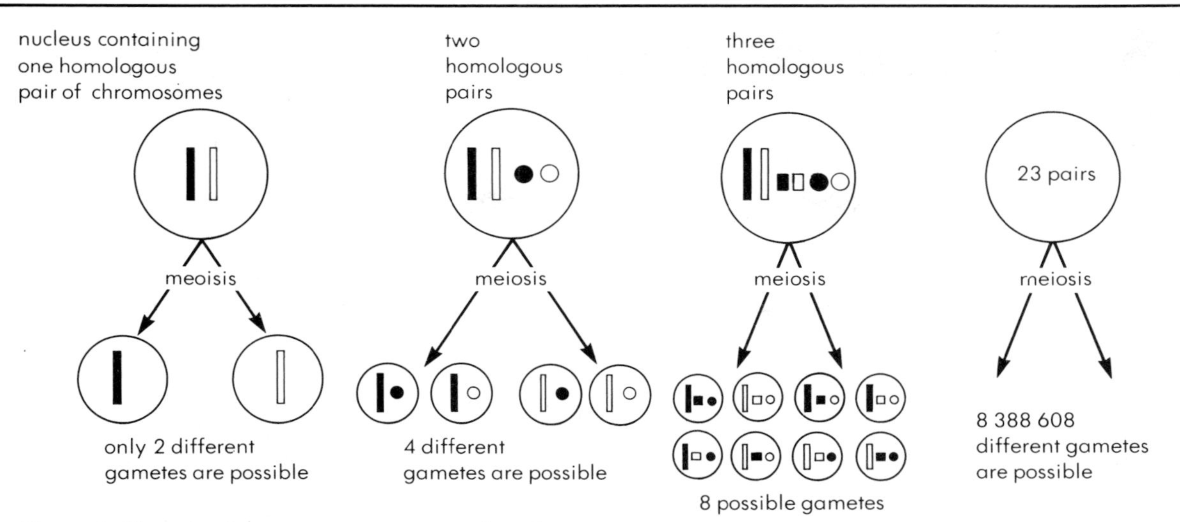

Figure 1 How the chromosomes segregate into the gametes. Notice that only one chromosome from each pair goes into any gamete

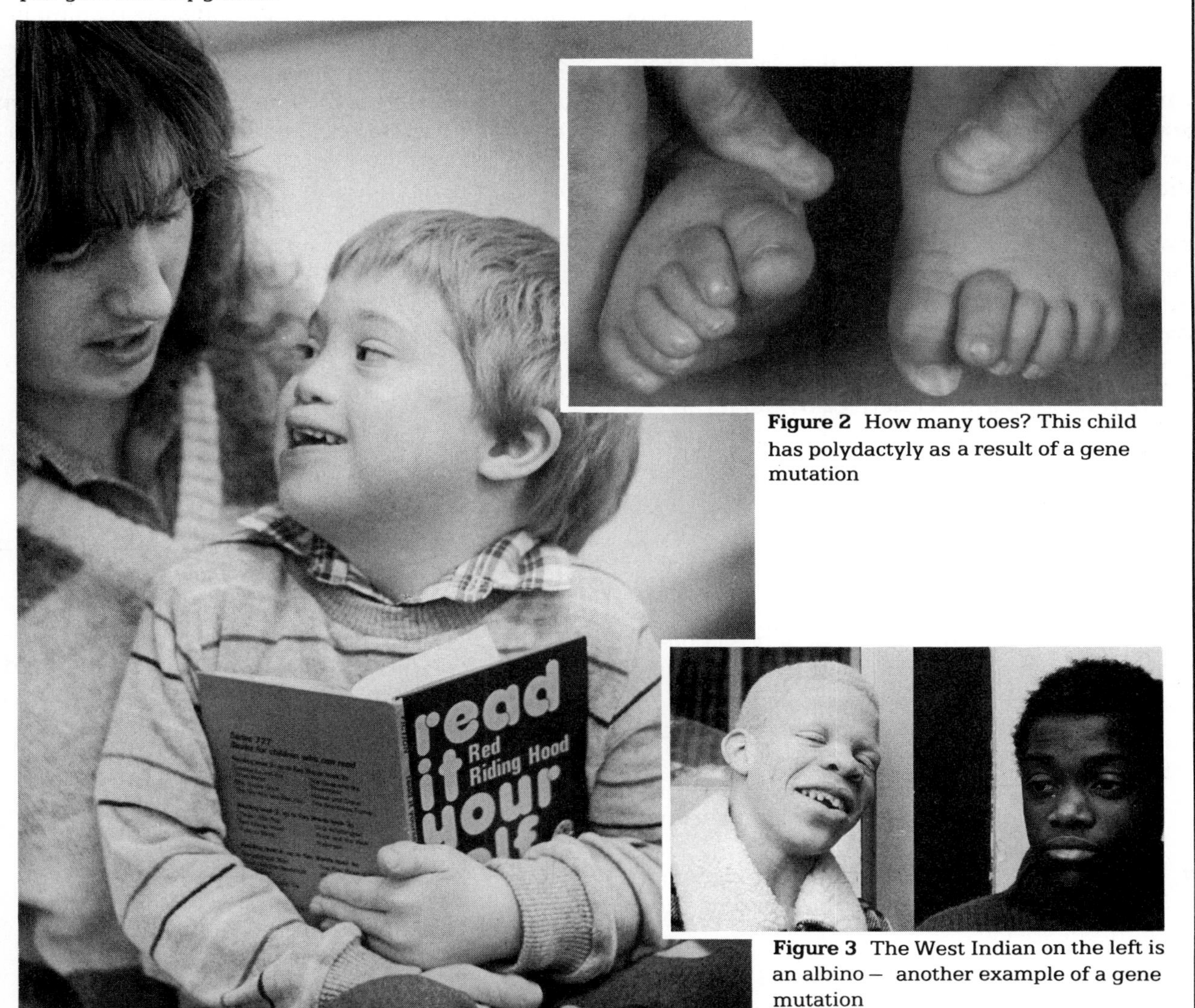

Figure 2 How many toes? This child has polydactyly as a result of a gene mutation

Figure 3 The West Indian on the left is an albino – another example of a gene mutation

Figure 4 A chromosome mutation leads to Down's syndrome

11.6 Evolution

What is artificial selection?

Breeding programmes are a common part of modern agricultural research. For example, it is now possible to produce low-fat lambs, pigs and beef cattle; rice and wheat plants that have a very high yield and dairy cattle which produce three times as much milk. Racehorse owners use breeding programmes to produce champion horses and horticulturists do it to produce new kinds of flowers. The term used to describe this kind of practice is **artificial selection**. It is thought that nature also selects; this is called **natural selection** and is the basis of evolution.

What is evolution?

The term evolution is used to describe very long series of gradual changes which have taken place to the simplest, earliest life forms on earth, resulting in the vast range of present day species. The idea that natural selection is a possible mechanism for evolution was first introduced by **Charles Darwin** in 1859. There have been many arguments for and against evolutionary theory since. Now, although it is largely accepted, recent versions also take into account our present knowledge of genetics. The main points are:

- Mendelian inheritance is correct – that is, characteristics are coded for by genes and these genes can be passed on to the offspring through the gametes.
- The process of sexual reproduction and especially the production of gametes by meiosis leads to considerable variation between the members of a population. More variation could result from mutations.
- Occasionally a combination of genes comes together which makes the organism better suited to its environment. This organism therefore survives better than the others and stands more chance of reproducing and passing its genes on. This is the basis of natural selection, sometimes called '**survival of the fittest**'.
- Over millions of years these selected changes add up, and eventually result in a completely different species.

What is a species?

A **species** is a group of only those organisms that can reproduce with one another to produce *fertile* offspring.

How many different species are alive today?

There are at present over one and a half million known species. These are often divided into 28 groups, the species in each group sharing similar features. Some of the main groups of organisms and how they could have shared ancestors are shown in figure 1.

What group do humans belong to?

We belong to the **mammals**. These are grouped together because they all have hairy skin, are **endothermic** (see page 114) give birth to live young (rather than eggs) and feed these on milk produced by mammary glands. Mammals are often thought of as the most advanced animals, but of the 3900 species in the mammal group only a few show any signs of intelligence. These intelligent mammals are often grouped together and called **primates**. Humans are thought to be the most intelligent of the primates (see figure 2).

Questions

1. Explain how artificial selection can be used to our advantage.
2. What is a species? In your own words explain how all the modern day species may have arisen from just one ancestral type.
3. Using diagrams explain the connection between modern monkeys, apes and humans.
4. How are mammals different to all other animals?
5. We are the most intelligent of all primates. Produce a table showing the similarities and differences between humans and other primates.

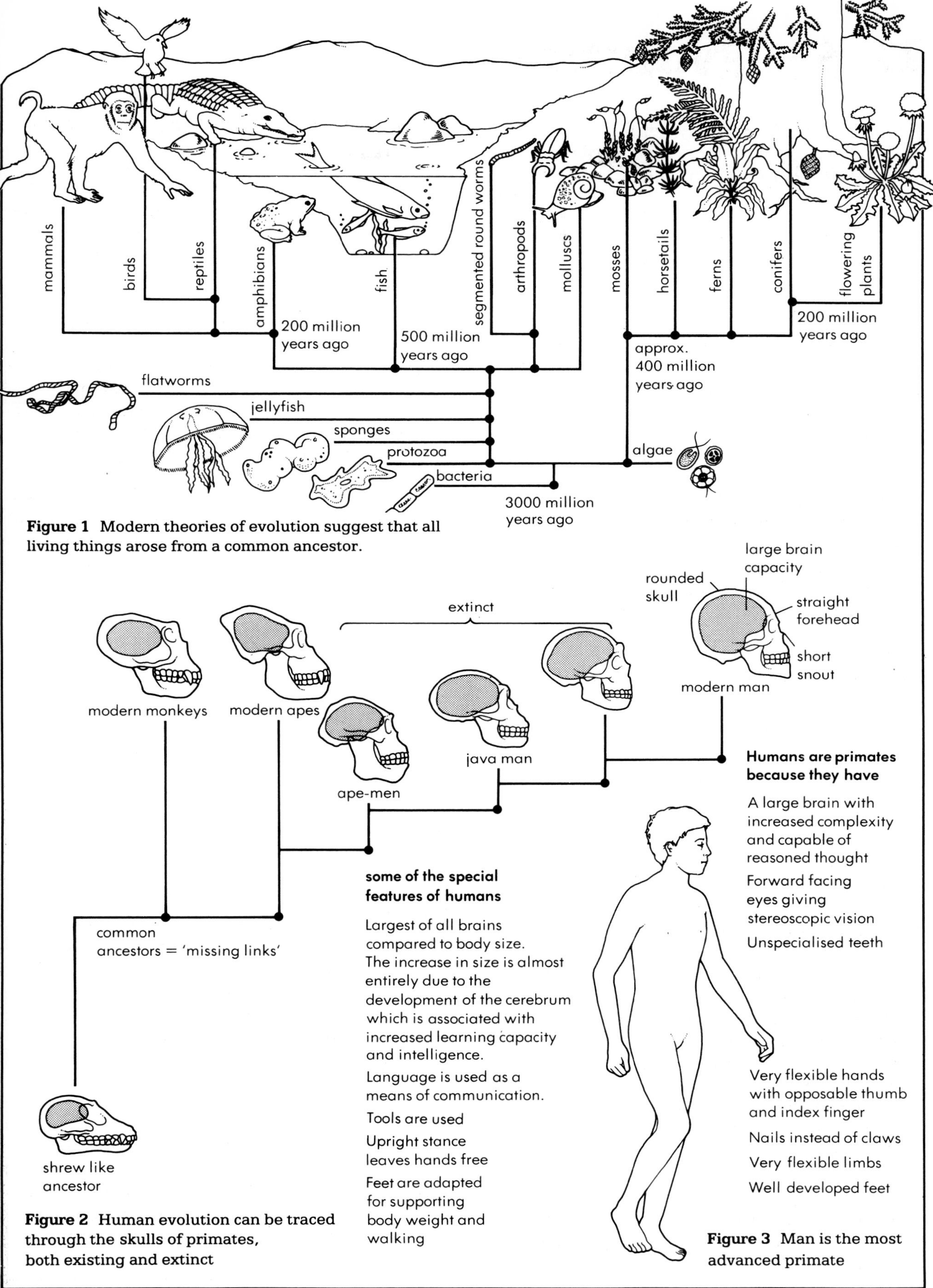

Figure 1 Modern theories of evolution suggest that all living things arose from a common ancestor.

Figure 2 Human evolution can be traced through the skulls of primates, both existing and extinct

Figure 3 Man is the most advanced primate

Questions

Recall and understanding

1 A gene is a:
A piece of clothing
B kind of cell
C unit of inheritance
D coin

2 A human diploid cell has:
A 23 chromosomes
B 42 chromosomes
C 92 chromosomes
D 46 chromosomes

3 The cell formed when two gametes fuse is called a:
A zygote
B clone
C embryo
D foetus

4 Alleles are:
A twins
B identical cells
C different forms of the same gene
D friends

5 When you describe someone by her colour you are describing her:
A genotype
B phenotype
C stereotype
D clothing

6 Meiosis is involved in:
A asexual reproduction
B growth
C growth of a tumour
D the production of gametes

7 When two parents of blood group genotypes AO and AO have children, the possible blood groups are:
A all AO
B A and O
C A only
D O only

8 Which of the following is not sex-linked?
A haemophilia
B red/green colour blindness
C German measles
D muscular dystrophy

9 Which of the following is not an example of continuous variation?
A foot size
B weight
C height
D blood group

10 A species is:
A group of similar organisms
B group of organisms that interbreed and produce fertile offspring
C group of organisms
D a type of animal

Interpretation and application of data

11 Two surveys were done to find out what percentage of the UK population had the various blood groups.

Survey 1 100 people

Number of people	Group
50	A
1	B
0	AB
49	O

Survey 2 10 000 people

Number of people	Group
4000	A
1000	B
500	AB
4500	O

a) The results of the two surveys disagree. Which would you believe and why?
b) Using blood groups AB and O show how blood group is inherited.
c) When Mendel did his experiments he got the following figures:

Characteristic	2nd generation results
tall and short	787 tall + 277 short
green and yellow	428 green + 152 yellow

i) What ratio would you have expected for the second generations?
ii) Why do you think they were not this ratio?
iii) How could Mendel have got nearer the expected ratio?

12 The table below shows the chances of producing a Down's Syndrome child.

Mother's age	Risk of occurrence
20–30	1 in 1500
30–35	1 in 750
35–40	1 in 600
40–45	1 in 300
over 45	1 in 60

a) Why are Down's Syndrome children produced?
b) Plot a graph of the increase in chance with age.
c) It is possible to detect Down's Syndrome by doing a test. What is this test and when is it done?
d) Another kind of mutation results in the disease Phenylketonuria (PKU).
i) What is PKU?
ii) Describe how two heterozygous parents can produce a baby with PKU.

13 The following results are for a class of 24 pupils who measured the length of their feet. The measurements are in mm.

250	262	237	231
258	250	249	243
236	237	251	251
247	243	252	247
253	252	260	248
249	247	256	245

a) Use the data to fill in the table below.

Length (mm)	231–235	236–240	241–245	246–250	251–255	256–260	261–265
Number of pupils							

b) Plot these figures on a graph.
c) What is the range of variation?
d) What is the mean foot size?
e) What kind of variation does this show?
f) What is the other kind of variation? Give three examples.

14 The effect of the environment on variation can be assessed by using identical twins.

Characteristic	Twins reared together	Twins reared apart
Height difference	1cm	1.2cm
Weight difference	1kg	7.4kg
IQ difference	10	30

a) Which of the characteristics are unaffected by the environment?
b) Which are affected?
c) For this study to be valid, another group of people should be used – who?
d) The results were obtained using only ten sets of twins. How could the experiment be improved so that more accurate results could be obtained?

15 Evolutionary relationships in the past have always been based on comparative anatomy. It is now possible to use genetic similarities.

	Anatomy									Genetic								
	1	2	3	4	5	6	7	8	9	1	2	3	4	5	6	7	8	9
Eskimos	✓	✓	✓				✓	✓			✓	✓				✓	✓	
Japanese		✓	✓	✓			✓					✓	✓		✓	✓		
Bushmen	✓					✓			✓	✓	✓	✓			✓	✓		
Indians		✓	✓				✓	✓			✓	✓	✓			✓		
Europeans	✓	✓	✓		✓		✓	✓		✓	✓	✓		✓		✓	✓	
Chinese			✓	✓		✓	✓			✓					✓	✓		✓
Aborigines	✓					✓	✓		✓	✓					✓			✓

A tick indicates features possessed.

a) Write out a list of races so that each one is next to those it is most related to, based on
i) comparative anatomy,
ii) genetic markers.
b) Do the lists for parts i) and ii) agree?
c) Which method do you think gives the most accurate evolutionary relationships? Explain why.

CHAPTER 12: *HEALTH*

How much are you responsible for your own health? If a member of your family has cancer, will you also get it? How can you improve your health? This chapter answers these questions and explains what *good health* is.

What effect is smoking having on their health?

Health and social class

What is social class?

Table 1 Social class is linked to occupation.

Social class	Examples of occupations
I Professional	Doctor, solicitor
II Managerial	Teacher, policeman, MP
III Skilled	
(a) non-manual	Secretary, draughtsman
(b) manual	Bricklayer, cook
IV Partly skilled	Barperson, postman
V Unskilled	Labourer, window cleaner

Social class is a classification of people based on their occupation. This arranges them into divisions often just called I, II, III, IV and V. See table 1.

How is health linked to social class?

In the United Kingdom health has always been linked to differences in social class. Statistics on mortality and disease show the link quite clearly.

- People in social class V are nearly twice as likely as people in social class I to die before they reach retirement age.

Table 2 The figures are as standard mortality ratios (SMR). The SMR is the ratio of the number of deaths in the group to the average number of deaths for all groups. Figures are percentages.

Social class	Men	Women
I	77	82
II	81	87
IIIa	99	92
IIIb	106	115
IV	114	119
V	137	135

- People in social class I live on average 5 years longer than those in social class V.
- More people in social classes IV and V suffer from long standing illness. This is called **chronic illness**.

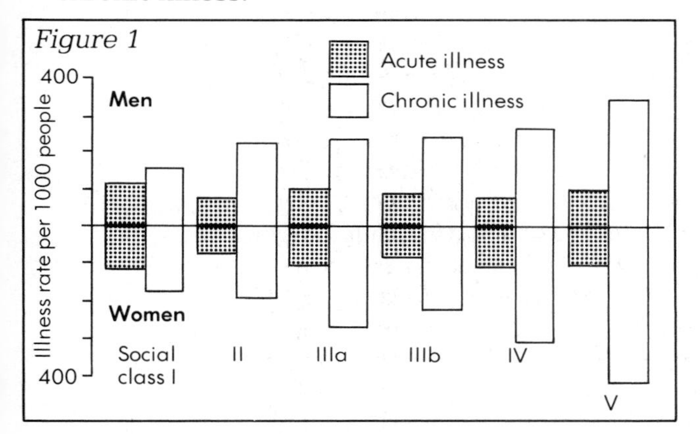

- Men in social class V are five times more likely to die of mental illness.
- A child born to parents in social class V is twice as likely to die between the ages of 1 and 14 as a child in social class I.

Why do these differences exist?

We can only guess although there is evidence that:

- Classes IV and V do not make as much use of the medical services, especially the preventive services such as dentists.

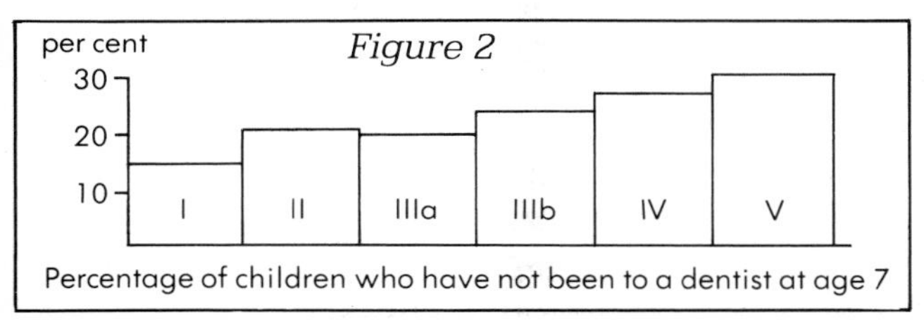

Percentage of children who have not been to a dentist at age 7

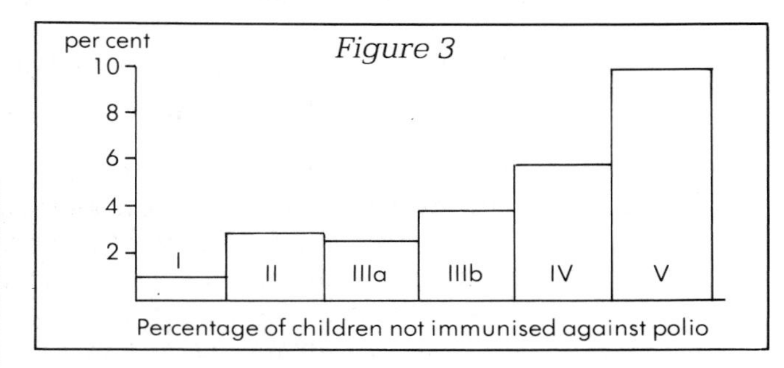

Percentage of children not immunised against polio

- Classes IV and V usually have low incomes and as such cannot afford to buy some of the necessities of a healthy life such as good food and good housing.
- Social conditions are poor for many people in classes IV and V. Houses are often overcrowded increasing the chances of diseases spreading.
- More people in classes IV and V smoke, increasing their chances of suffering certain diseases and creating an unpleasant environment for their children.

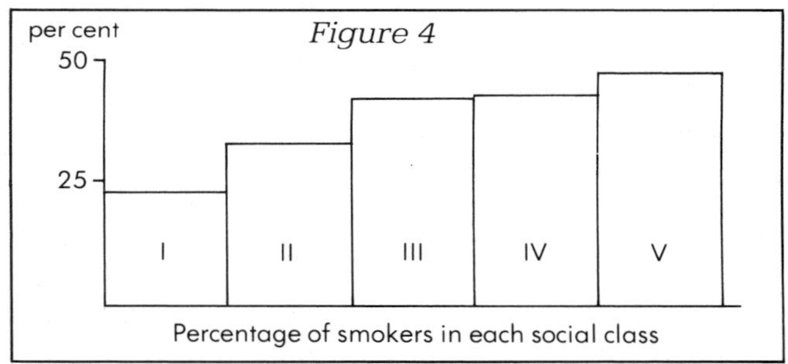

Percentage of smokers in each social class

What is the solution to these inequalities?

In 1948 the government thought it had the solution. It introduced the National Health Service. The NHS made good medical treatment and facilities available to every member of a community, whether rich or poor. Its success to some extent can be judged by the findings of a government working party in 1980. These can be summarised as follows:

- A higher percentage of the population now make use of the health services
- Generally, the health of the population is improving
- There has been a greater improvement in the health of social classes I and II than of social classes III, IV, and V

In other words, health is improving generally, but class differences are becoming greater.

12.1 Influences on health

How healthy are you?

Everybody at some time experiences poor health, but just how healthy are we the rest of the time? If you are in perfect health this usually means your body is working at peak efficiency both physically and mentally. Anything less than this and you are unhealthy in some way or another. So not many of us can say we have perfect health! But clearly there are degrees of unhealthiness. A person could just be a bit unfit. At the other extreme they could be seriously ill with cancer.

What sort of things can affect your health?

Your health is affected by almost everything you do. But it is also affected by what you are. These **inbuilt factors** are things such as genotype, age and sex:

- **Genotype** Every child inherits from its parents a blueprint for its future growth and development. Unfortunately if the parents had certain disorders their children inherit these. For example **haemophilia**, or **red green colour blindness**.
- **Age** Many diseases take time to develop and are therefore associated with old age. **Coronary heart disease** and **cancer** are examples. Some other diseases or disorders occur because your body is wearing out. It is thought that the mental illness **dementia** is caused by the progressive loss of brain cells as you age.
  Figure 1 shows deaths from various causes at different ages for women and men.
- **Sex** Many illnesses are linked to your sex either directly or indirectly because of lifestyles. **Pre-menstrual tension** clearly only affects women. **Lung cancer** is three times more common in men, but it is not clear why.

What else affects your health?

Everything that happens to you – what you do as a job; how you live; where you live – can affect your health for good or bad. Here are some of the main **environmental factors** that may influence your health:

- **Climate** Conditions which are either too cold or too hot can be harmful (see page 116). Even too much sunbathing has a strong link with **skin cancer**.
- **Pollution** In an industrial society such as our own, pollution is inevitable. The amount of pollution a person has to suffer depends to a large extent on her lifestyle and in particular, her occupation. This is constantly being looked at and checked by the work of the **environmental health services** (page 218).
- **Other organisms** Many organisms obtain their food from other organisms. In this way diseases are spread. There are two main groups that affect health; **parasites** and **saprophytes**. Parasites feed off other living organisms and in doing so damage them. Examples are many of the **microbes**, **insects** and **worms**.
  Saprophytes feed off dead organic matter. **Ringworm** and **thrush** are diseases caused by **fungi** feeding on the dead layer of skin covering your body (see chapter 13).
- **Education** To live a healthy life, you must be aware of what is good and bad for you. In many countries there is now a carefully planned programme of health education (figure 2). In the USA this has contributed to a 50 per cent fall in the number of deaths from coronary heart disease.
- **Occupation** A person's occupation exposes him to many potentially harmful situations and substances. In the UK occupation is generally linked to social class (page 186).
- **Political state of the country** A recent report on health by the World Health Organisation (WHO) stated that good health can only be achieved in peaceful, prosperous and socially sound societies. Do you think your society fits in with this?

Questions

1. How would you describe perfect health?
2. There are some influences on our health which we cannot at present do anything about. Explain why.
3. The text outlines six of the major environmental influences on health. Explain how each of these has or is affecting your health.
4. Health has generally been improving over the last 100 years (figure 3). List as many reasons as you can think of for this improvement.
5. Use figure 1 to work out the main causes of death in (a) males and (b) females at the ages of 5, 25 and 65. Write a list of these from highest to lowest.

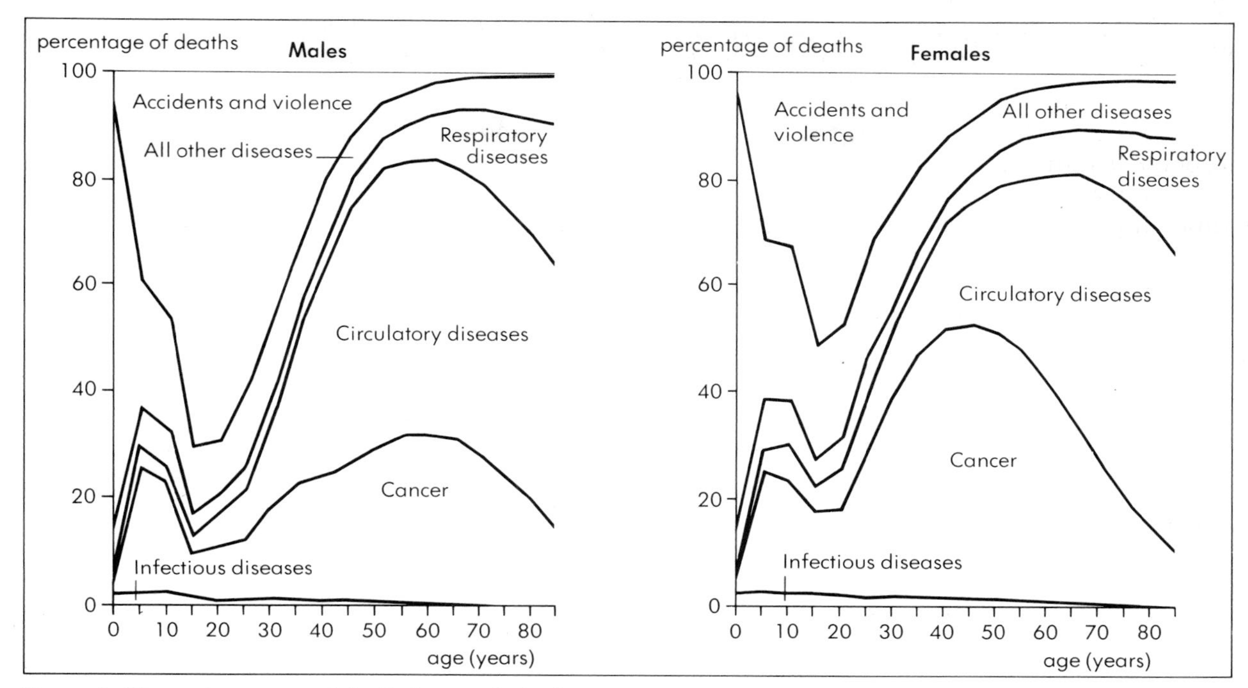

Figure 1 The major causes of death for people in the UK in 1982

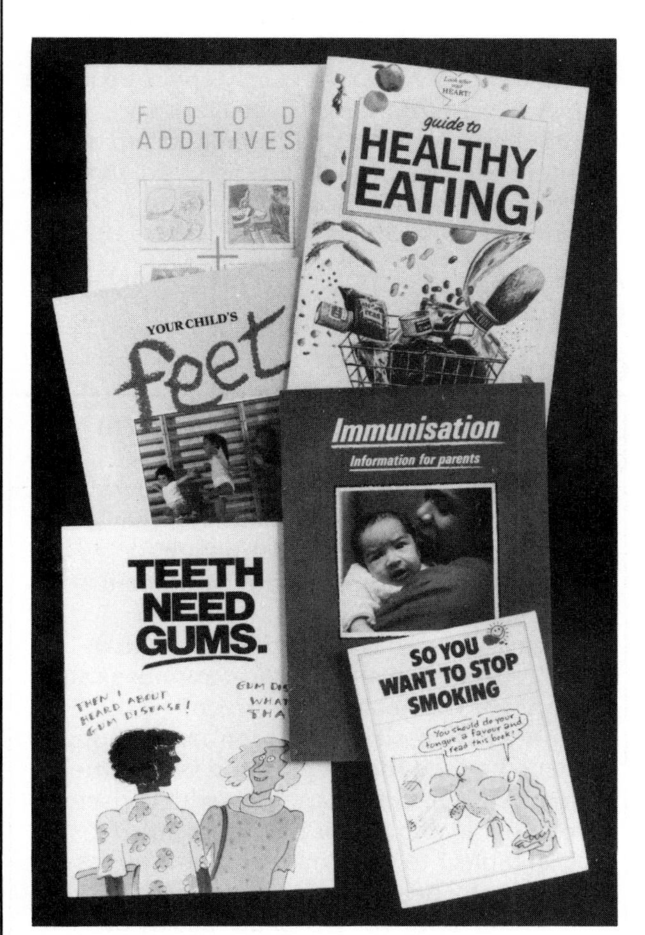

Figure 2 The Health Education Authority produce a variety of leaflets and posters

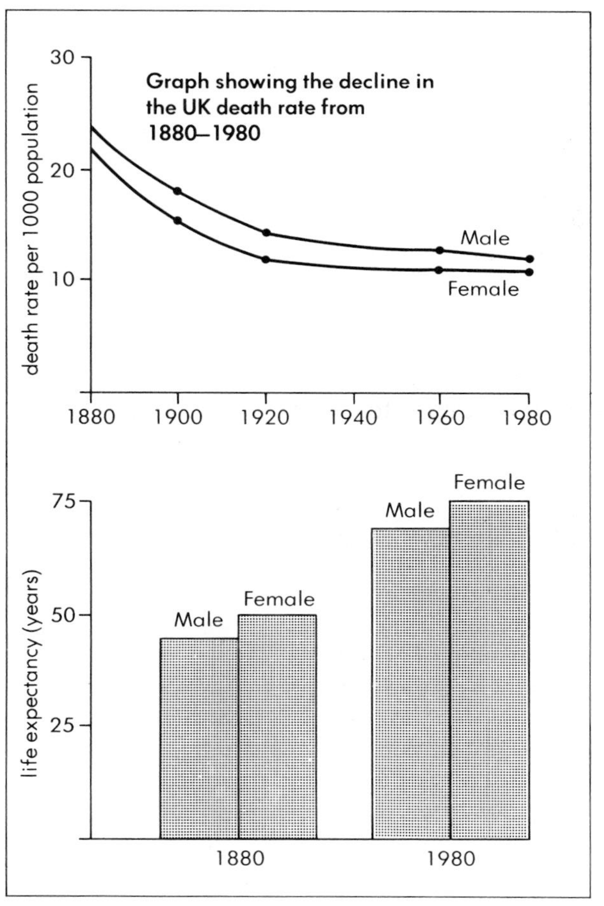

Figure 3 These graphs show how health has improved over the last 100 years

12.2 You and your health

What about behavioural factors?

Your health is not only influenced by the characteristics you are born with and by the environment in which you grow up but also by what you do to your body. A person usually has more control over these **behavioural** factors than any of the others. It is often these that make the difference between good and bad health. So it is very important to take care. You can influence your health, and therefore your lifespan, a great deal. Here are some of the main behavioural factors that should concern you:

- **Diet** There is a lot of truth in the old saying we are what we eat (figure 1). When you are run down, your poor general health is nearly always due to poor nutrition. Most of us would benefit from an improved diet, but for many people it is not just a question of likes and dislikes. Some people cannot get a better diet because of the poor social and economic conditions in which they live.
- **Smoking** Smoking-related diseases such as **bronchitis** use up more than their fair share of doctors' time and national health service resources (£111 million in 1984). In the UK the number of men and women who smoke is falling. The average number of cigarettes smoked per person is also falling.
- **Drinking** Drinking alcohol can become a problem when it is overdone or is a regular habit. Drinking any amount during pregnancy can affect the health of the child. Teenagers who drink also risk their health. It also endangers your safety and the safety of other people if you drink and drive.
- **Drug-taking** Drugs can be very helpful in overcoming illness in the short term. Many however can be very damaging if taken for long periods or for the wrong reasons. Drug abuse is one sure way of finding an early grave. Table 2 shows how it is increasing from year to year.
- **Personal hygiene** A clean body is obviously essential to good health. It is something that everybody can achieve but again, unfortunately, it is related to education. If you know why you should do something, you are more likely to do it. Many parasites are more likely to find a home on dirty people. Many are actually transmitted from person to person by bad habits and poor sanitation.
- **Accidents** Obviously, no-one has an accident on purpose, but many people do take unnecessary risks. In 1980 there were about one and a half million reported accidents, 12 000 of which were fatal. In children under fifteen years old, accidents were the biggest single cause of death (table 1).
- **Exercise and rest** A person who does not exercise will be unfit and therefore more likely to break down when her defences are threatened or extra demands are put on her body. The number of people who have a heart attack whilst running for a bus is very high. Even regular short periods of exercise would prevent this by raising the overall fitness. It is also important to get enough rest. This is the period when your body does its running repairs.
- **Use of medical services** The health of the average American adult is generally better than the average English person. One reason could be that to get health insurance each American must have a yearly check up. Their health is therefore monitored more frequently and minor problems sorted out before they become major. So it is a good idea to use the medical services for *prevention* as well as for cure (figure 2).

All these factors are discussed more fully in the relevant sections.

Questions

1. We can all improve our general health by altering our behaviour. Discuss this statement with reference to your own health.
2. Is there any evidence that health education can improve the health of a population? Explain.
3. What evidence is there to suggest that more regular use of medical services can improve the health of a person?
4. Preventive medicine has for many people gone beyond the six monthly visit to the dentist. What else can the health conscious person do to monitor his/her health?
5. Use table 1 to work out the main causes of accidental death in these age groups: a) under 1, b) 10–14, c) 25–29, d) 65–74. Write these out in a list, highest to lowest for each age group.

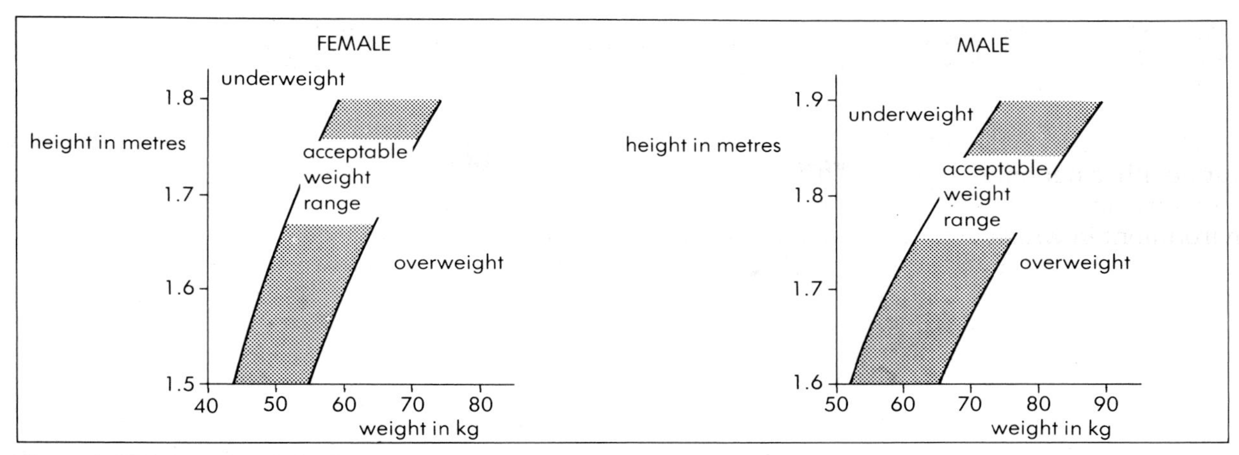

Figure 1 Height and weight charts

Accident classification	Under 1	1–4	5–9	10–14	15–19	20–24	25–29	30–44	45–64	65–74	75 and over	All ages
	AGE GROUP (YEARS)											
Railway accidents	–	1	–	1	14	10	8	21	21	3	2	86
Road transport accidents	9	80	160	188	817	736	392	671	764	714	675	5012
Water transport accidents	–	2	2	1	6	9	7	14	11	2	–	54
Air and space transport accidents	–	–	–	–	2	1	5	15	6	–	–	29
Poisoning from drugs and medicaments	1	7	–	1	19	32	27	101	121	34	33	376
Poisoning from other solids, liquids and gases	1	2	3	14	32	17	17	73	57	24	26	266
Falls	6	20	13	19	28	39	33	144	357	502	2767	3928
Fires and flames	6	47	22	14	11	21	15	52	115	89	249	644
Natural and environmental factors	19	1	3	5	6	6	1	8	52	68	148	317
Drowning and suffocation	55	53	20	38	62	55	42	112	174	103	141	855
Surgical and medical misadventures	1	–	–	–	1	–	–	5	23	58	81	169
Other accidents	3	15	14	18	34	50	38	88	136	57	99	553
Late effects of accidents	–	–	–	1	3	1	4	12	23	9	19	72
Total accidents	101	233	237	300	1035	977	589	1316	1860	1663	4240	12551

Table 1 The main causes of accidental death in the UK in 1984

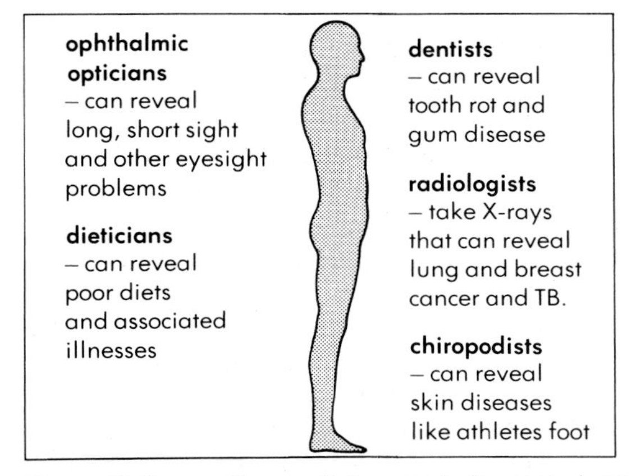

Figure 2 Preventive medicine can help you to keep healthy

Year	Male	Female
1973	644	163
1976	745	239
1981	1607	641
1982	1976	817
1983	2979	1207
1984	3840	1575

Table 2 Drug addiction in the UK is on the increase

Questions

Recall and understanding

1 Which of the following diseases cannot be passed on to the offspring through the genes?

A haemophilia   C phenylketonuria
B malaria   D spina bifida

2 Which of the following illnesses is age related?

A dementia   C polio
B measles   D malaria

3 An organism that gets its food from another living organism is a:

A germ   C parasite
B saprophyte   D worm

4 Which of the following is not a saprophyte?

A bread mould   C flea
B *Candida albicans*   D mushroom

5 A substance that alters the working of your mind or body is a:

A poison   C antigen
B drug   D food

6 A radiologist:

A takes X-rays
B works with radiation
C repairs radios
D measures radiation

7 The biggest killer disease in the western world is:

A malaria   C cancer
B gonorrhoea   D coronary heart disease

8 Most accidents happen:

A at work   C in the home
B on the roads   D on holidays

9 Chiropodists examine:

A hands   C tonsils
B chairs   D feet

10 The United Nations organisation set up in 1948 to monitor health is:

A FAO   C WHO
B UNESCO   D Oxfam

Application and interpretation of data

11 The data below refers to deaths from various accidents in children.

Place		Age 0–4	Age 5–14
Road	Boys	78	325
	Girls	55	150
Home	Boys	141	44
	Girls	123	31
Others	Boys	75	119
	Girls	31	30

a) Draw a bar chart with these figures.
b) Where do most accidents happen?
c) Name some of the places in the 'other' category.
d) The figures quite clearly show that more boys have accidents than girls. Why do you think this is so?
e) The graph below refers to home accidents:

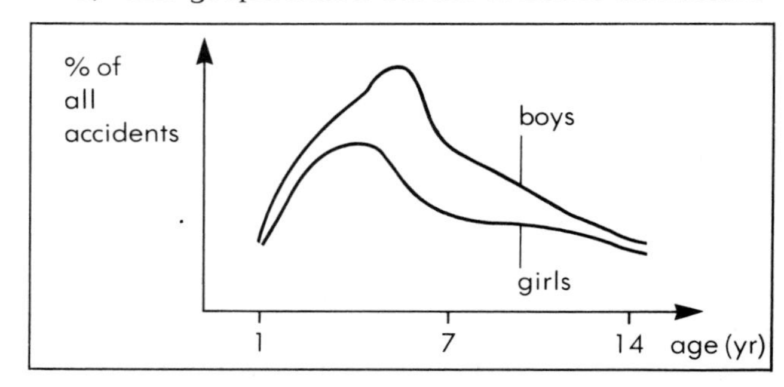

i) At what age is the proportion of accidents greatest?
ii) List some of the accidents which may result in death.
iii) Why are there very few accidents before the age of one?

12 The graph below shows the risk of being fat.

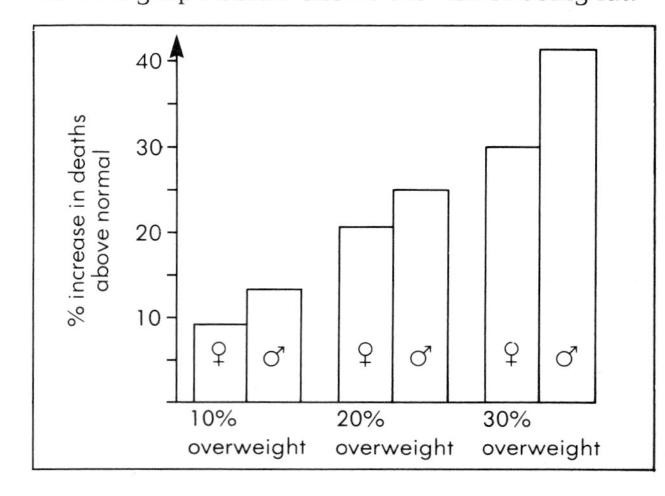

a) What precisely does the graph show?
b) Most of the increase in deaths is due to heart disease. State precisely how an increase in fatness can cause an increase in heart disease.
c) The figures below show the percentage of a person's daily energy obtained from three nutrients.

	UK diet (typical)	WHO recommendations
Fat	43	30
Protein	13	11
Carbohydrate	45	59

i) How must we alter our diet to meet the WHO recommendations?
ii) Calculate how much less fat we must eat (as a percentage).

13 The following shows the amount of sleep 24 ten year old pupils think they need. The measurements are in hours.

10	9	8	10	6
10	9	8	8	8
12	12	8	9	10
8	6	9	9	8
10	6	9	8	10
9	8	8	7	12

a) Calculate the average.
b) Fill in the gap in this table:

Age	1 week	6m	1yr	4yr	10yr	20yr	60yr
Hours sleep needed	16–18	13	11	10		8	7

c) What does this show you about the amount of sleep needed as you grow older?
d) Suggest why this is.
e) Predict how much sleep a 100 year old would need.

14 The data below is about injuries received at work (UK, 1984).

a) In which industry were there most injuries?
b) In which industry were there most deaths from injury?
c) Which is the most dangerous industry?
d) If you were head of an insurance company, which industries would you give a premium discount to?

	Agriculture and forestry	Mining and quarrying	Manufacturing	Construction	Transport and communication
Injuries to employees (numbers)					
Fatal	29	52	123	100	44
Fatal and major	297	532	4,729	2,369	545
Incidence of injuries to employees (rates per 100 000 at risk)					
Fatal	8.5	17.5	2.3	9.8	3.3
Fatal and major	86.9	178.7	87.4	232.6	41.2

15 A survey involving one million people done in 1959 in the USA came up with the following results.

Problem	Smokers (%)	Non smokers (%)
Cough	33.2	5.6
Poor appetite	3.3	0.9
Short of breath	16.3	4.7
Chest pains	7.0	3.7
Diarrhoea	3.3	1.7
Easily tired	26.1	14.9
Abdominal pain	6.7	3.8
Hoarse voice	4.8	2.6
Weight loss	7.3	4.5
Stomach pain	6.0	3.8
Insomnia	10.2	6.8

a) Suggest with reasons whether the survey is valid or not.
b) The incidence of at least three health problems is clearly related to smoking. Use the data to find these.
c) Another study was also carried out at the same time on 68 000 smokers.

Disease	Expected death rate/100 000	Actual death rate/100 000
CHD	1973	3361
Lung cancer	37	397
Other cancers	651	902
Ulcers	25	100
Other diseases	453	486

i) Is this study valid or not? Explain why.
ii) How could the study be improved?

CHAPTER 13: *DISEASE*

Many of the diseases we suffer from are caused by micro-organisms. This chapter is all about these micro-organisms. It also looks at how the diseases they cause can be spread and ways of preventing this.

AIDS – a disease for which we have no cure, yet.

Bugs can be useful!

Not all micro-organisms cause disease. In fact the vast majority of them are harmless to us. Some are even very useful and much of the flourishing biotechnology industry has developed around them. This section explains how.

What is biotechnology?

Biotechnology makes use of natural organisms or parts of them to make the substances we need. For example, micro-organisms are used to produce antibiotics.

The products of micro-organisms

Table 1 is a list of some more useful substances that we can now use micro-organisms to make.

Table 1 All these substances are produced in ***fermenters****.*

Bacteria	Yeasts	Other fungi
Protein (SCP) Enzymes Organic acids Antibiotics Food additives Hormones Gum Plastics Pesticides	Protein (SCP) Enzymes Alcohol Anti-viral agents Single cell fat Hormones	Mycoprotein Enzymes Organic acids Antibiotics Single cell fat Food additives

What is a fermenter?

A fermenter is any suitable container in which the micro-organisms can grow and make their products.

Figure 1 The ICI pruteen fermenter

Industrial fermenters

Industrial fermenters are run by a controller who can alter the internal environment (pH, temperature, oxygen concentration etc.) to produce the *optimum* conditions for particular substances to be produced.

The largest industrial fermenter in the world belongs to ICI. It has a capacity of 1500 m^3. They use it to produce single cell protein (SCP).

Why does biotechnology use micro-organisms so much?

The use of micro-organisms in industrial processes has developed for several reasons:

- They have a relatively simple structure consisting of one cell only (except some fungi).
- They have a very rapid reproductive rate. Table 2 compares their doubling time (the time it takes a group of them to double in number) with some other common organisms.

Table 2

Organism	Doubling time
Bacteria and yeasts	minimum 20 mins
Other fungi	2–6 hours
Algae	2–6 hours
Cereals	1–2 weeks
Poultry	2–4 weeks
Pigs	4–6 weeks
Beef cattle	1–2 months

- They are easily grown in easily maintained conditions.
- They have simple food requirements and some will even feed on our waste.
- They have a relatively high protein content.
- Their chemistry is well understood.
- They contain some useful enzyme systems.

Micro-organisms and genetic engineering

Many of the micro-organisms used in industry have been genetically altered in some way. The most common technique involves inserting genes so that they can make substances they would not normally make. This is covered more fully in the sections on genetic engineering (page 171) and insulin (page 113).

13.1 Organisms that cause disease

What is disease?

A **disease** is a condition that prevents your body, or part of it, working properly. There are many different kinds, each having its own particular cause (table 1). Your doctor can tell these diseases apart because each one has its own **symptoms** (signs).

Most diseases are actually caused by other organisms entering your body then feeding and reproducing in it. In doing so, they may damage your body, or poison it by excreting toxic waste substances (**toxins**). These organisms can be very tiny (**micro-organisms**), or quite large as are some of the worms. The main kinds of organisms that cause disease are: **viruses**, **bacteria**, **protozoa**, **fungi**, **insects** and **worms**.

What are viruses?

Viruses are very simple organisms consisting of two parts only: a protein coat and a strand of genetic material (figure 1). They are the smallest of all living things and can only be seen by using an electron microscope. Most viruses are less than one thousand times smaller than a full stop on this page.

All viruses are parasites and can only reproduce inside the cells of another living body. This can be a plant, animal or bacterium. Most viruses cause disease.

How do viruses cause disease?

When a virus enters a cell in your body it takes over control of it. The cell is made to stop all its normal work and start making more viruses. Eventually the cell bursts, releasing the new viruses. Each of these new viruses will enter another body cell and reproduce itself (figure 2). In a matter of hours, tens of thousands of cells can be destroyed. This destruction of cells usually results in disease.

Viral diseases can often be very serious because there are no drugs that a doctor can give you to fight them. It is up to your own body to fight them by making **antibodies**, and if you are incapable of doing this, you will probably die. Table 2 shows some of the main human viral diseases. Viral diseases also do a lot of damage to our crops and domestic animals.

Are there any useful viruses?

Viruses have no natural uses, but as we learn more about them we are beginning to make them work for us. We have started to use them to control pests such as the alfalfa caterpillar. As few as six viruses released per acre of land is sufficient to keep the size of the caterpillar population at an acceptable level. A virus used in this way is called a **biological control agent**.

More recent research has made use of them in genetic engineering experiments. For example, bacteria can be 'told' to make human insulin by inserting the correct genes into their DNA. Viruses can be used to transfer these genes into the bacteria (see page 170).

Can viruses be grown artificially?

It is useful to be able to grow organisms like viruses so that we can study them and learn how to prevent diseases. Viruses can be grown, but the procedure is very complicated. It involves using **tissue cultures**. These are living cells taken out of an organism and kept alive in special sterile containers. The viruses are then grown in these cells.

Questions

1 Sort the following diseases into groups based on what causes them:
malnutrition, alcoholism, cancer, stress, dementia, colour blindness, phenylketonuria, obesity, drug addiction, malaria, diabetes, asbestosis.

2 With the help of a diagram, describe how a virus reproduces.

3 Explain why viral diseases are often more serious than bacterial diseases.

4 Which of the following are viral diseases?
mumps, measles, syphilis, food poisoning, rabies, AIDS, herpes, gonorrhoea, thrush, influenza, German measles, common cold, tuberculosis, whooping cough.

5 Viruses have always been very difficult to study. Suggest two reasons why.

Cause of disease	Example of disease
Nutritional deficiencies	Malnutrition, starvation, obesity
Infection by other organisms	Food poisoning, influenza, malaria
Inherited in the genes	Haemophilia, colour blindness, muscular dystrophy
Occupation (as one factor)	Asbestosis, stress, cancer
Social behaviour	Alcoholism, drug addiction
Ageing (as one factor)	Arthritis, dementia, cancer
Metabolic disorders	Diabetes, phenylketonuria, cancer

Table 1 Diseases can be caused by a variety of things

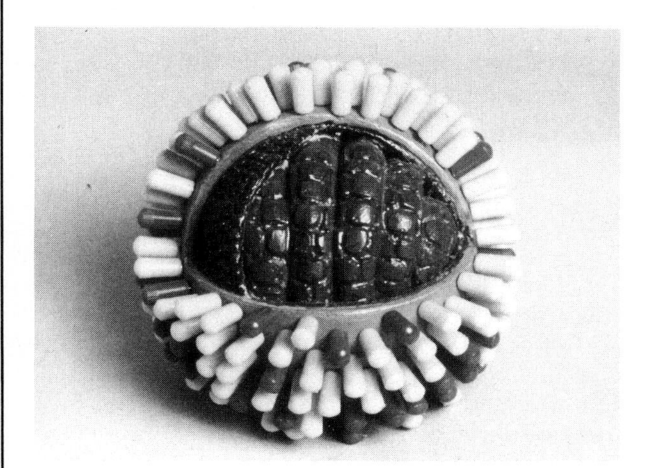

Figure 1 A model of the influenza virus

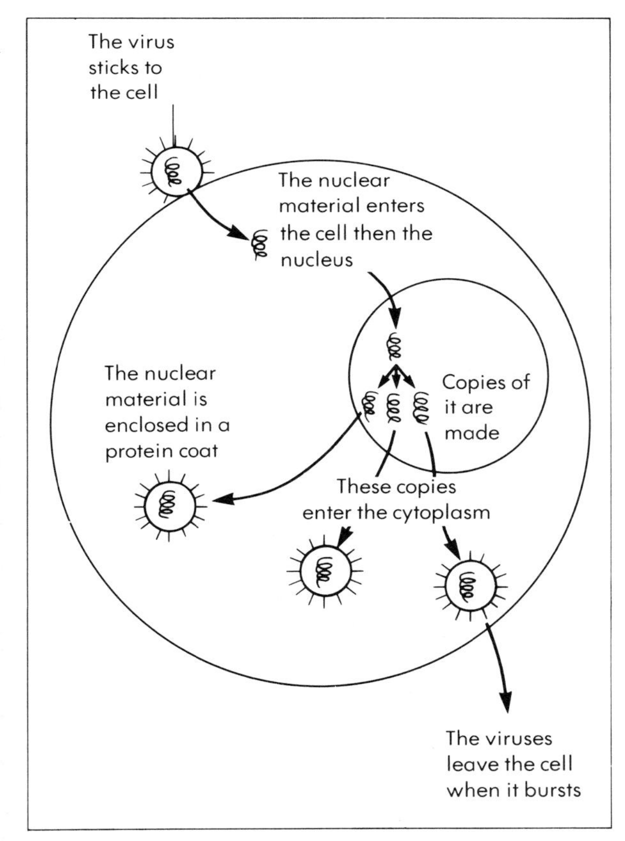

Figure 2 How a virus takes over a cell

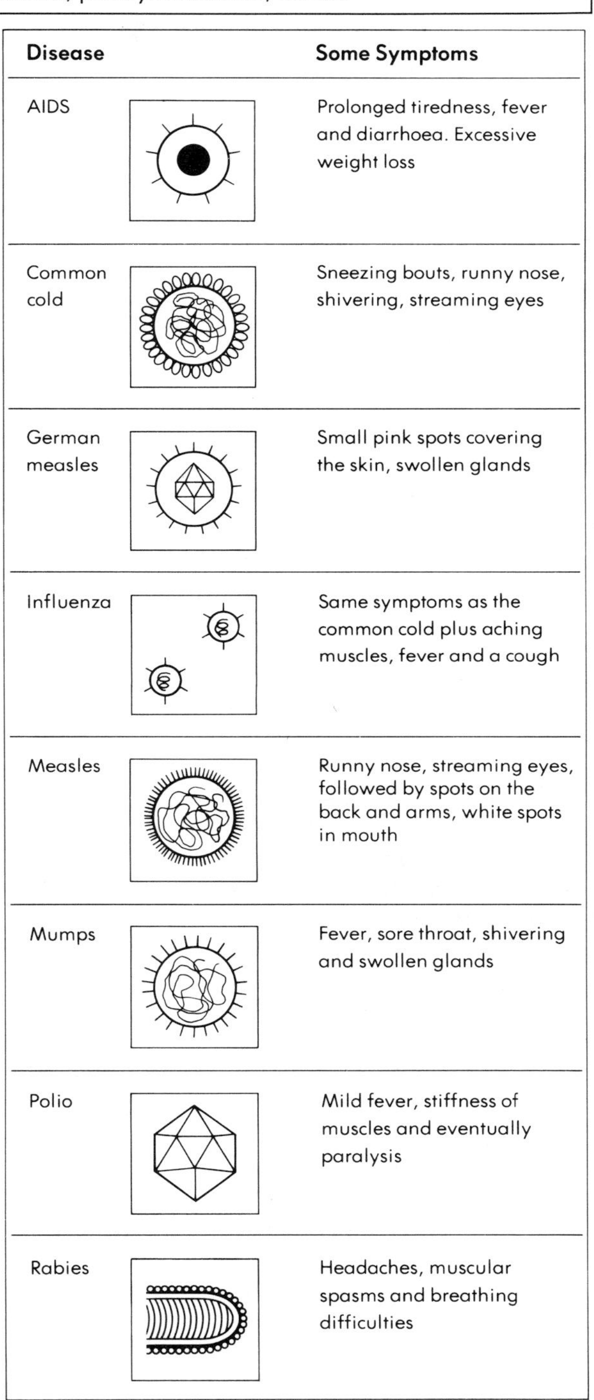

Disease	Some Symptoms
AIDS	Prolonged tiredness, fever and diarrhoea. Excessive weight loss
Common cold	Sneezing bouts, runny nose, shivering, streaming eyes
German measles	Small pink spots covering the skin, swollen glands
Influenza	Same symptoms as the common cold plus aching muscles, fever and a cough
Measles	Runny nose, streaming eyes, followed by spots on the back and arms, white spots in mouth
Mumps	Fever, sore throat, shivering and swollen glands
Polio	Mild fever, stiffness of muscles and eventually paralysis
Rabies	Headaches, muscular spasms and breathing difficulties

Table 2 Some common viral diseases and their symptoms

13.2 Bacteria

What are bacteria?

Bacteria are very small and primitive single celled organisms (figure 1). They were first seen by a Dutchman called Van Leeuwenhoek when he examined scrapings from his teeth through a microscope. We now know that bacteria are found almost everywhere. Most are very useful and have an important role in nature but a few also cause serious diseases in humans, in the animals we breed and in the plants we use for food.

How are bacteria useful?

Bacteria have many uses, some of which are:

- The recycling of nutrients e.g. nitrogen (see page 24).
- The treatment of sewage (see page 210).
- The production of vitamin K in our intestines.
- The digestion of cellulose in herbivores' guts.
- The production of cheese, yoghurt and vinegar.
- The production of drugs.
- The production of silage.
- As a high protein food source (see page 16).

How are bacteria harmful?

Bacteria are harmful to us in two ways:

1. Some of them cause diseases (table 1). These are referred to as **pathogenic** bacteria.
2. Some of them feed on our food, making it go bad. Organisms that feed on dead plants and animals are called **saprophytes**. The main saprophytes are bacteria and fungi.

Which diseases do bacteria cause?

The link between bacteria and disease was first discovered by Louis Pasteur when he was working with diseased silkmoths. Twenty years later, Robert Koch managed to show that a killer disease of cattle (anthrax) was also caused by bacteria. The technique he used to do this is still used today in pathology laboratories. We now know of hundreds of diseases caused by bacteria, some of which are shown in table 1.

What happens if you have a bacterial disease?

If your doctor suspects that you have a bacterial disease, he or she will take an infected sample from your body and send it to the local **pathology** or **public health laboratory**. Here the sample will be transferred onto the surface of a culture plate. The culture plate is then incubated for 24–48 hours at 37 °C. Any bacteria present will grow and produce colonies. These bacteria are then examined through a microscope. If any unusual bacteria are found your doctor is told. You will then be treated.

Bacteria and food poisoning

Some of the bacteria that can live on our food can also cause disease. They do this by:

- Excreting toxins into the food which will poison your body when you eat the food.
- Infecting your body after being eaten with the food.

Food poisoning is the term used to describe the group of illnesses these bacteria cause. These can range from sickness and diarrhoea to severe intestinal bleeding, and even death. A recent estimate by the World Health Organisation suggested that as many as 1000 million people suffer some form of food poisoning every year. Figure 2 shows some of the main types of bacteria which cause this food poisoning.

Questions

1. Imagine you are a doctor. A patient comes to you complaining of headache, sickness and abdominal pain. You know that this could be one of several diseases. How would you confirm which one it was?
2. Your local MP is campaigning to rid the world of bacteria. Write a letter to him suggesting that he changes his mind, stating your reasons.
3. Food poisoning is one of the commonest diseases, yet it is easy to prevent. List five things that you do to make sure that you do not catch food poisoning from the food you eat.

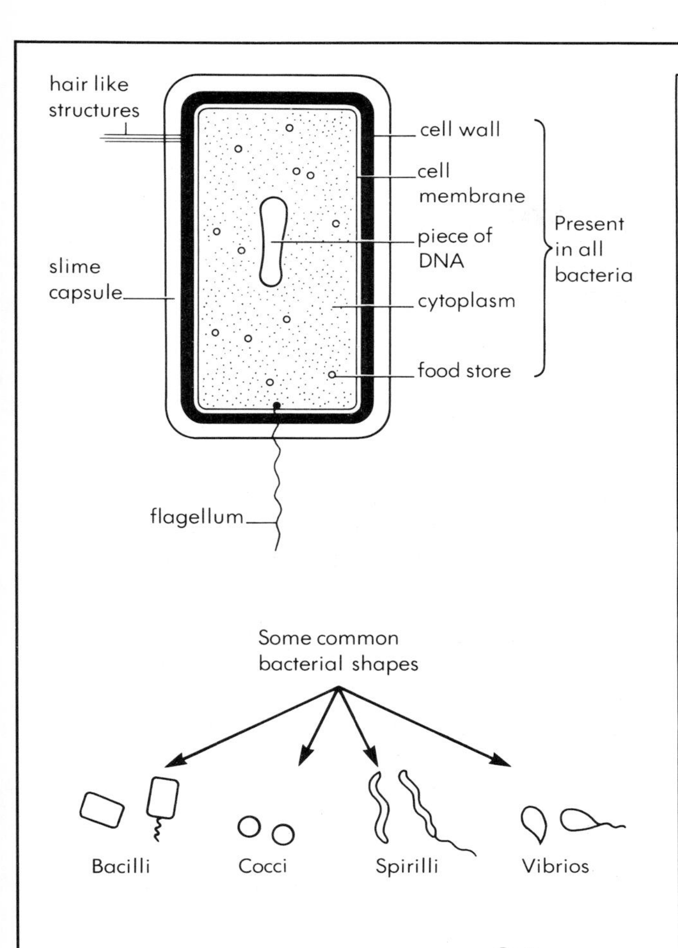

Figure 1 The structure of a typical bacterium and some common bacterial shapes

Disease	Some symptoms
Cholera	Prolonged sickness, diarrhoea and abdominal pain
Diphtheria	Sore throat with swollen glands and fever
Food poisoning	Sickness, diarrhoea and abdominal pain
Gonorrhoea	Painful urinating and a milky white discharge (p.156)
Meningitis	Severe headache with vomiting, fever and neckache
Syphilis	Painless sore replaced by body rash which also soon disappears (p.156)
Tetanus	Headache, fever, stiff muscles followed by spasms
Tuberculosis	Persistent cough bringing up blood - stained sputum. Fever
Whooping cough	Frequent coughing bouts which interfere with breathing often followed by vomiting
Typhoid	Sickness, diarrhoea, abdominal pain and fever

Table 1 Some common bacterial diseases and their symptoms

Salmonella species
The commonest cause of food poisoning particularly from poultry and other meats

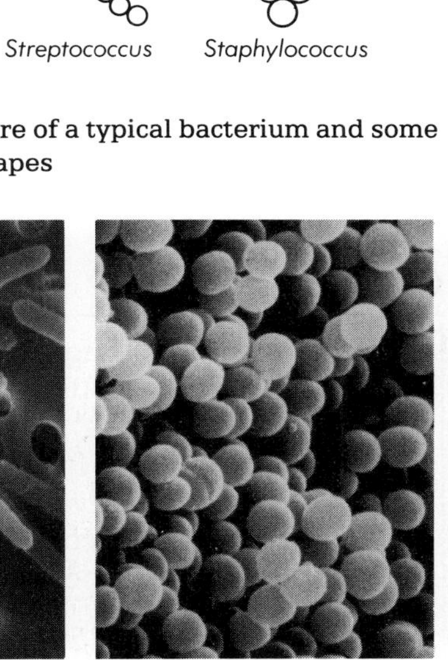

Staphylococcus aureus
Always present in your nose and on your skin. Multiply rapidly in food

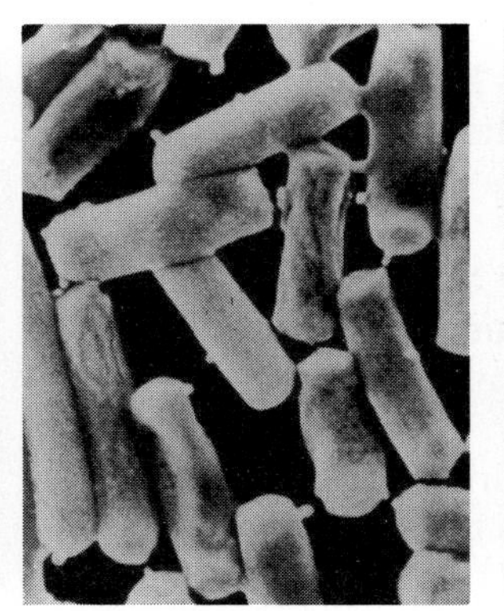

Clostridium botulinum
Produces the most deadly known toxin. Usually caught from canned fish

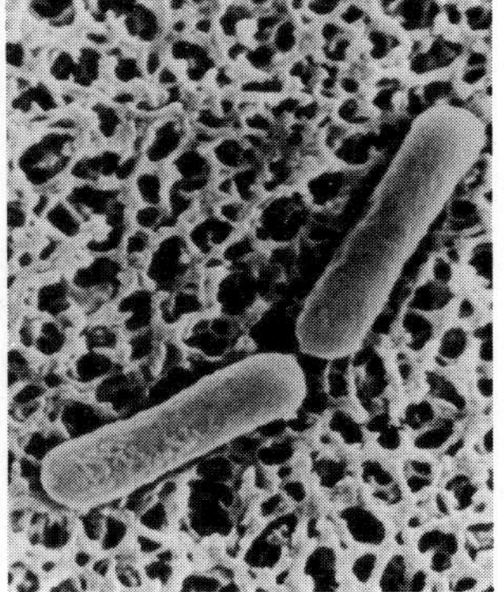

Clostridium welchii
Produces spores which resist cooking. Usually caught from canned meats.

Figure 2 These bacteria can all cause food poisoning. (All are magnified about 10000 times)

13.3 Protozoa

What are protozoa?

Protozoa are single celled animals, slightly larger than bacteria but still microscopic. Most are harmless and some are even useful. For example, some protozoa live in the alimentary canals of animals and help them digest cellulose. In return they get food and shelter. This kind of relationship where both organisms benefit is called a **symbiotic** relationship. Some kinds of protozoa are used to make sewage harmless.

A few protozoa cause very serious diseases, the most widespread being **malaria**, **sleeping sickness**, and **dysentery**.

Malaria

About 500 million people a year catch malaria and about three million die as a result. Many of the rest continue to suffer from the illness.

The protozoan that causes malaria is called **plasmodium**. This has a complex life cycle involving both mammals (including humans) and the mosquito. The mosquito transmits the plasmodium from one mammal to another. We call it the **vector** (carrier).

The symptoms of malaria are closely linked with the parasite's life cycle (figure 1). In humans it first makes its way to your liver where it feeds and reproduces for about two weeks. During this time, you start to feel ill and your temperature gradually rises. Eventually you develop a fever, just as the parasites start to leave the liver and enter your bloodstream. The fever subsides as each parasite enters a red blood cell. A couple of days later the fever returns as the red blood cells burst, releasing lots more parasites and their waste products (toxins) into your blood.

Very few people actually die from malaria. Most are killed by a **secondary infection** such as bronchial pneumonia caught while their body is in such a weakened state.

Female mosquitoes take in the protozoan parasite when they remove blood from an infected animal (figure 2). They need this blood for the normal development of their eggs. The parasite undergoes sexual reproduction in the mosquito's stomach and the offspring make their way to the salivary glands. The next time the mosquito feeds on blood, the parasites are injected into the new host, along with its saliva. This link between mosquitoes and malaria was discovered by Ronald Ross.

Sleeping sickness

Sleeping sickness is widespread in parts of tropical Africa. It is caused by a protozoan called **trypanosome** (figure 5) and the vector is the **tsetse fly**. In the human, the parasite enters the nervous system where it feeds and reproduces. Its waste products cause sleepiness and lethargy. One form of sleeping sickness can even kill you.

Dysentery

Dysentery is very common in countries where there is poor sanitation. It has many different causes, one of which is the protozoan **entamoeba**. This lives in your intestine, feeding on bacteria and occasionally on the lining cells. The symptoms of dysentery are severe diarrhoea, sickness and ulcers. People die from dysentery.

The vector is often the **housefly**, but it is sometimes passed on by humans. The parasites are passed out in human faeces. If they get into drinking water, because of bad sanitation, or food, because of poor personal cleanliness, the disease can rapidly spread to many more people.

Questions

1 Malaria and sleeping sickness are usually referred to as serious economic diseases. Can you think why?

2 A person suffering from malaria has periodic bouts of fever. Explain why.

3 People who live in countries where malaria is endemic sleep under netting (figure 5). Why?

4 If you wander round many African villages you will see many of the inhabitants apparently asleep against the walls of their houses. You could be forgiven for thinking they are lazy, but what is more likely to be the real reason for their inactivity?

5 We get very few cases of dysentery in this country. Why do you think this is?

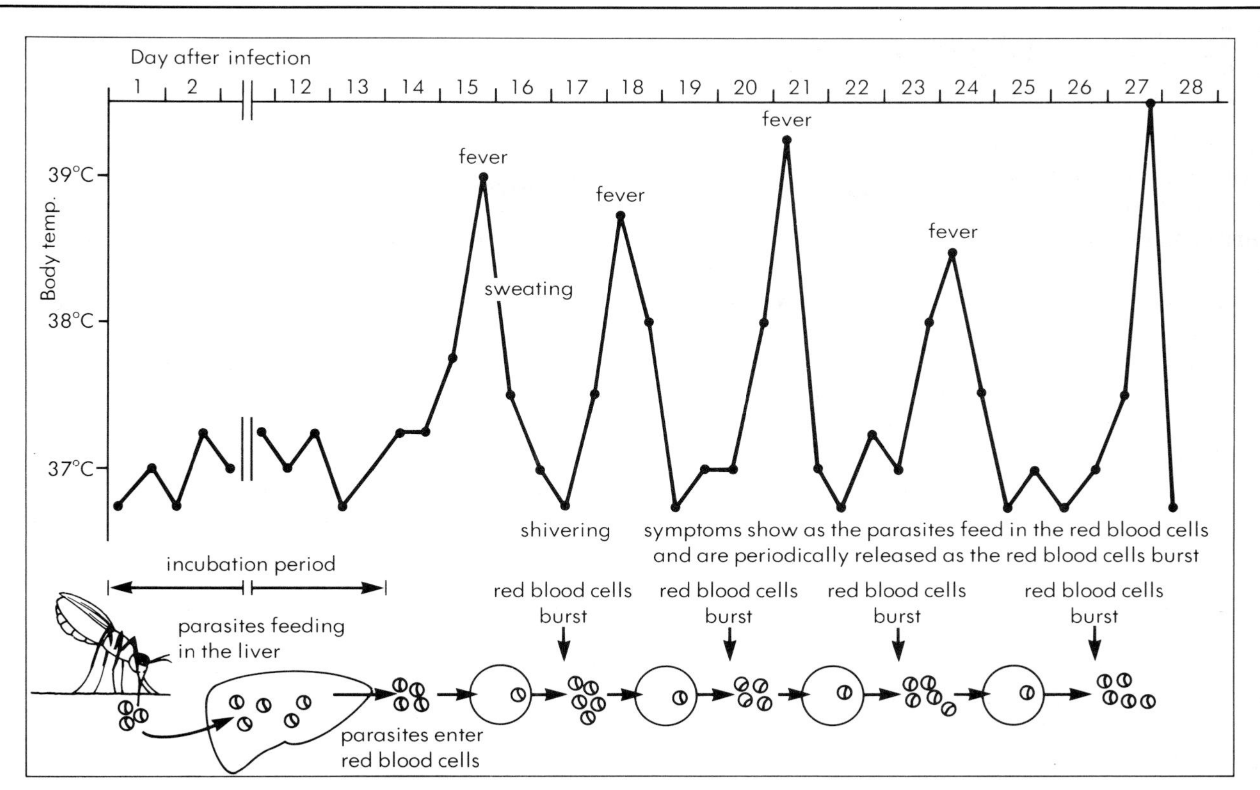

Figure 1 Infection with the malaria parasite causes a series of high fevers to develop

Figure 2 A mosquito takes blood from an arm

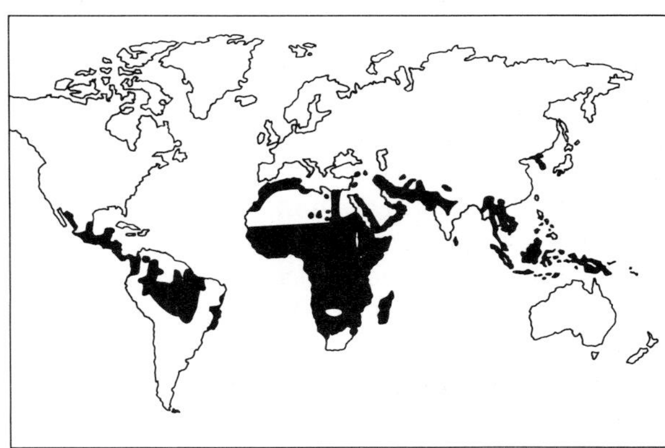

Figure 3 Malaria is usually confined to the tropical areas of the world

Figure 4 When visiting areas where malaria is endemic, it is best to take suitable precautions such as sleeping under a mosquito net

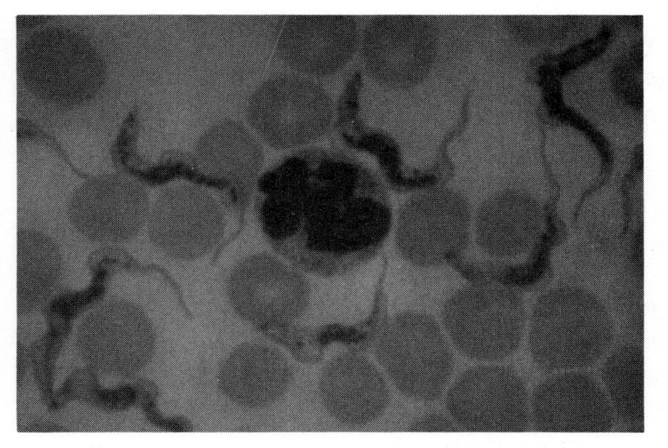

Figure 5 How many trypanosomes can you see among the blood cells here?

13.4 Fungi

What are fungi?

Fungi are plants, but unlike most plants they are not green. Most feed as **saprophytes**, that is, they feed on dead plant and animal matter, and are quite harmless. Some are even useful, but a few cause disease.

Yeast, **mushrooms** and **moulds** are all types of fungi (figure 1).

How are fungi useful?

It is well documented that the Chinese were using yeast for bread making and a mould to produce soya sauce over 2000 years ago. These processes have changed very little.

Some present day uses of fungi are:

- Yeast (figure 1) is used by the baking industry to make dough rise.
- Yeast is used by the brewing industry to produce alcohol from sugars. This process is called **fermentation**.
- Moulds are used to produce antibiotics, e.g. *Penicillium* (figure 1) is used to produce penicillin.
- Moulds are used by the cheese manufacturers to ripen cheeses and give them their individual flavours, e.g. one kind of *Penicillium* is used in Camembert cheese.
- Mushrooms (figure 1) are used as food.
- Various fungi are used to produce protein concentrates (Single-Cell Protein, see page 16).
- Various fungi are used for the large scale production of enzymes, e.g. one kind of *Aspergillus* is used to produce amylase. Amylase is used by the brewing industry for malting, see page 2.

How are fungi harmful?

Fungi are harmful to us in four ways:

- Some of them spoil our food.
- Some destroy our crops, e.g. potato blight.
- Some destroy our houses and furniture, e.g. dry rot.
- Some cause diseases (figure 2) for example:

1 **Athlete's foot** This disease is caused by a kind of **ringworm fungus** feeding on the dead, sweaty skin between your toes. This feeding damages the living skin underneath, causing soreness and inflammation.

The same fungus can also feed in the groin, causing the disease called **dhobi itch**, or on the scalp causing **ringworm**. All three diseases are extremely contagious and are usually passed on in public places such as swimming pools and communal changing rooms.

Ringworm, athlete's foot and dhobi itch can be cured by **antibiotics** or by using fungicide lotions on the skin. If you think you may have one of these diseases, you should go to see your doctor as soon as possible.

2 **Thrush** This is caused by a yeast-like fungus called *Candida*. It lives and feeds in the delicate mucous membranes lining your body's natural openings, such as the mouth and vagina, and can become a problem when you are 'run down' and unwell.

Questions

1. There are three different kinds of fungi. State one commercial use of each kind.
2. Fungi will feed on most things. How can this be used to our advantage?
3. Many of the fungal diseases are caught from public places such as swimming pools. In an attempt to prevent this happening, many of these places display signs saying 'Please make sure you dry yourself thoroughly before getting dressed'. Why would this help stop the spread of fungal diseases?
4. Write a list of as many other ways as you can think of, in which fungi are harmful to us.

Figure 1 Some of the many useful fungi

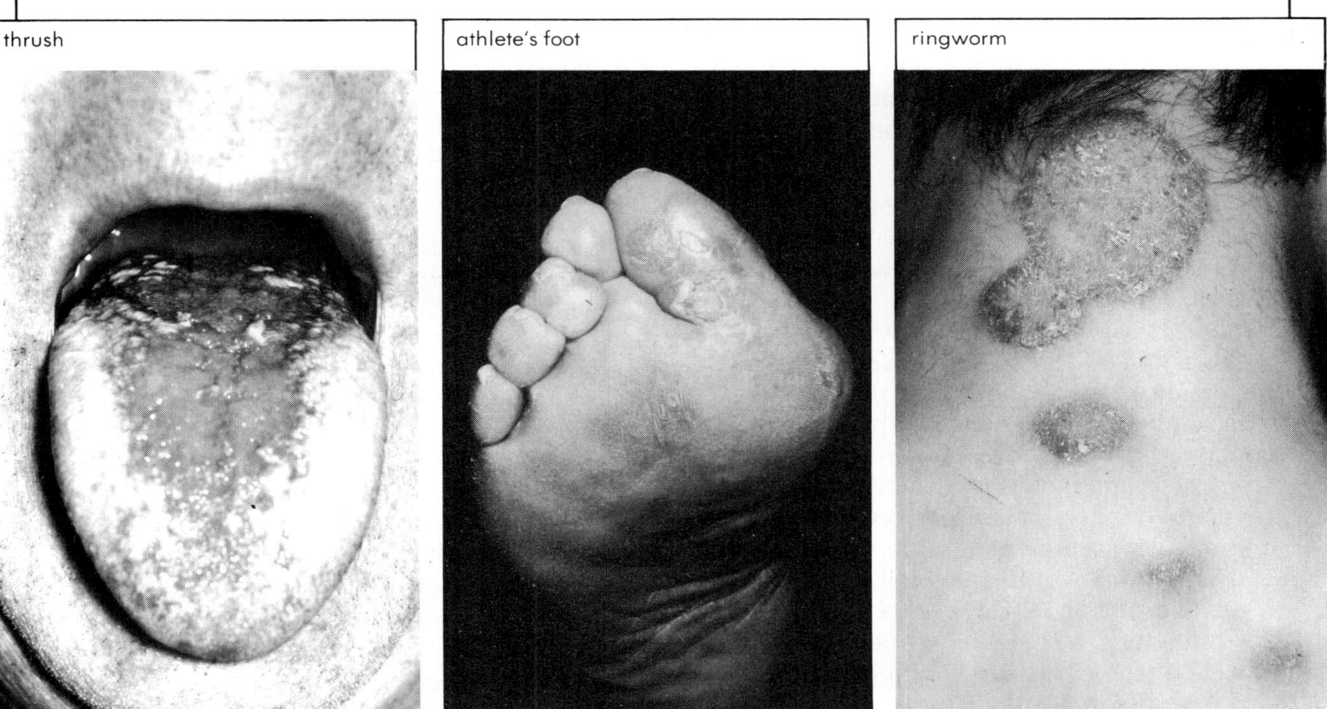

Figure 2 Some of the diseases fungi cause

13.5 Insects and worms

Insects

Insects are the most abundant of all animals, but surprisingly, very few are actually directly responsible for disease. Many, however, are vectors of disease and are harmful in other ways.

How are insects harmful?

Insects are harmful to us in the following ways:

- Some are vectors for human disease-causing organisms (see page 206).
- Some are vectors for diseases of our crops, e.g. aphids carry plant viruses.
- Some are vectors of diseases that affect our domestic animals, e.g. fleas transmit myxomatosis.
- Some cause disease by removing blood, e.g. bedbugs can cause anaemia.
- Some destroy our crops, e.g. locusts and cabbage white butterfly caterpillars eat the leaves.
- Some spoil our food, e.g. weevils taint flour.
- Some damage our buildings and furniture, e.g. termites damage wood; clothes moths damage clothing.

Some insects that spread disease are shown in figure 1.

Which worms cause disease?

The two groups of worm that cause disease are the **roundworms** and the **flatworms**. The names of these worms describe the shape of their bodies.

- **Roundworms** There are over forty kinds of roundworm that cause disease. The two most common of these are the **hookworms** and the **filarial** worms.

  Hookworms live for most of their larval stage in water. They enter a human by burrowing through the skin as the person walks or bathes in the water. The adult hookworms live in your intestine damaging lining cells and feeding on blood. This causes severe abdominal pain, sickness, diarrhoea, and sometimes anaemia.

  Adult filarial worms (also called threadworms) usually live in the lymphatic system often causing a blockage. This blockage stops the flow of lymph, and it builds up in the tissues. This is how the condition known as **elephantiasis** comes about.

- **Flatworms** Pathogenic flatworms include the **blood fluke**, and **tapeworms**. The blood fluke called **shistosoma** lives in the blood causing a disease called **bilharzia**. Bilharzia affects over a quarter of a billion people in the developing countries. It is often classed as a serious economic disease because it makes people too weak to work.

  Tapeworms are the largest of all disease-causing organisms. Sometimes they may reach lengths of thirty metres or more. They have two hosts: the one in which they reach sexual maturity is called the **primary host** and the other is called the **secondary host**. The fish, beef (figure 2), and pork tapeworms all have humans as their primary host.

  All tapeworm infestations can be prevented by cooking food properly and disposing of sewage effectively.

Questions

1. Insects are vectors for some of the world's deadliest diseases. It has been suggested that the easiest way to stop these diseases spreading is to destroy all insects. In your opinion would this be wise?
2. The following statements all refer to the life cycle of the beef tapeworm. Place them in the correct sequence:
   the young worms burrow into the muscles of the cow
   eggs pass out with the faeces
   the eggs hatch
   the worms reproduce in the gut
   the eggs are eaten by a cow
   the young worms are eaten in poorly cooked beef
   human faeces are used as manure
3. Suggest two ways by which tapeworm infestations can be avoided.
4. Filarial worms often block lymph channels. What are the consequences of this?

Flea (actual size: 2 mm)
Fleas suck blood and can transmit the bacteria which cause typhus and plague

Body louse (actual size: 2 mm)
Lives in clothing and sucks blood. Transmits typhus and impetigo

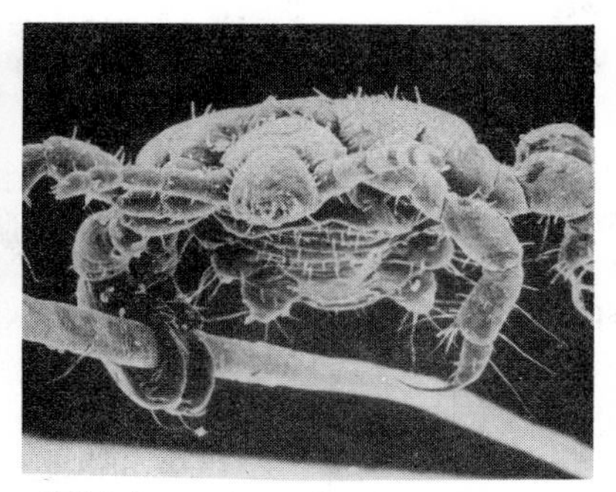

Pubic louse (actual size: 2 mm)
Sucks blood. Transmits impetigo

Cockroach (actual size: 30 mm)
Cockroaches are scavengers and will feed on anything. They spread food poisoning

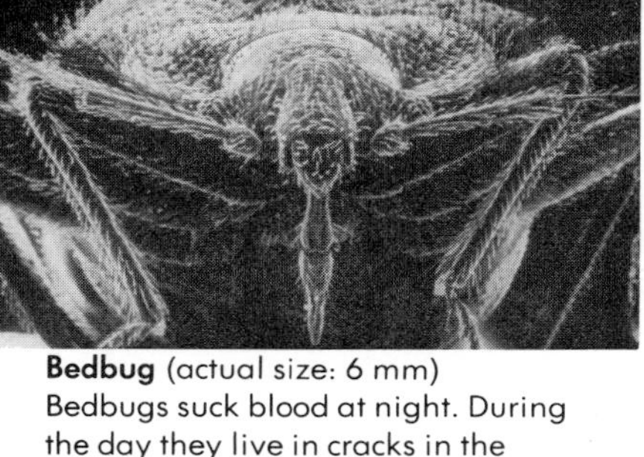

Bedbug (actual size: 6 mm)
Bedbugs suck blood at night. During the day they live in cracks in the floor.

Housefly (actual size: 10 mm)
Houseflies spread food poisoning in the UK but abroad they may spread cholera, dysentery and typhoid

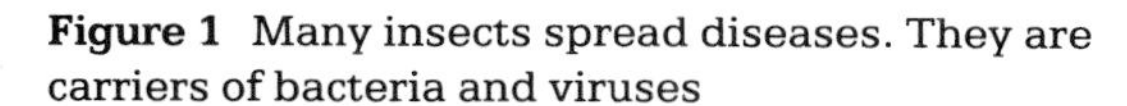

Figure 1 Many insects spread diseases. They are carriers of bacteria and viruses

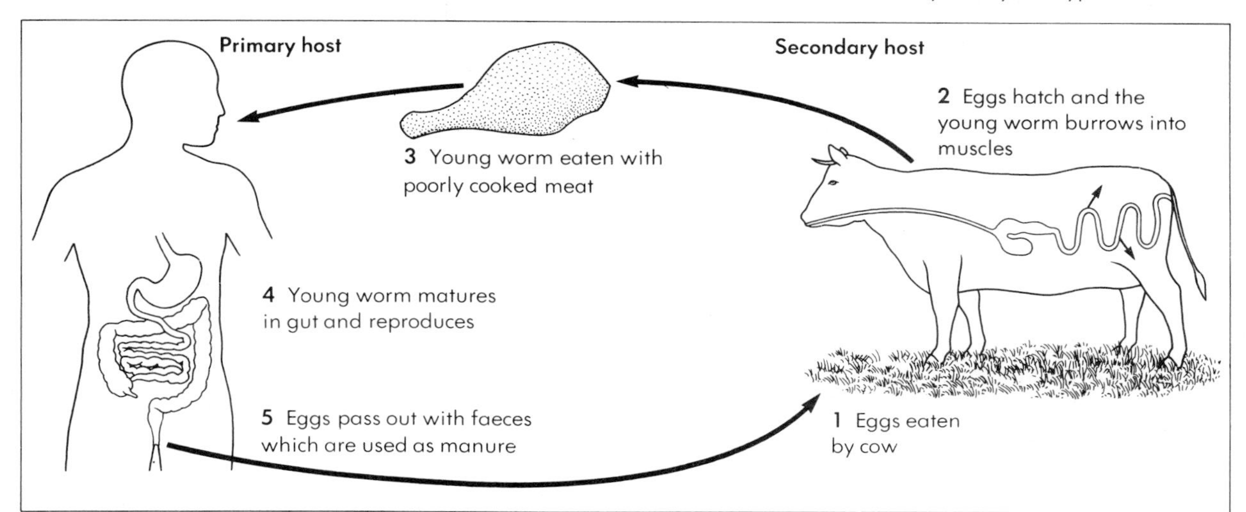

Figure 2 The life cycle of the beef tapeworm

13.6 The spread of disease

What are infectious diseases?

Diseases which can be passed on from person to person are called **infectious** diseases. In a few cases the organisms causing these diseases can move by themselves, but in most cases a third agent must be involved.

How are diseases spread?

Organisms causing disease can be spread:

- **By the air** Coughing, sneezing, breathing and even talking release tiny droplets of moisture into the air. Microbes stick to these droplets and may enter another person when he or she breathes in. This is how many diseases are spread in crowded, stuffy places like buses, trains, pubs and shops.
- **By infected food and water** Food can be contaminated in many ways. For example it may have been handled with dirty hands or carved up with a dirty knife. Water is usually contaminated by passing raw sewage into it.
- **By objects** Books, toys, clothes, towels and money can all carry micro-organisms.
- **By direct contact** Some pathogens can only be passed on by direct contact with an infected person. These diseases are said to be **contagious**. The organisms involved can only live inside a body.
- **By insects and other animals** Organisms that spread disease from one individual to another are called **vectors**.

Table 1 shows how some common and not so common diseases are spread.

Which vectors spread which diseases?

A lot of the world's most serious diseases are spread by insects:

- malaria is spread by the **mosquito**,
- sleeping sickness is spread by the **tsetse** fly,
- cholera, polio and dysentery are spread by the **housefly** (figure 1),
- plague is spread by the **flea**.

Rats and **mice** are not only vectors of disease but also do serious damage to property. The diseases they have been known to spread are **food poisoning**, **bubonic plague** and **infective jaundice**.

Dogs, **cats** and **foxes** are vectors of the disease called **rabies**. **Rabies** is a very serious disease, **endemic** in most countries of the world, but so far it has not even entered the UK mainly due to the strict **quarantine** regulations (figure 2).

What is quarantine?

Quarantine is a period of isolation which all animals must go through before they can enter the UK and many other countries. The length of this period of isolation depends on the country the animal has just travelled from (or even through). If, for example, it has come from a country where there have in the past been cases of rabies, then the period of isolation will be six months. This is much longer than the incubation period of the disease and if no signs have been shown by the animal within this period, it is assumed that it does not have rabies and is allowed to enter the country.

Questions

1. What is an infectious disease? List the main ways by which infectious diseases can be spread.
2. How are the following diseases spread: influenza, tuberculosis, typhoid, cholera, salmonella food poisoning, gonorrhoea, syphilis, athletes foot and rabies?
3. Name three insect vectors and the diseases they spread.
4. Are we vectors for any diseases? If so name them.
5. One of the major worries after an earthquake is that epidemics of typhoid, cholera and dysentery will develop. Why will these suddenly develop?
6. Why has rabies so far not entered the UK?
7. Many people are very worried that a channel tunnel will allow rabies into this country. What do you think?

This is what happens when a fly lands on your food.

Flies can't eat solid food, so to soften it up they vomit on it.

Then they stamp the vomit in until it's a liquid, usually stamping in a few germs for good measure.

Then when it's good and runny they suck it all back again, probably dropping some excrement at the same time.

And then, when they've finished eating, it's your turn.

Cover food. Cover eating and drinking utensils. Cover dustbins.

Figure 1 Would you eat this food now?

Disease	Main method of spread
Common cold Diphtheria German measles Influenza Measles Mumps Polio Rabies Smallpox Tuberculosis Whooping cough	On droplets of moisture produced by coughing, sneezing and even breathing
AIDS* Gonorrhoea* Ringworm Syphilis* Thrush	By direct contact and in many cases sexual contact (*)
Bilharzia Cholera Dysentery Food poisoning Typhoid	On objects or in contamined food and water
Malaria Sleeping sickness Typhus	By blood sucking insect vectors

Table 1 How some common diseases are spread

Figure 2 Part of a campaign to stop rabies reaching this country

13.7 Fighting disease

How do we fight disease?

You could say that in the world today we carry out the fight against disease on three levels: the personal level, the community level and the worldwide level.

How can an individual fight disease?

A person can reduce his or her chances of getting some diseases by:

- maintaining good personal hygiene (figure 1),
- eating a good balanced diet,
- taking regular exercise,
- getting enough rest,
- not smoking or drinking alcohol in excess,
- using the preventive medical services (figure 2),
- behaving sensibly.

What can a community do to prevent disease?

Because we often live close together, we share a lot of facilities which affect our health – water, sanitation, rubbish disposal etc. So it is important that we share the responsibilities of good healthy conditions between us, in our local community.

Community health in England is the responsibility of each local government. Each local authority has to provide by law certain services for the community. They must:

- **Provide safe water** for drinking, washing etc.
- **Deal with the sewage and refuse** made by the community.
- **Provide medical care** for those who require it.
- **Monitor standards** of health and hygiene within the community and implement the **Food and Drugs Act**, the **Health and Safety Act** and the **Factories Act**.

Most local governments also provide many additional services which are not required of them by law.

What about the worldwide level?

There are many organisations concerned with world health, but the largest and most influential is the **World Health Organisation** (WHO). This was set up by the United Nations Assembly in Geneva in 1948 and still retains its headquarters there. The WHO long term aim is to raise the level of health of all the citizens of the world so that by the year 2000 they will all be able to lead socially productive lives. At present there are 166 Member States involved and with their co-operation the WHO is active in the following areas:

- the promotion and development of comprehensive health services,
- the monitoring of health and disease,
- the prevention and control of disease,
- the planning and implementation of health programmes,
- the improvement of environmental conditions,
- establishing standards for drugs, pesticides etc.
- the co-ordination of research programmes,
- the training and provision of health personnel.

There have already been some successes in some of these areas. For example, the infant mortality rate in many countries has fallen dramatically due to work done on behalf of the health of mothers and babies – in particular education and diet. The promotion of mass immunisation has virtually wiped out smallpox all over the world. Malaria has been eradicated from many areas of the world and safe water is now available to millions more people.

The WHO also works in conjunction with other UN agencies such as the **Federal Agricultural Organisations** (FAO) and **UNESCO** to tackle problems like starvation and malnutrition in developing countries. There are many more organisations such as **The International Red Cross**, **Christian Aid**, **Live Aid** and **Oxfam** which are also deeply involved in all aspects of world health.

Questions

1. Write a list of everything you do every day/week to maintain your own personal hygiene. Compare this list with figure 1.
2. Regular checks on your body can reveal many problems well before they get out of hand. List some of these problems.
3. By monitoring new outbreaks of disease, the WHO is able to predict future epidemics. Suggest why this information would be useful.
4. One of the main roles of the WHO is the implementation of health education programmes. Why is this so important to the long term aim of the WHO?

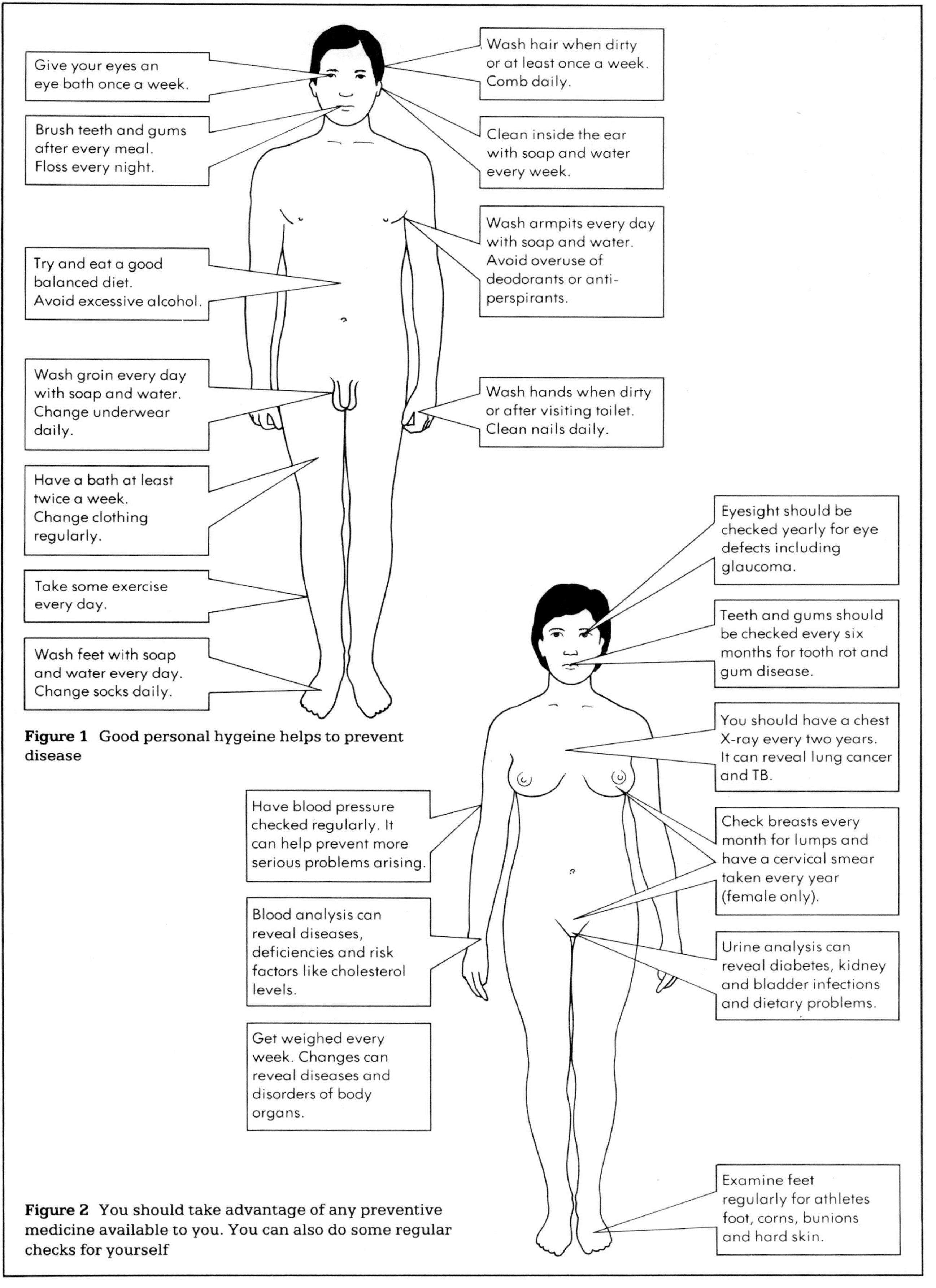

Figure 1 Good personal hygeine helps to prevent disease

Figure 2 You should take advantage of any preventive medicine available to you. You can also do some regular checks for yourself

13.8 Safe water and sewage disposal

Where does our water come from?

The average person in the UK uses about 120 litres of water every day. This means an average sized city (300 000 people) will need to find 36 million litres of water every day. Most of this comes from fresh water rivers, lakes and reservoirs. Sea water can be used, but it is very expensive to process. The lack of water is a major problem in some parts of the world.

How is water made safe to drink?

Very few sources of water provide clean, safe water suitable for drinking straight away. To make it safe to drink it must undergo a series of treatments. The extent of these will depend upon where the water came from. Figure 1 shows how river water is treated. Water taken from underground sources or mountain springs is already fairly clean and does not undergo so many processes.

What is sewage?

Sewage is the wet rubbish from houses, factories and the streets. **Domestic sewage** consists of human urine and faeces, waste water from the kitchen and bathroom and rainwater from roads and roofs. **Industrial sewage** mainly consists of waste water and unpleasant liquids left over from the manufacturing processes. These two kinds of sewage are often dealt with separately.

What happens to domestic sewage?

In industrial countries, that have the equipment and the money, sewage is taken from houses to treatment works in underground pipes called **sewers**. Nowadays most houses in the UK are connected to the sewers, but the few that are not have to use a **septic tank**. This consists of two concrete lined containers at a suitable distance from the house. Sewage is discharged into the first container and allowed to stand for a while. Any solids settle out and form a sludge at the bottom. Special **bacteria** in the tank start to eat this. The liquid part overflows into the second container where it trickles through small stones. The surfaces of the stones are covered by **protozoa** which feed on the organic chemicals in the liquid. Eventually all that is left is water which can easily be disposed of.

Sewage that goes to the sewage works is treated in a similar way (figure 2).

What is refuse?

Refuse is the rubbish that we all produce during our day to day living. It includes things like empty cans, plastic containers, potato peelings, newspapers and wine bottles. In the UK 15 million tonnes of it are produced every year. This is usually collected on a weekly basis by council workers and taken to areas where it can be dealt with.

At present most refuse is either burnt in large incinerators or covered in disinfectant and buried. Most of the buried refuse eventually decomposes, but substances like plastics that are not **bio-degradable** do not. A better way of disposing of these is by recycling them. Some success in recycling has been achieved with metals, glass and paper. Recycled paper is used to make newspaper, cardboard, toilet paper and tissues (see page 232).

Questions

1 The following processes describe how river water is made safe:

chlorination
sedimentation
screening
filtration.

Place these in sequence and outline why each is carried out.

2 Many years ago (and still in some parts of the world), people used to take their drinking water out of the same river as they dumped their sewage in. Explain the connection between this and the high death rate from disease during these times.

3 Our waste is food for other organisms. Explain this statement with respect to sewage.

4 Imagine civilisation as we know it broke down and you had the task of getting rid of the family sewage. Explain how you would do this.

5 Explain what happens to refuse from the moment it is collected.

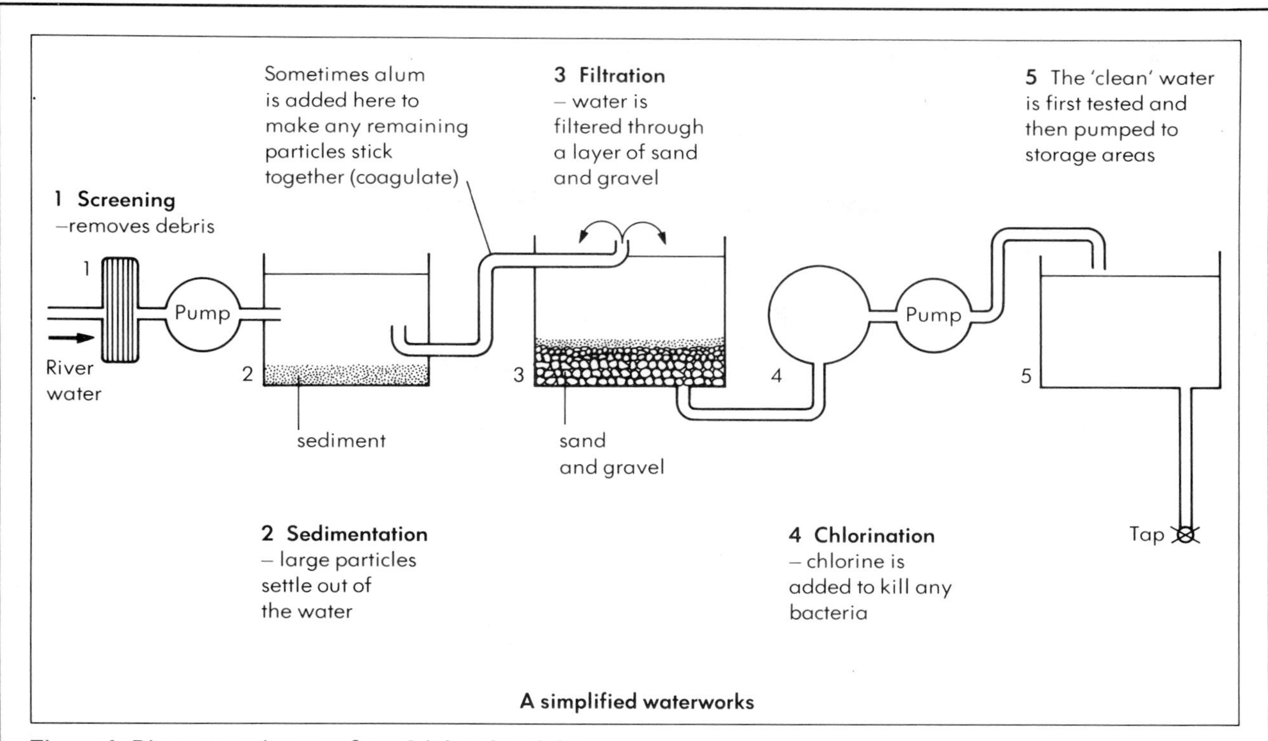

Figure 1 River water is not safe to drink unless it has been treated

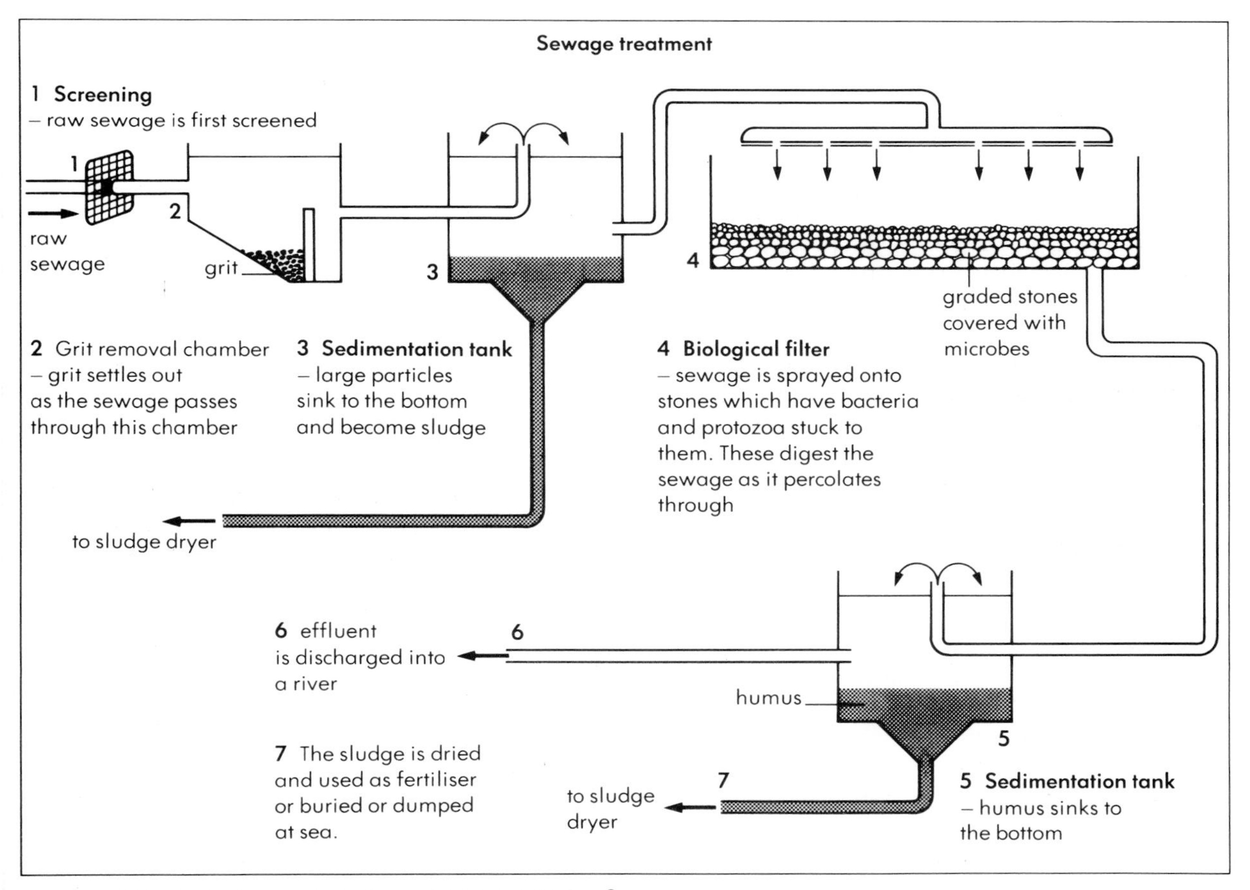

Figure 2 Why is the treatment of sewage an example of biotechnology?

13.9 Safe food

What precautions can you take to make food safe?

Food poisoning is the second most common cause of illness in the UK, yet it can so easily be avoided by taking a few sensible precautions.

The bacteria that cause food poisoning are almost everywhere, even in some of the foods themselves. Keeping it safe, therefore is not just a question of preventing contamination, but also of making sure that you do not give the bacteria the conditions they need to thrive. The three main danger areas are:

- **During handling and preparation** Bacteria can be transferred into food from:
  - dirty hands, especially finger nails,
  - dirty work surfaces,
  - dirty utensils such as knives and chopping boards,
  - other foods via utensils, work surfaces or hands,
  - flies and other animals.
- **During the cooking and serving** Bacteria will thrive in warm food. This can result from:
  - undercooking due to not properly defrosting the food or failing to pre-warm the oven,
  - allowing food to cool too slowly or too long before serving,
  - not reheating the food quickly enough,
  - not reheating the food sufficiently.
- **During the storage** Contaminated food stored in warm, damp conditions can be lethal because the bacteria will quickly build up their numbers. Situations to avoid are:
  - leaving food at room temperature for a few hours,
  - leaving food uncovered,
  - covering the food with dirty cloths or container lids,
  - keeping food warm in warmed containers for long periods.

  The last of these conditions is often seen in some pubs, wine bars and restaurants, where food is prepared in advance and put on display. Be careful about buying food from these places and from shops which do not store and display their food properly. Food should only be bought from shops which:
  - have clean, well-ventilated premises,
  - do not allow pets in,
  - keep all food covered,
  - keep cooked and uncooked meats separate and in chilled cabinets,
  - keep dairy produce chilled,
  - do not allow the assistants to handle the food,
  - label the fresh foods with sell-by dates.

How can the natural life of food be extended?

All foods go off for two reasons:

- Because of the actions of saprophytic organisms (some of which cause food poisoning).
- Because enzymes within the food catalyse unwanted chemical reactions.

Trying to make food last longer has now developed into a large industry. It is because of this industry that we can now enjoy some foods out of season and make better use of surpluses. We can store them to be used later.

All micro-organisms require food, water and warmth to live and reproduce. The methods of preserving food make use of this by denying them water or warmth and so they are killed off or prevented from reproducing. In addition, chemicals can be used to kill the micro-organisms.

Enzymes cannot work without water and are destroyed by heat. Cold inactivates them.

Questions

1 A good friend of yours is very proud of the sausage rolls he has just made in home economics. He said you can have one later on, but he must first warm them up. He places them on top of a radiator, being careful to cover them with his handkerchief so that no dust will settle on them. During the next two hours he tells you why he was late for school that morning. He had been milking the cows. He shows you the milk stains on his hands to prove it. When the rolls are warm he offers you one but you say no. Explain why.

2 Design a poster to promote safe food.

3 Look at figure 1. Make a list of reasons why you would not shop there.

Figure 1 Would you shop here?

Figure 2 The foods in this photograph have all been treated to make them last longer

Figure 3 Cooked and uncooked foods should always be stored separately and in a chilled compartment

13.10 Preserving food

What is food preservation?

For thousands of years people have used many different methods to make food last longer, such as treating food with salt, cooking and pickling food, preserving fruit etc. Now with new technology and the widespread industrial processing of food large scale methods are used to preserve it.

The main methods of preserving food

The most important commercial methods of preserving food are by **sterilisation**, **freezing** (figure 1) and **dehydration**.

- **Sterilisation** involves heating the food to high temperatures. This kills most of the microbes and also destroys the enzymes. The heat treatment is usually carried out after the food has been sealed in suitable containers (cans, bottles or polythene bags). The commonest method of heating is by using steam.

  Unfortunately, sterilisation does not always work as some microbes can produce heat resistant forms called **spores**. As the food cools, these spores grow into new microbes and re-contaminate it. Many of the **toxins** which some microbes produce are also unaffected by heat.

  Sterilising food also alters its flavour. You may know this if you have tried sterilised milk. Often, in an attempt to get over this problem, a milder form of heat treatment called **pasteurisation** is used. This involves heating the food quickly to 72 °C and holding it there for 15 seconds before cooling rapidly. This treatment is used for the cow's milk and much of the wine we drink.
- **Freezing** of food works because most microbes are incapable of reproducing at very low temperatures and the action of enzymes is slowed down or stopped.

  The most useful method of freezing is known as **quick freezing**. This involves bringing the overall temperature of the food to below −18° C very quickly. The speed of the process is essential to prevent large ice crystals forming which could damage the food and lead to the loss of nutrients. Blanching is also used as a treatment for some foods, before they are frozen. To blanch food, it must be put into hot water (90–100 °C) for 1 to 5 minutes. This destroys many of the enzymes.
- **Dehydration** is the removal of water. This is usually done by passing hot air over the food. The more quickly the water is removed, the less damage there is to the food. The fastest method currently used is called **Accelerated Freeze Drying (AFD)**. AFD involves first freezing the food and then heating it inside a vacuum. The ice turns directly into steam, leaving the food in the process.

  Dehydrated food has many advantages, not least that if all the water is removed, it will last forever.

Other methods of preserving food

There are other methods used to preserve food commercially such as:

- **Pickling** is soaking in acid solutions to reduce the pH of the food to a level at which most microbes and enzymes are inactive.
- **Curing** is a kind of dehydration. Addition of salt or sugar to food removes water from it.
- **Adding preservatives** such as sulphur dioxide inhibits the growth of microbes. Some of these additives are harmful.
- **Smoking** food covers it with chemicals which prevent bacteria growing.
- **Vacuum packing** removes oxygen and this prevents many microbes feeding.
- **Irradiation** with alpha-, beta-, or X-rays sterilises food. However, it also alters the texture and flavour of the food, and as yet no one is sure of the long term effects on the consumer.

How does preserving food affect its nutrients?

Heat obviously alters some of the nutrients but not to any great extent. The main problem with preserving food is the nutrient losses, especially vitamins. Some foods can lose up to 25 per cent of their water soluble vitamins (table 1).

Questions

1. What conditions are required for microbes to thrive in food?
2. Use the methods of preserving food as headings. Underneath each heading write down as many foods as you can think of which are preserved by that method.
3. Explain how the various methods of food preservation can affect the flavour, texture and nutritive value of a food.

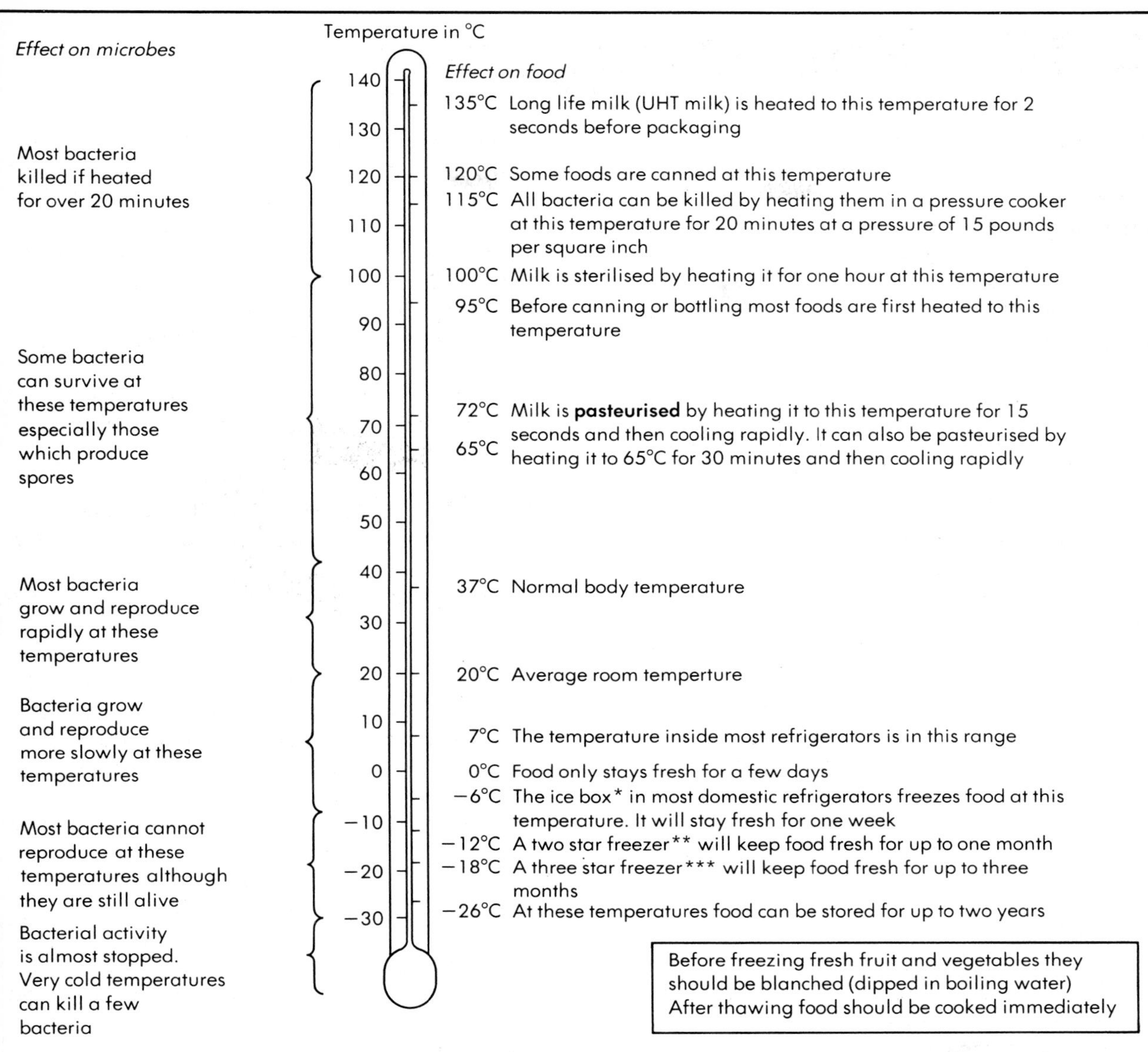

Figure 1 Microbes are affected by temperature; they can be killed by high temperatures and inactivated by low temperatures

	Using heat	Freezing	Dehydration	Altering the pH
Proteins	may alter the structure	may alter the structure		may alter the structure
Fats	may cause rancidity	may cause rancidity		
Vitamins	causes losses of A and C	blanching results in loss of C	causes losses of C and D	causes losses of C
Minerals		causes losses on thawing		

Table 1 Preserving a food can alter the nutrients it contains

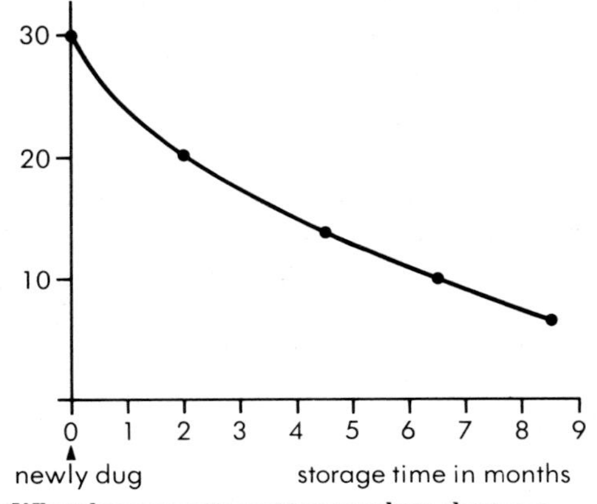

Figure 2 What happens to potatoes when they are stored?

13.11 Health care

The health services

Community health care in the UK is provided by the **National Health Service (NHS)**. This was set up by the Government in 1948, with the intention of making good medical treatment and facilities available to every member of a community, whether rich or poor. It was hoped that in the long term this would improve general health standards and, in turn, this would reduce the demand for the services.

The services are provided and administered by the local government authorities, but overall control and financing comes from the central government through the Department of Health and Social Security. The DHSS also controls the support services to the NHS (figure 1).

What services does the NHS provide?

The NHS provides three main types of medical care:

- **The hospital service** You go to hospital when you need continuous medical treatment by qualified doctors and nurses (figure 2). You may also need to go to a hospital for routine tests, checks or X-rays and not need to stay in. Large hospitals may also have research departments where many important medical advances are being worked on. There are three main types of hospital:
  1. **General hospitals** which can treat most illnesses (table 1).
  2. **Specialist hospitals** which provide expert help for one or two kinds of illness, e.g. psychiatric hospitals provide expert help for the mentally ill.
  3. **Teaching hospitals** which are large general hospitals attached to a medical school or university. These train new doctors and nurses.

- **The General Practitioners** (GPs). A GP is a doctor who works within the community and is usually the first person you go to if there is something wrong. She may treat you or refer you to a hospital. Other health care services available locally are those of the dentist, optician and pharmacist (chemist).

- **The community health services.** These services are mainly concerned with monitoring health standards and preventing illness. The NHS also works closely with the **Personal Social Services** to provide non-medical aids such as home helps. The full range of services both of these provide is:

  The Environmental Health Service
  The School Health Service
  The Child Welfare Service
  Care of the Elderly and Home Help Service
  Midwifery and Maternity Services
  The Mental Health Services
  The Ambulance Services
  The Health Education Services

Who pays for all this?

The NHS cost about £15 billion to run in 1985/6, which is about £250 per person. Ninety per cent of this money comes from direct **taxation** which all working adults pay. The remaining 10 per cent comes from national insurance contributions (deducted from working people), prescription and dental charges.

The personal social services are paid for by another form of tax called the rates. This is a local tax paid by every homeowner and most tenants.

Questions

1. When and why was the National Health Service set up?
2. Who provides and administers the NHS?
3. How is the NHS financed?
4. In most other countries of the world, health care is only available to those people who have taken out medical insurance. Which do you think is the best system? State your reasons.
5. Two people, A and B visit their respective doctors on the same night with the same problem – haemorrhoids. A pays medical insurance, B does not. A is in hospital the next day whereas B has to wait six months. Comment on this.
6. a) What are the Community Health Services? List them.
   b) List the Personal Social Services.
   c) Which of these services have you or members of your family used in the last year?
7. Design a general hospital.
8. Explain the staffing structure of a general hospital.

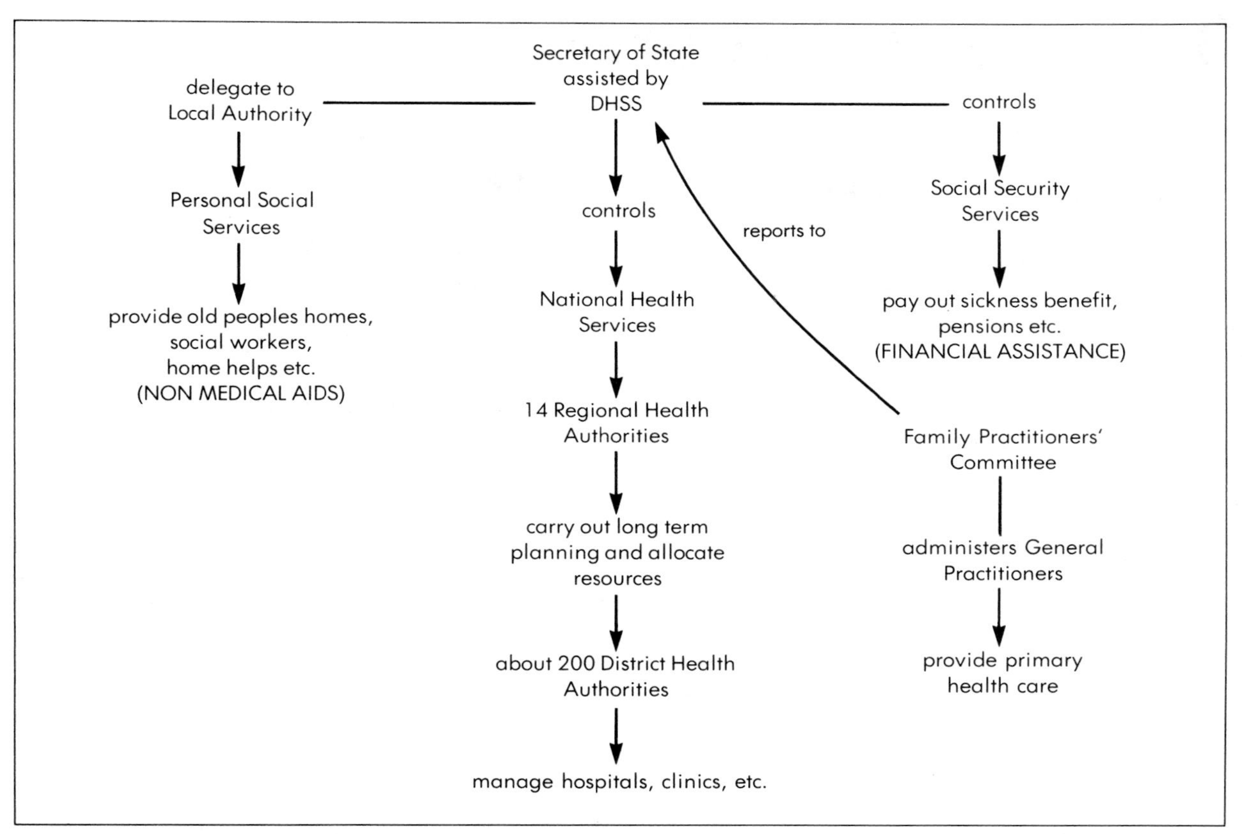

Figure 1 The organisation of the Health Service in England, 1986

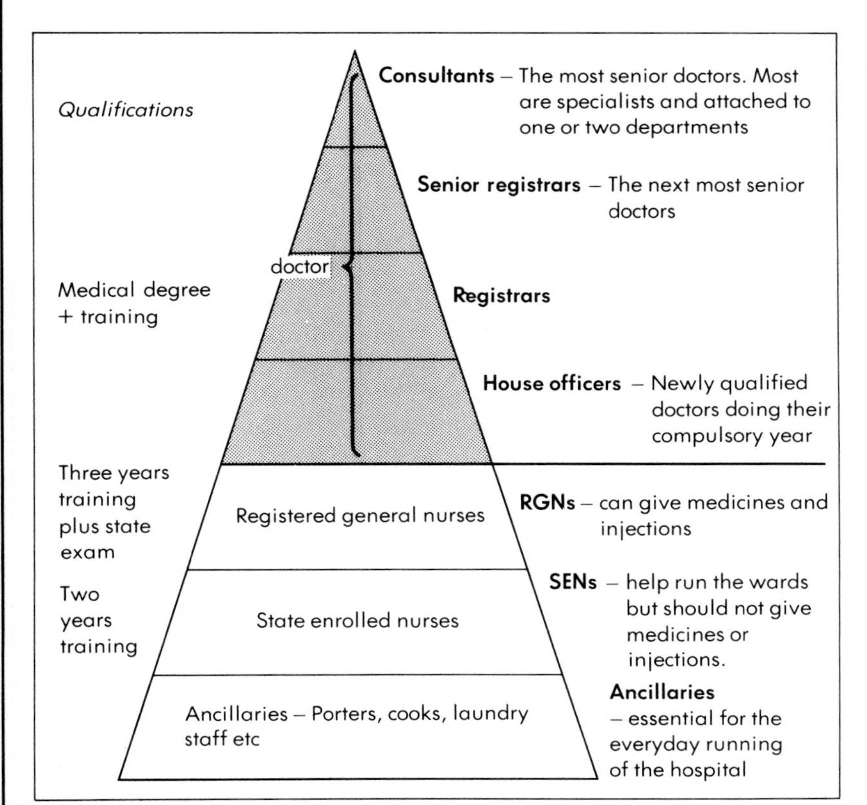

Figure 2 The staffing of a typical general hospital

Department	
Out patients	→
Casualty	↑
Radiology	↑
Blood transfusion unit	→
Surgical ward	→
Medical ward	→
Childrens ward	→
Maternity	←
Geriatrics	←
Physiotherapy	↑
Isolation ward	↑
Intensive care unit	←
Operating theatre	←
Pathology laboratory	←
Sterilisation unit	↑
Incinerator	↑
Pharmacy	→
Mortuary	←

Table 1 The departments in a typical general hospital

13.12 Environmental health

What is the Environmental Health Department?

To a large extent good health in a community relies upon good hygiene. With this in mind, successive governments in the UK have passed laws which provide rules and regulations to protect the health of the community. Individual people, shopkeepers, restaurants etc. must live and work within these laws. To make sure they do, the **Environmental Health Department** was set up. This employs officers (Environmental Health Officers) to keep checks on every aspect of community health, with the aim of improving it and maintaining safety.

What do the Environmental Health Officers do?

EHOs in the majority of local communities carry out the following duties:

- **Housing** EHOs make sure that houses are fit to live in (figure 1). If they are not, they can order their demolition or repair.
- **Food** EHOs make sure that food sold in restaurants and shops is fit to eat. They regularly inspect kitchens, slaughterhouses (figure 2) and shops for cleanliness and if necessary have food samples analysed at the **Public Health Laboratories** (table 1). They also check all imported food.
- **Air** EHOs make sure that air is fit to breathe. They regularly check it for **pollutants**, especially in industrial areas and can if necessary order factories to close down.
- **Water** EHOs make sure water in the mains is fit to drink (figure 3).
- **Waste disposal** EHOs check that refuse is collected and disposed of hygienically. They also carry out regular checks on drains (figure 4) and the effluent discharged from sewage works.
- **Infectious diseases** EHOs investigate all cases of **notifiable diseases** and if necessary take action to prevent an **epidemic**. The notifiable diseases in the UK are **rabies**, **cholera**, **smallpox**, **diphtheria**, **typhoid**, **scarlet fever** and **the plague** (table 2). Many of these are endemic (always present) in other countries, but have been kept out of the UK by regular health checks on people at ports.
- **Health and safety at work** EHOs make sure that working conditions are of a minimum standard required by law. They investigate all accidents at work.
- **Pest control** EHOs make sure that their area is kept as free as possible from pests such as rats, mice and insects (table 3).
- **Other duties** EHOs make regular checks on pet shops, riding schools, farms, kennels, hotels, and so on. They investigate complaints from the public and are involved in health education.

What is health education?

Any long term strategy aimed at preventing diseases and improving health must include health education. If people know why they get diseases, they will take precautions to avoid them. Health education is all about improving your awareness about health and related factors so that you can make informed decisions about anything that might affect your health, now or in the future. It is also about developing good positive attitudes to health so that the health of the community will improve.

Health education in England and Wales is now the responsibility of the Health Education Authority. This has recently taken over from the Health Education Council. Its aim is to help local authorities to implement a programme of health education in their community by providing support, advice, materials and training. The Health Education Authority also runs the AIDS campaign.

Questions

1. What is the overall aim of the Environmental Health Department?
2. Why would an EHO order the demolition of a house?
3. Why would an EHO send a food sample to a public health laboratory?
4. Many notifiable diseases in the UK never reach epidemic proportions. Why do you think this is?
5. Health education is now part of the curriculum in most schools. Do you think people really need to be taught about what is good and what is bad for them?
6. Imagine you were put in charge of informing the general public about AIDS. How would you go about it?
7. In what ways do you find out about health related factors?

Figure 1 An Environmental Health Officer checking whether a house is fit to live in

Bacterial sampling of food – Hull area 1980		
	Number taken	Unsatisfactory
Milk	132	9
Cream	43	3
Ice cream	47	1
Fresh meat	89	2
Prepared meat	40	4
Fish	87	8
Others	127	3

Table 1 Results from the random sampling of foods for harmful bacteria

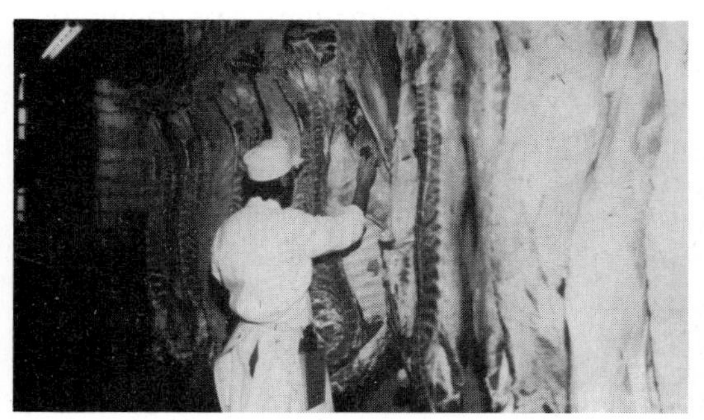

Figure 2 Slaughterhouses are inspected regularly for cleanliness

Figure 3 This water sample from a stream will be taken back to a laboratory to check it for pollution

Notification of infectious diseases – Hull 1980	
Meningitis	56
Dysentery	6
Food poisoning	122
Infective jaundice	183
Malaria	1
Measles	728
Scarlet fever	34
Tuberculosis	17
Whooping cough	223
Others	2

Table 2 A record is kept of some important infectious diseases, including all notifiable diseases, so that people can be warned of an epidemic

Premises treated against insect pests Hull area 1980	
Ants	76
Bees	26
Bugs	45
Cockroaches	1277
Earwigs	22
Fleas	442
Houseflies	12
Silver fish	5
Wasps	580

Table 3 Cockroaches were the most common insect pest in Hull, 1980

Figure 4 EHOs take regular samples from drains

13.13 Preventing disease

What can we do about the spread of disease?

Because we know that many human diseases are spread by other animals (the vectors) we can try to limit these and stop them getting around. To do this successfully first we have to study the vector thoroughly and find out where the weak link in its life cycle is before applying the method of control.

What methods are used to control these vectors?

- **Chemicals** Chemicals which kill animal pests are called **pesticides**. Pesticides which are used against insects only are sometimes called **insecticides**. The main advantages of using chemicals are that firstly, they usually act very quickly and secondly, they can often be used to kill more than one type of pest. Unfortunately, they also have side effects.

  DDT is an insecticide which in the past has been used against almost every known insect vector and now many insects are **resistant** to it. Even more serious is the fact that the DDT has been found to accumulate (build up) in the bodies of other organisms including humans, with serious side effects to health (figure 1). So while trying to solve one health problem we've created a new one! Many of these organisms eventually die.

  Discoveries like this have resulted in the Government restricting the sale and use of pesticides. They also have to undergo strict tests before being allowed on the market and this makes them very expensive.

  Examples of pesticides currently being used are **dieldrin**, **lindane (BHC)** and **malathion**.
- **Other organisms** This is often called **biological control** because it makes use of the vector's natural enemies such as its predators. It has the advantage that it only kills the organism you intend it to. It doesn't upset the balance of nature too much and it does not cause pollution. Unfortunately though biological control takes a long time to have an effect.

Controlling some important vectors

The mosquito and the housefly are both very worrying vectors. They carry some fatal diseases to humans in many parts of the world.

- **Mosquitoes** The mosquito is the vector for the malaria parasite (figure 2). The main methods used to try and control it are:
  1. The adult's resting and breeding places are sprayed with insecticides such as malathion.
  2. The mosquito larvae predators are encouraged to live and breed in the waters where the mosquito larvae live.
  3. Static water in the areas where the mosquitoes live is covered by a layer of oil. This prevents the adult mosquitoes getting out and about!
  4. Water is not allowed to collect in areas where the mosquitoes live. This deprives them of suitable places to leave their eggs.

- **Houseflies** Houseflies are vectors for more than one hundred disease causing organisms, including those which cause cholera, polio, dysentery and typhoid (figure 2). The main methods used as attempts to control it are:
  1. The breeding places are sprayed with insecticides.
  2. Refuse and sewage are disposed of hygienically so that houseflies can't get to it.
  3. Food for human consumption is always covered when left in an area where houseflies live.
  4. Sterile males are released into areas where houseflies breed. Females who mate with these males will lay unfertilised eggs.

Questions

1. Indiscriminate use of pesticides has in the past resulted in the short term control of the pest, but has also created many additional problems. What are these?
2. Before applying a pesticide it is very important that you research the life cycle of the pest it is intended to control. Why?
3. Look at the life cycle of the mosquito (figure 2). There are several weak links in it which can be exploited. What are these?
4. Even if you use a very dilute concentration of a pesticide you could still be endangering your health. Explain this.
5. Why is biological control so much better than chemical control?
6. Houseflies have been a problem for centuries. How is this problem being tackled?

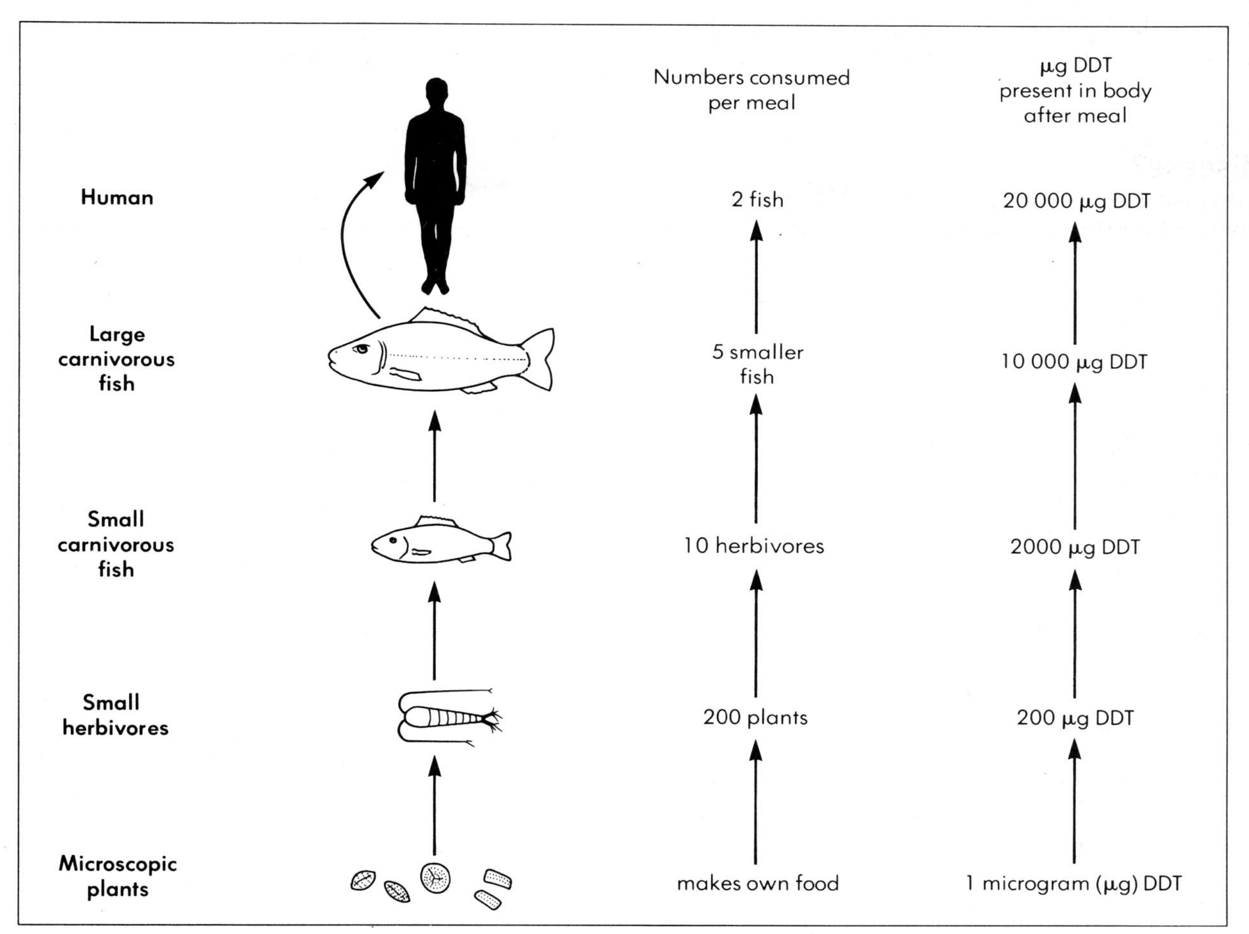

Figure 1 How chemicals are concentrated along a food chain

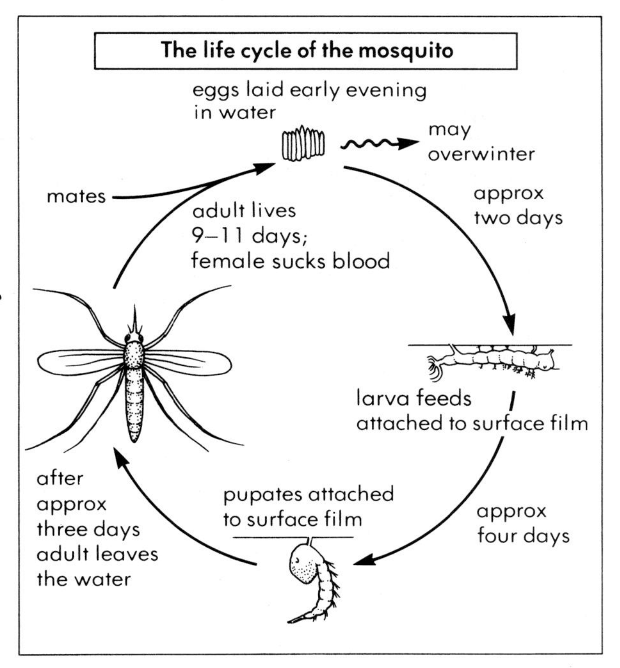

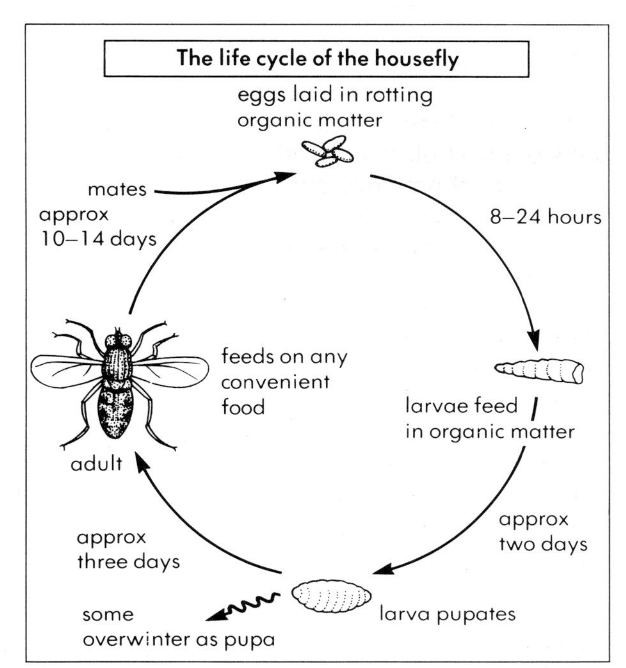

Figure 2 Control of these insects is all about exploiting the weaknesses in their life cycles

13.14 Antiseptics, disinfectants and antibiotics

Using chemicals to kill microbes

The discovery that microbes and disease were linked was very important in the development of medical science. Scientists now started looking for ways of killing these microbes. **Louis Pasteur**, a French scientist (1822-1895), was the first to be successful. He showed that heat would kill microbes and developed the process of **pasteurisation**.

The next major success was by an English surgeon called **Joseph Lister**. He was very concerned by the number of deaths from infections following surgery. He soon realised that the source of infections was the bacteria in the air in the operating theatre. He tried chemicals to kill these, eventually testing **carbolic acid**. He soaked the dressings for the wounds in it and sprayed it around the operating theatre. This seemed to work.

After Lister's success, progress was rapid. We now know of many chemicals which can be used against microbes. Some, like carbolic acid merely stop microbes reproducing and are called **antiseptics**. Chemicals, such as bleach, kill microbes and are known as **disinfectants**.

Can these chemicals be used inside the body?

Antiseptics and disinfectants can only be used outside the body. It was not until 1910 that a chemical was found that could be used to kill microbes inside the body. This chemical was called **salvarsan** and was effective against **syphilis bacteria** and **trypanosomes**. The scientist who discovered it was **Paul Ehrlich**. Unfortunately, salvarsan produced horrific side effects.

Research continued and in 1935 another group of chemicals was discovered. These chemicals called **sulphonamides** were effective against many kinds of bacteria.

Possibly the most useful of all chemicals was discovered by accident. In 1928, **Alexander Fleming** was examining some culture plates on which he had grown staphylococci bacteria. He noticed that one of the plates was contaminated with a mould and that no bacteria had grown anywhere near this mould. After further investigations he concluded that the mould must be producing chemicals which stop the bacteria growing. He went on to extract a juice from the mould which he called **penicillin**. Fleming saw the enormous potential penicillin might have, but it was left to **Florey** and **Chain**, twelve years later, to purify penicillin and turn it into a chemical which could be used medically. Penicillin is now used as one of the most common **antibiotics** to treat bacterial infections. Since then many more antibiotics have been isolated from various moulds and bacteria. Some of the main ones are shown in table 1.

Can we now destroy all known microbes?

Unfortunately it is not yet possible to destroy all microbes inside your body simply by using chemicals. The biggest difficulty are the viruses. These have resisted nearly all attempts to destroy them although research continues all the time. At the moment the only way you can recover from a viral disease is for your body's own defences to do it for you.

Questions

1. What is the difference between an antiseptic and a disinfectant? Write out two lists, one of antiseptics and the other of disinfectants.
2. Look at figure 3. Which of the antibiotics is the most effective against the bacteria on the plate?
3. Scientists are always discovering new chemicals which appear to be effective against microbes, but before they can be used in humans they must first be tested in other animals. Draw up a table of arguments for and against this practice.
4. An alternative to using other animals would be to pay people to undergo trials or to use desperately sick people. What do you think of this suggestion?
5. If you catch influenza, your doctor will usually tell you to go to bed and let the disease run its natural course. Why is this the only help she can give?

Figure 1 The pharmaceutical industry in the UK has an annual turnover in excess of £800 million

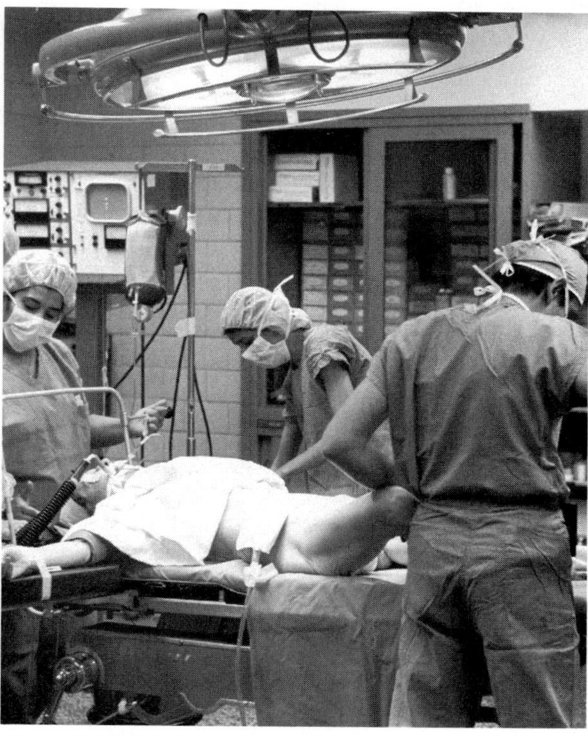

Figure 2 Modern operating theatres are as sterile as possible

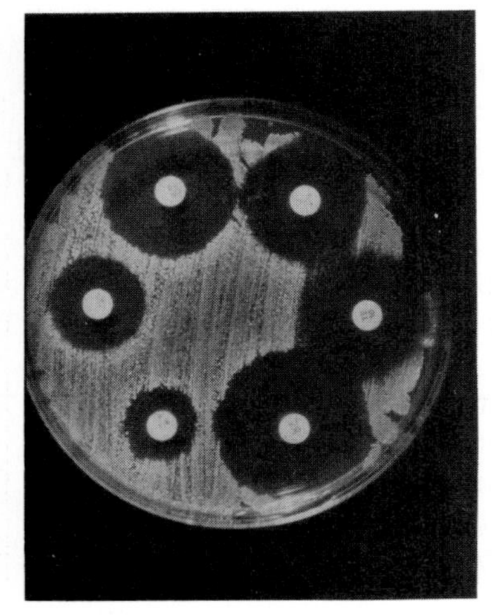

Antibiotics are placed on here:
CB = methicillin
P = penicillin
E = erythromycin
TE = Tetracycline
S = Streptomycin
C = chloramphenicol

Figure 3 A multodisk is used to discover which antibiotic is best to use

Group and examples	Isolated from	Good for
Penicillins e.g. Benzylpenicillin Ampicillin	Fungus (mould)	Blood poisoning Throat infections Gonorrhoea and syphilis Bronchitis Pneumonia
Cephalosporins e.g. Cephalexin	Fungus (mould)	Food poisoning Urinary infections
Macrolides e.g. Erythromycin	Bacteria (streptomycetes)	Used instead of penicillin if person is allergic to it.
Aminoglycosides e.g. Streptomycin Gentamycin	Bacteria (streptomycetes)	Used in conjunction with penicillin to kill bacteria in the gut and urinary tract
Tetracyclines e.g. Tetracycline	Bacteria (streptomycetes)	A broad spectrum drug especially good for fevers and diarrhoea
Antifungals e.g Nystatin Griseofulvin	Bacteria (streptomycetes)	Thrush Ringworm Fungal disease of skin
Others e.g. Bacitracins Gramicidin	Bacteria	Skin infections resulting from burns Urinary infections

Table 1 There are many different kinds of antibiotic

13.15 The body fights back

What do our bodies do?

You may think that if you are careful about the food you eat, the water you drink, washing your hands etc. that you will be safe from disease. But despite all our precautions, many micro-organisms do actually get into our bodies. When this happens, most are quickly devoured by the **phagocytic white blood cells** and those that survive are attacked by **antibodies**.

What are antibodies?

Antibodies are special chemicals which can destroy or lead to the destruction of foreign substances which enter the body. They are produced by the **lymphocytes** in the blood and lymphatic systems. The foreign substances that cause the production of antibodies are called **antigens**. Figure 1 shows how they do this. Most antigens are chemicals attached to the surface of invading microbes, but the cells of **transplanted organs** also have antigens attached to their surfaces. This is why they are often rejected by the body they have been put into (see page 142).

Antibodies are very specific chemicals: that is, each antibody will only destroy one type of antigen. For example, the antibodies that are made to destroy diphtheria antigens will not destroy tetanus or any other antigens. This has very important implications for the body.

Antibodies and immunity

There are many diseases which, once you have had them, you can never catch again, i.e. you are said to be **immune** to them. For example, let us look at what happens if you get measles.

When the measles viruses first enter your body, the antigens on their surface cause the production of antibodies. To start with, this is a slow process because the antigen has first to be identified (**typed**) before antibodies can be produced. It is during this initial period that the measles viruses multiply and cause the symptoms of the disease we call measles. The time from infection to the onset of the first symptoms is called the **incubation period**. This is different for every disease (table 1). Eventually, you recover from measles as the viruses are destroyed by the antibodies. This typical course that a disease follows is shown in more detail in figure 2.

The next time measles viruses enter your body, antibodies are already present and circulating in the blood. The viruses are therefore quickly destroyed before they can build up in large enough numbers to cause any symptoms. It is the presence of circulating antibodies that makes you immune. Unfortunately, immunity does not always last forever because the antibodies die.

Does immunity cause any problems?

The immune response can cause many problems, three of which are:

- **Tissue or organ rejection** after a transplant. An individual's cells have their own particular antigens attached to their surface. When put into a strange body, these antigens cause the production of antibodies, which then start destroying healthy cells. Now special drugs are given to transplant patients to prevent this (see page 142). These are called **immunosuppressant** drugs.
- **Blood transfusions** There are four blood groups; A, B, AB and O. These groups are determined by the antigens (A or B) attached to the red blood cells and only some groups can be mixed (see page 48). If you are given the wrong blood during a transfusion your body will start to make antibodies to destroy the antigen.
- **Allergies** Sometimes, even a harmless substance can cause the production of antibodies. If this happens, the person is said to be allergic to that substance.

Questions

1 Explain the connection between the words in the following pairs:
antibody/antigen
lymphocyte/antibody
virus/antigen
antigen/rejection.

2 Describe the course of a typical infectious disease making sure you use the following words:
fever, recovery, infective, symptoms, incubation, antigens, antibodies.

3 People who have had organs transplanted into them have to take immunosuppressant drugs for the rest of their lives. Explain why.

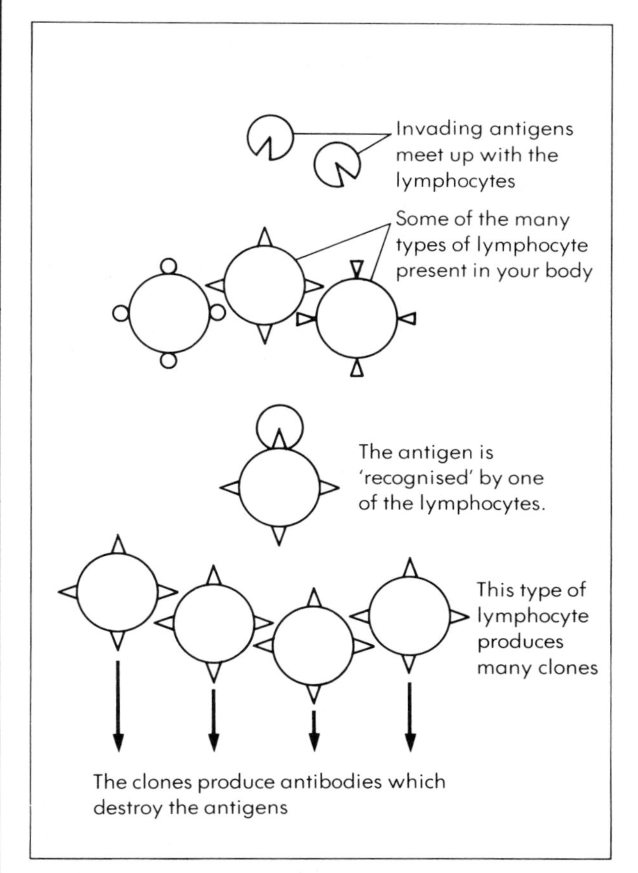

Figure 1 The relationship between antibody and antigen

Disease	Incubation period
AIDS	6 months to 6 years
Chicken pox	17 days
Common cold	average 4 days
Dysentery	average 3 days
Food poisoning	average 1 day
Gonorrhoea	5 to 10 days
Influenza	4 days
Measles	10 days
Mumps	21 days
Rabies	6 months
Syphilis	2 to 6 weeks
Whooping cough	14 days

Table 1 The incubation period of common diseases varies from one day to six years

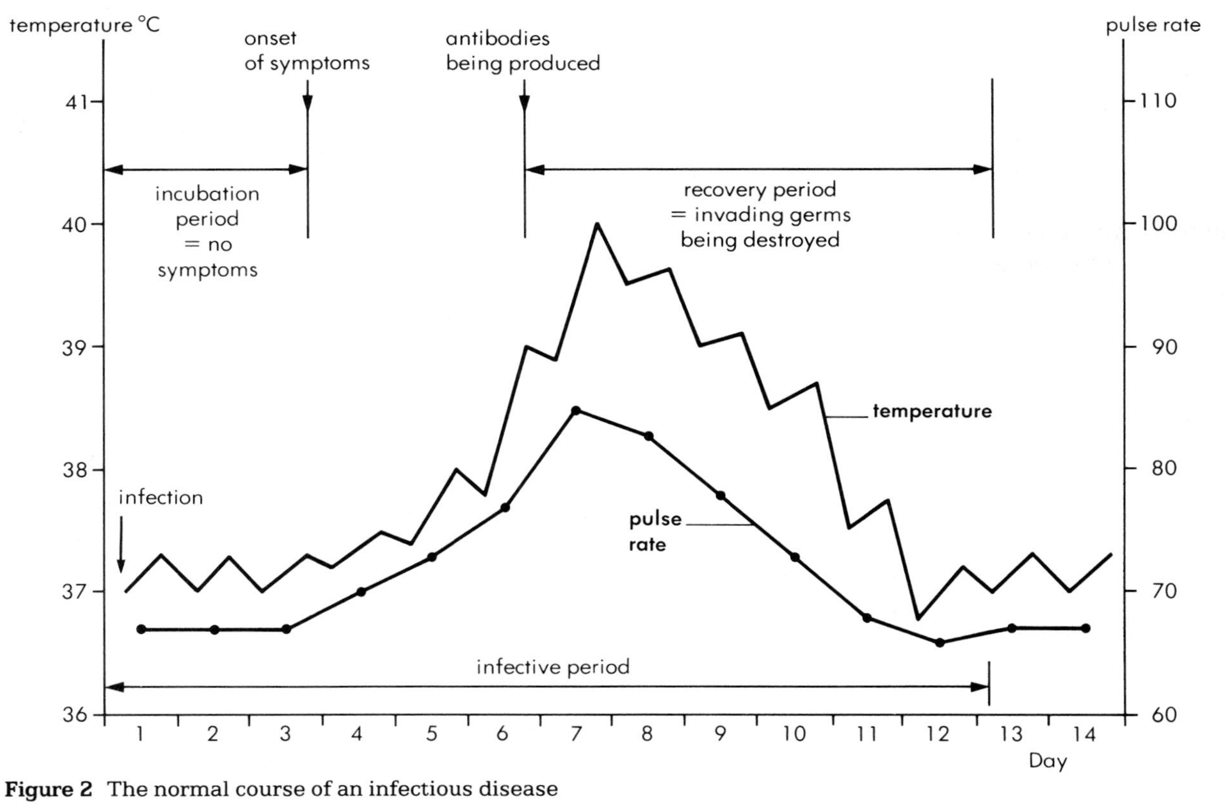

Figure 2 The normal course of an infectious disease

13.16 Helping the body fight back

Can you be given immunity to a disease?

You can be given immunity to many diseases. There are two forms this can take: **active immunisation** and **passive immunisation**.

What is active immunisation?

Active immunisation is when your body is artificially persuaded to make its own antibodies to a particular antigen. It will do this if it is injected with a **vaccine**, a process called **vaccination** or **immunisation**. The vaccine can be made from:

- **Dead organisms** which still have antigens on their surfaces yet cannot grow and reproduce. The antigens cause the production of antibodies. This is the basis of the polio and cholera vaccines.
- **Attenuated organisms** which are less **virulent** forms of the organisms. They are still living, but have lost the ability to reproduce. Attenuated vaccines are used for German measles and tuberculosis (figure 1).
- **Toxoids** are mild forms of the toxins produced by some bacteria. They are not poisonous, yet contain antigens which can stimulate the production of antibodies. Toxoids form the basis of the tetanus and diphtheria vaccines.

The first person to use a vaccine was **Edward Jenner**. In 1796 he successfully immunised a small boy against smallpox by injecting him with a very similar but less dangerous germ which causes a disease called cowpox. Unusually, the antibodies which your body makes to fight this are also effective against smallpox.

How long does active immunity last?

Once acquired, active immunity can last a lifetime, but not always. For example, you must be revaccinated against diphtheria every five years to maintain immunity. This revaccination takes the form of a **booster**. Active immunity can also develop naturally if you catch a particular disease.

What is passive immunisation?

Passive immunisation is when you are injected with a serum. The **serum** contains **ready-made antibodies**. In the past this was taken from another animal such as a horse, but now it is produced in culture vessels. Passive immunity is also acquired by a baby as it takes breast milk from its mother. This milk contains some of the mother's antibodies. It is also thought that a foetus receives antibodies from its mother during its development.

Passive immunisation is given to people who are going to parts of the world where certain diseases are endemic. It only lasts until the antibodies die, usually a maximum of six weeks and so if immunity is needed for longer periods a vaccine is given well before it is required. The injection against hepatitis is an example of passive immunisation. Table 1 shows the immunisations you need when visiting other countries.

Which diseases will you have been vaccinated against?

Everybody in this country can be vaccinated against polio, tetanus, whooping cough, diphtheria and tuberculosis. Girls can also be vaccinated against German measles (rubella). The programme of vaccinations starts when you are about three months old (table 2) and continues into adulthood. Some parents choose not to have their children vaccinated because not all vaccines are 100 per cent safe. Parents have to weigh up the risk of suffering from a particular disease against the risk from the vaccine.

The ability of mass vaccination to reduce the death rate from a disease is clearly shown in figure 2.

Questions

1 Distinguish between:
a vaccine and a serum
active and passive immunisation

2 A friend of yours has just had the six needles test for tuberculosis to see if she needs the main jab. She is determined not to have the larger jab because she has been told it hurts. She said she intends to spend all week gently scratching the six pin pricks so they will turn red and swell. Persuade her how foolish this would be.

3 Edward Jenner is usually credited with discovering immunity. Explain why he was very lucky in his choice of disease to work on.

4 Many diseases can now be controlled by mass vaccinations. Explain how the data for whooping cough supports this.

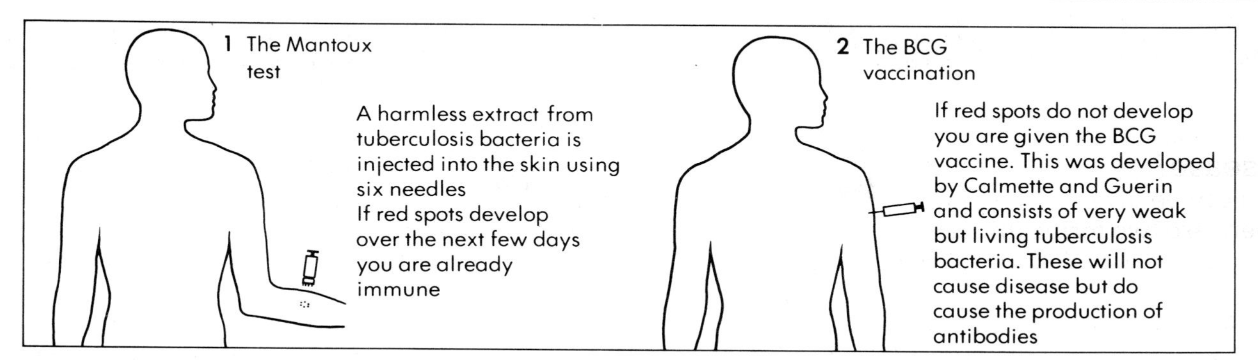

Figure 1 Inducing active immunity to tuberculosis

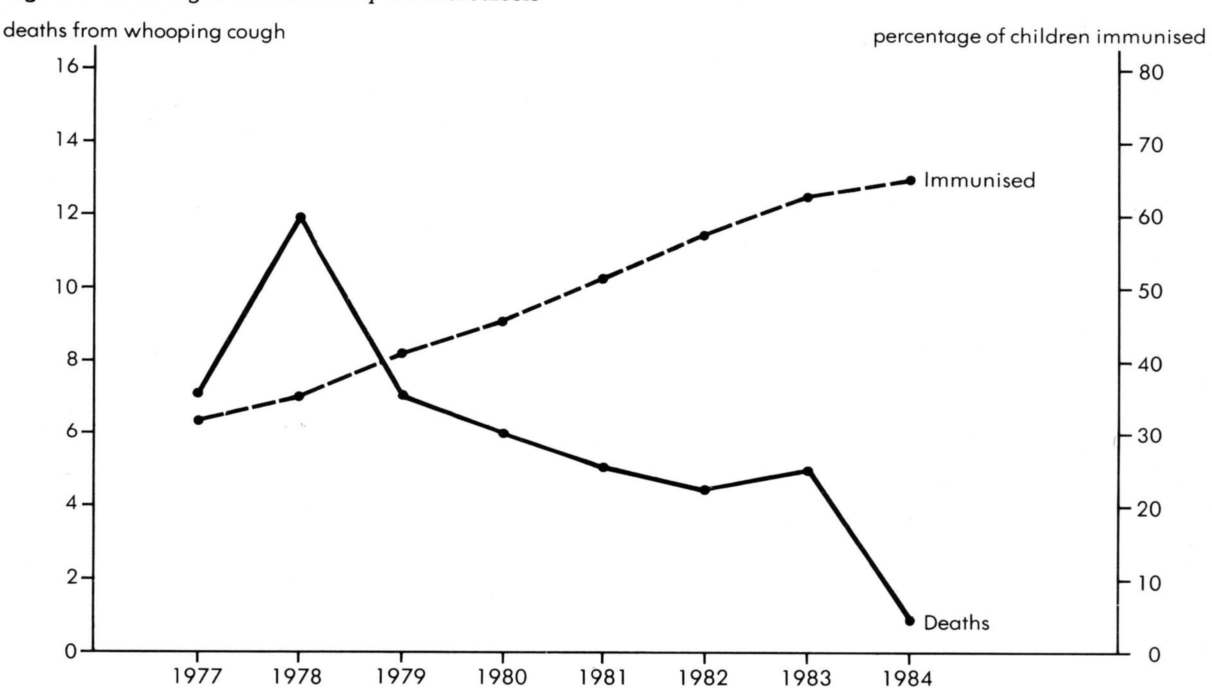

Figure 2 Is mass immunisation effective in reducing the number of deaths from whooping cough?

Recommended (WHO) vaccinations for travellers abroad

Area	CHOLERA	TYPHOID	POLIO	HEPATITIS	MALARIA	YELLOW FEVER
Asia	✓	✓	✓		✓	✓
Europe			✓			
Africa	✓	✓	✓	✓	✓	✓
N. America			✓			
C. America		✓	✓		✓	
S. America		✓	✓	✓	✓	✓
Middle east		✓	✓			
India		✓	✓	✓	✓	✓
Australasia		✓	✓			

Table 1 When you travel abroad it is important to have the right vaccinations

Age	Vaccines to be given
from 3 months	First dose of diphtheria, tetanus, whooping cough and polio vaccine
5–6 months	Second dose of diphtheria, tetanus, whooping cough and polio vaccine
9–11 months	Third dose of diphtheria, tetanus, whooping cough and polio vaccine
12–24 months	Measles vaccine
4½ – 5 years	Booster dose for diphtheria, tetanus and polio
11–13 years	Tuberculosis if needed
10–14 years	German measles (girls only)
15–19 years	Booster dose for tetanus and polio

Table 2 A programme of vaccinations is available for all children in the UK

13.17 AIDS

What is AIDS?

AIDS stands for **Acquired Immune Deficiency Syndrome**. It is a condition which develops through your body's defences not working properly. The condition is characterised by a particular pattern of illnesses, two of the most serious being a rare form of skin cancer and a form of pneumonia. People who die from AIDS usually die from one of these illnesses or a combination of things which their body just cannot fight.

What causes AIDS?

AIDS is caused by a virus (called HIV) which takes over your body's lymphocytes eventually destroying them. This reduces your body's ability to make antibodies to combat other diseases.

What are the symptoms of AIDS?

AIDS has a variety of symptoms many of which are very common symptoms of other diseases. When attempting to diagnose AIDS, a doctor will therefore look for a *pattern* to the symptoms.

Some of the symptoms which may suggest AIDS are:

- **profound fatigue**, which lasts for weeks,
- **swollen lymph glands**,
- **rapid and excessive weight loss**,
- **persistent fever and night sweats** lasting for many weeks,
- **persistent shortness of breath**,
- **a persistent dry cough**,
- **diarrhoea** which lasts for more than a week,
- **pink/purple patches on the skin**,
- **lethargy and depression**.

People with the HIV virus are particularly susceptible to other infections, especially thrush, warts, eczema and pneumonia.

How long does it take for symptoms to show?

It is thought that some people who catch the HIV virus may not actually develop AIDS. But at the moment this is difficult to judge because so little is known about the HIV virus and how long it takes to develop. Those who do get AIDS may develop it in as little as six months or it may take six years.

How is the HIV virus passed on?

The HIV virus can only be transmitted in body fluids such as blood and semen. Ninety percent of all transmissions in the UK are during sexual activity. Homosexual men are the highest risk group at the moment although greater awareness has lowered the rate of transmission in this group considerably. Remember anyone who has unprotected sex could be at risk – not just gay men. When little was known about AIDS, some people were infected through blood transfusions, but now all blood in the UK is heat treated to destroy the virus. Similarly, haemophiliacs caught it through contaminated blood clotting agent, but again this is now safe. Intravenous drug users are also a high risk group. They can pass on infected blood through sharing needles (see tables 1 and 2).

Can AIDS be cured?

Unfortunately, there is as yet no cure for AIDS, but it can be controlled. We know what causes it and how it is transmitted, so we should be able to stop it spreading. Transmission of it through blood transfusions and blood clotting agents has been stopped in the UK. Most of the other transmissions can be avoided by safe sex. The Government has run a campaign to educate people about AIDS. But it is a worldwide disease and needs the care and attention of everyone to prevent it from reaching epidemic proportions. It is already a huge problem in parts of Africa, USA and elsewhere (figure 2).

Is there any hope of a cure?

Research has been going on frantically for the last five years or so in an attempt to find a cure. There are some promising signs, but we must remember that most new drugs and vaccines take ten to twenty years to perfect (see page 74).

Questions

1. There may not be a *cure* for AIDS for some time but what can be done now to try to control it?
2. Design a poster which explains the seriousness of AIDS to the general public.
3. Look at tables 1 and 2. What are the high risk groups? Are these the same for the UK and the USA?

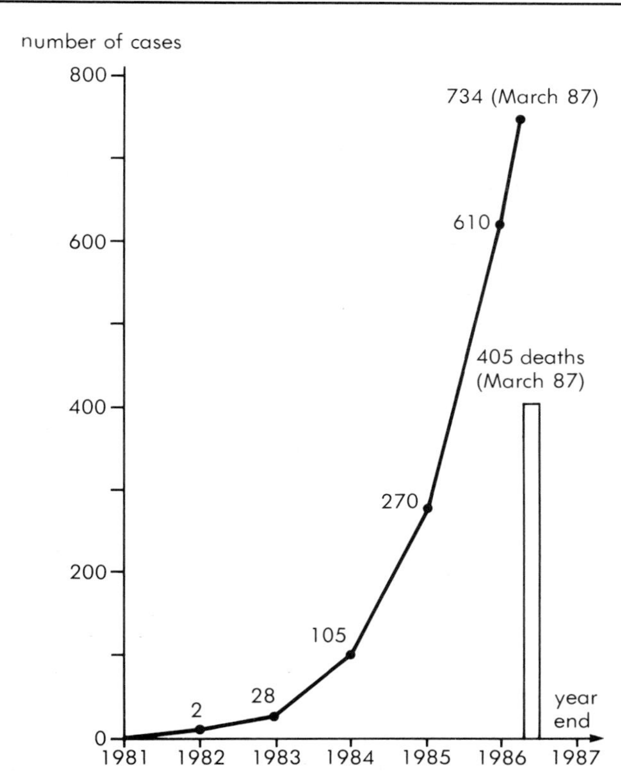

Figure 1 The number of AIDS cases in the UK has been increasing exponentially

Risk group/ Patient characteristics	Number of people with positive antibodies	Number of AIDS cases	Number of deaths
Homosexual/ bisexual	2546	640	342
Intra-venous drug addicts (IVDA)	939	10	5
Homosexual and IVDA	30	7	4
Haemophiliac	1028	31	23
Blood recipient	109	13	11
Heterosexual	181	25	15
Other	807	8	5
TOTAL	5571	734	405

Table 1 The table shows the groups of people at risk from AIDS in the UK (1987)

Risk group/ patient characteristics	Number of AIDS cases	Number of deaths
Homosexual/bisexual	27 483	15 323
Intra-venous drug addicts (IVDA)	6853	4156
Homosexual and IVDA	3129	1890
Haemophiliac	410	245
Blood recipient	952	654
Heterosexual	1644	937
Other	1711	1040
TOTAL	42 182	24 245

Table 2 The table shows the groups of people at risk from AIDS in the USA (September 1987)

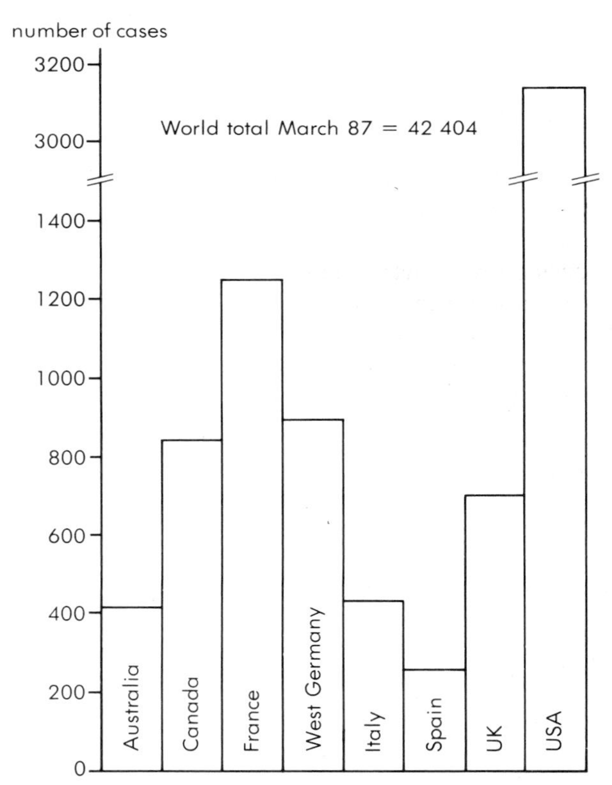

Figure 2 In less than six years AIDS has spread to over 170 different countries

Questions

Recall and understanding

1 Which of the following is not a type of micro-organism?
A salmonella bacterium
B influenza virus
C malaria protozoan
D beef tapeworm

2 Phenylketonuria is an example of a:
A metabolic disorder C viral disease
B bacterial disease D social disease

3 Malaria is caused by a:
A bacterium C warmth
B virus D protozoan

4 Ringworm fungus is responsible for the disease:
A thrush C anthrax
B acne D athlete's foot

5 Which of the following diseases is not spread on droplets of moisture?
A AIDS C polio
B influenza D pneumonia

6 Micro-organisms are used to treat:
A water C detergents
B raw sewage D plastics

7 The correct sequence for treating water is:
A filtration settlement chlorination
B filtration screening chlorination
C screening settlement filtration chlorination
D chlorination filtration settlement

8 Which of the following is not a bacterial disease?
A measles C tetanus
B cholera D syphilis

9 Antibiotics are only effective against:
A bacterial infections C worm infestations
B viral infections D insects

10 During the first few days of feeding, a breast fed baby develops passive immunity to many diseases because the milk contains:
A calcium C antibodies
B antigens D antibiotics

Interpretation and application of data

11 The table below shows the average cost to the NHS of various age groups:

Age (yr)	Average cost (£/yr)	
	Male	**Female**
less than 1	554	452
1–4	205	183
5–15	130	117
16–24	134	130
25–44	148	153
45–64	226	215
65–74	477	434
greater than 75	904	1109

a) What are the average costs of men and women to the NHS?
b) Write the age ranges in order of most to least cost.
c) The average working person pays £10 a week for 40 years for health services. How much is this?
d) Do you consider this is good value for money?
e) Cervical smears have been available on the NHS for many years now. The table below shows the number of females using this service.

Year	Number examined (1000's)	Detection rate (per 1000)	Death rate (per million)
1971	2205	4.5	91
1976	2803	5.1	86
1979	3002	6.2	82
1980	3211	6.6	80
1981	3293	6.9	79
1982	3241	7.3	76

i) Plot on a graph the number examined and the death rate.
ii) What does this graph tell you about the value of using preventive medical services?

12 The data below refers to the disease AIDS in the UK:

Year	Total number of cases
1982	2
1983	28
1984	105
1985	270
1986	610

a) Plot this data on a graph.
b) Predict the number of cases in 1990.
c) By March 1987 there were 4,471 people known to be infected by the AIDS virus – 734 of these had developed the full syndrome. What percentage is this?
d) Of the 734, 405 had died. What percentage is this of the
i) total with AIDS infection,
ii) total with full AIDS syndrome?

13 The table below shows the percentage of children immunised against various diseases:

Disease	Percentage immunised in Europe	Percentage immunised in UK
Diptheria	84	84
Measles	63	63
TB	76	86
Tetanus	91	84
Whooping cough	81	64

a) Which diseases are the children in the UK less well protected against?
b) The number of vaccinations in the UK for whooping cough fell dramatically after 1974. Explain why.
c) To eradicate all these diseases from Europe, the WHO has suggested that 95 per cent immunisation is necessary. Predict for which diseases we might achieve this first.
d) The graph below relates to diphtheria in the UK:

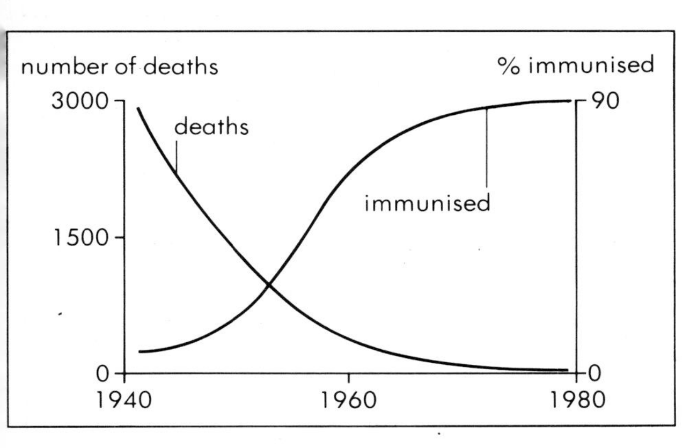

Does the information suggest mass immunisation is or is not effective?

14 Notifications of infectious diseases in England and Wales 1972–1982:

	1972	1976	1980	1982
Meningitis	98	89	102	67
Dysentery	9048	6217	2709	2850
Food poisoning	5494	9226	10318	9954
Infective jaundice	12269	5963	5143	10602
Malaria	363	1174	1296	1180
Measles	145916	55502	139487	94195
Scarlet fever	11211	9712	11118	7601
TB	11100	10102	9145	7406
Whooping cough	2069	3907	21131	65810
Tetanus	20	15	18	12
Rabies	0	0	0	0

a) Re-write the diseases out under the following headings:

On the increase	Static	On the decrease

b) Why have there been no notifications of rabies?
c) Why have the number of cases of malaria increased?
d) Suggest why the number of cases of whooping cough have increased so dramatically.
e) Which department must be informed of outbreaks of these diseases?

15 A multodisk experiment was set up to find out which antibiotic was the most effective against a bacterium.

Antibiotic	Diameter of clear zone
* Chloramphenicol	8mm
Erythromycin	2mm
Methicillin	-
Penicillin	10mm
Streptomycin	4mm
* Tetracycline	20mm

a) Which antibiotic is the most effective against the bacterium?
b) Which antibiotic is the least effective?
c) The patient with the disease is allergic to the antibiotics marked*. Which antibiotic was therefore prescribed?
d) Another experiment must be done to find out which dose to use. Describe this experiment.

CHAPTER 14: *THE ENVIRONMENT*

The earth was formed about 5000 million years ago. The first life forms appeared about 3000 million years ago. Humans only appeared 35 000 years ago.

Two hundred years or so ago, the age of 'modern' man started with the industrial revolution. Look at some of our achievements since then:

- Destruction of nearly half of all rain forests.
- Holes in the ozone layer.
- Poisonous air.
- Acid rain.
- Expanding deserts.
- Extinction of many life forms; estimates suggest one species/hour.

The list is almost endless. This chapter explains how and why these things are happening.

Fortunately it is not all doom and gloom. There have been notable successes of which we can be proud. Many animals and plants have been saved from extinction. Rivers have been cleaned up and restored to their former beauty. International agreements control fishing and so on. This chapter also explains what else we are doing to protect the earth and its remaining resources.

Recycling is becoming big business

A banner from the environmental group Greenpeace on a barge loaded with refuse in New York.

The problem of waste

This section is about how we are trying to cope in this country with one major problem, that of household waste.

Every year over 18 million tonnes of refuse is generated by the households in the UK. In addition 38 million tonnes are produced by industry. In the past the majority of this waste has been buried in large holes or burnt. However, more and more of it is now being reclaimed and recycled. We even have a government minister in charge of waste and reclamation.

Is recycling worthwhile?

Recycling is worthwhile simply because it makes resources last longer. Unfortunately this is not enough reason for many governments. It also has to be *cost effective* and *energy efficient*.

At present recycling may not be the cheapest method of disposing of waste, but as raw materials become scarce and their market price goes up, recycling will become more attractive.

Even now most recycling saves energy. For example the recycling of glass saves the equivalent of 30 gallons of oil per tonne.

What sort of things can be recycled?

The recycled materials in waste can be divided into several categories:

- **Metals** Over 9% of the average dustbin consists of waste metal cans. These are either tin-plated steel or aluminium. The aluminium is easy to recycle and maintains its quality. In addition, the recycling only requires 3% of the energy needed to produce the original product.

  Tin-plated steel is more difficult as it has first to be separated into its components. Even so, half of the UK's steel now comes from recycled cans.
- **Glass** Glass is one of the few substances that can be completely recycled without loss of quality.

  In 1976, the UK glass industry introduced a bottle bank scheme. In its first year this collected 25 000 tonnes of re-usable glass. By 1986 this had risen to 227 500 tonnes. This is about 13% of the total product. Many European countries recycle much more than this (table 1).

Unless bottle banks are conveniently positioned in the street or car parks, like this one, people won't use them.

Table 1

Country	Glass recycled (kg/head) 1984
Switzerland	21.7
Germany	15.0
Netherlands	13.9
France	9.4
Italy	5.5
Great Britain	2.2

- **Paper** Total consumption of paper in 1985 was about 7.7 million tonnes of which 2.1 million tonnes (27%) was reclaimed.

  Recycled paper is no longer just suitable for the production of newsprint, toilet tissue and cardboard. Many major charities use it as stationery.
- **Plastics** Plastics have always presented a special problem because many are not biodegradable. However, most can now be used to recover energy by a process called **pyrolysis**. This involves heating without air at a high temperature. The products are gases, including methane, an 'oil' and carbon. All these are useful fuels.

Table 2 The extent of recycling in the UK, 1986.

Material	Percent recycled 1986
Iron and steel*	49
Aluminium	21
Paper and board*	27
Glass	14
Plastics	6
*1985 figures	

- **Textiles** Many blankets are made from recycled textiles.
- **Food** Food and animal dung can be used to produce methane and fertiliser by a fermentation process (page 26). Waste plant material is being used in countries such as America and Brazil to produce ethanol. Most of Brazil's cars now run on a 50/50 mixture of ethanol and petrol.

Careless disposal of waste can cost lives. In 1966 a slag heap demolished a school in Aberfan killing nearly 200 people, many of them children.

14.1 Living with the environment

What on earth do we do!

We are in a privileged position on the earth, in that we are the only organism that can:

- extensively alter the environment to suit our needs,
- produce our own food almost anywhere on the earth,
- make use of the earth's resources for our own benefit,
- cure human diseases and control the organisms responsible.

All this needs to be done with great skill and care so as not to upset the natural balance that exists between other organisms and the environment (figure 1).

How can our activities affect the environment?

Many of the effects we have on the environment are the result of our attempts to cope with an ever increasing population. This has created demands for more living space, more food production, more energy and more manufactured goods. In meeting these demands, many forests have been destroyed, resources have been over exploited and more pollution has been created.

How does removal of trees affect the environment?

In recent years trees have been removed in vast quantities to provide: space for housing; more land in which to grow food; wood to burn; materials for building. At first little thought was given to the possible effects on the environment because little was known about it. However, we are now experiencing some of these bad effects at first hand.

For example, the concentration of carbon dioxide in the atmosphere has been steadily rising because there are fewer trees to use it (figure 2). Now, the more carbon dioxide there is in the air, the warmer the average air and land temperatures get. This is called the **greenhouse effect**. At the present rate of increase, the temperature will have risen by 1°C by the turn of the century. This could have substantial effects on the climate. One predicted effect is that the polar ice caps will melt. If this happens, the sea level will rise by several metres, enough to flood much of the earth's lowland!

One irony of removing vast quantities of trees is that the land is of little value for growing crops after the first few growing seasons. Forest land is already very low in nutrients and this is made worse by the extra rain which now reaches it. This extra water flowing through the soil **leaches** out more of the vital nutrients eventually leaving it completely infertile.

How many trees are being removed?

Figure 3 shows the extent of the world's tropical rain forests. This is being removed at the rate of 100 000 km^2 (1.5 per cent) per year. If this continues, by 1990 the forests in Australia, Bangladesh, India, Malaysia, Sri Lanka, Vietnam, Central America, Madagascar and East Africa will have virtually been destroyed.

Nearer home, a similar situation exists with the natural woodland in the UK (figure 4).

Questions

1. Many of our activities affect the environment (figure 1). Choose any six of these and explain how.
2. The Brazilians have been exploiting the Amazon rain forest for many years now. Their original idea was to clear large areas, making use of the trees and leaving the land for cultivation of crops. All they have achieved is the production of more deserts. Explain why.
3. The Amazon rain forest covers about 500 million hectares. It is being destroyed at the rate of 10 million hectares per year. Calculate how long the forest will last at this rate.
4. The removal of trees in vast quantities could result in large areas of land being flooded. What is the connection?
5. The removal of trees also affects the other organisms that normally live there (figure 2). Destruction of their habitats will lead to a reduction in their numbers. What is the ecological significance of this?

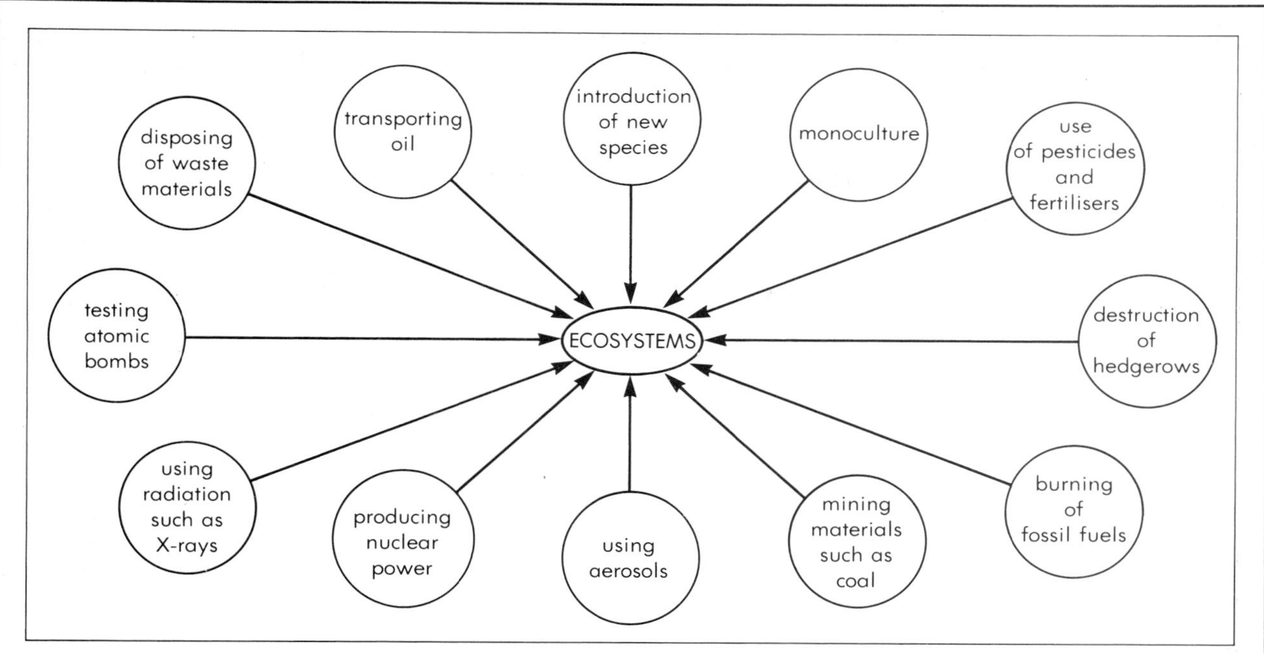

Figure 1 Many of our activities affect ecosystems

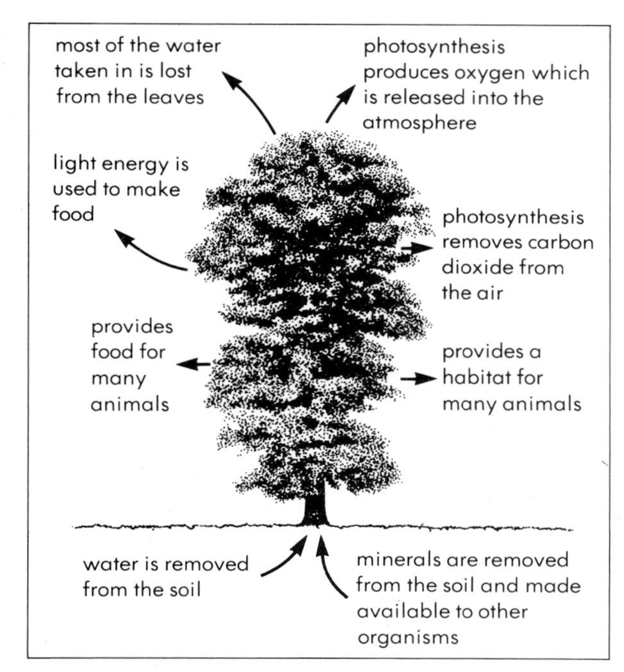

Figure 2 Trees are a vital part of many natural cycles

Area	% loss of natural woodland
Northern England	46 (8412 Ha)
Midlands	46 (20823 Ha)
Southern England	37 (12471 Ha)
Wales	55 (12430 Ha)

Figure 4 Britains natural woodlands have declined over the last 50 years

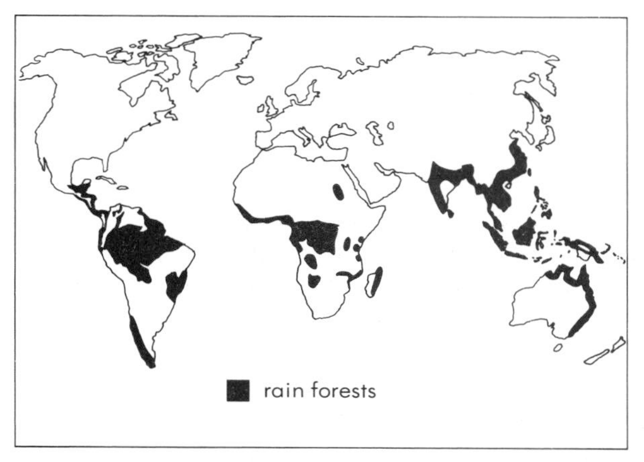

Figure 3 The World's tropical rain forests

Figure 5 In the last 30 years about 150 000 miles of hedgerow have been removed to create larger more manageable fields

14.2 Management of resources

What have we done to the earth's resources?

One of the biggest problems facing modern society is 'where are we going to get our energy from?' in future. For years coal, oil and natural gas have been the main suppliers of our energy, but they are non-renewable. At the present rate of consumption, oil and natural gas will only last another 40 to 50 years and coal a little bit longer (figure 1). Once the earth runs out – that's it. It took thousands and thousands of years for these resources to form in the earth and it now looks as if our careless management of them is catching up on us.

Is nuclear energy a good alternative?

There are of course, alternative forms of energy. **Nuclear energy** is probably the most likely form to take over (figure 2). It already accounts for about 18 per cent of the electricity produced in the UK and is in fact cheaper than conventional forms to produce. But do we really know enough about it for it to be a safe alternative? While nuclear power can provide us with energy for another 2000 years or so, if anything goes wrong, it can also go a long way to destroying the planet. Even so, for many people the biggest worry when dealing with radiation is that its effects do not show immediately, but take time to develop, sometimes a generation. Our mistakes now could cost our children dearly in the future. The Chernobyl disaster has virtually been forgotten, but many scientists are convinced it will return to haunt us ten, twenty or even fifty years from now.

What about other natural forms of energy?

The other choices are expensive, but if we can use them they will be virtually inexhaustible. Some of these are:

- **Using water** to generate electricity. This can be achieved in three ways:
  1. By using the potential energy in water on high ground. As the water falls to lower ground, it can be used to rotate turbines and thereby generate electricity.
  2. By using the tidal movement of sea water. High tides occur twice a day. If the water can be cut off and stored during one of these high tides and then released at low tide, it again can be used to rotate turbines.
  3. By using the energy in waves. The up and down actions of waves contain a lot of potential energy.
- **Using the sun's energy** (solar energy). The amount of solar energy falling on the UK in a typical year is over 80 times more than the energy we need! In some places solar panels are used to heat domestic water supplies.
- **Using waste** produced by us all. Our refuse can be used to produce more oil and gas by a process call **pyrolysis**. Organic wastes such as sewage can be used to produce **bio-gas**, a mixture of methane and carbon dioxide. The methane can replace petrol and diesel fuels.
- **Using the wind** to rotate turbines and thereby generate electricity.
- **Using the earth's internal heat** (geothermal power). This energy source looks the most rewarding as it is by far the largest. It has been estimated that if we could tap this we would have sufficient energy for many thousands more years (figure 7).

At the moment it is expensive to use these alternative energy sources compared with present resources. But as these run out their cost rises and these other sources become more attractive. Perhaps we should be spending more time and money on them now, before it is too late.

Questions

1. There is a law of physics that states that energy cannot be created or destroyed, yet we are told that we are rapidly running out of energy. Explain.
2. At present the best solution to the energy crisis seems to be nuclear power. Why do you think there is so much opposition to this?
3. The chances of dying in a road accident are about 8000 to 1. The chances of dying due to an accident at a nuclear power plant are about 1 million to 1. Are people's fears therefore justified?
4. Look at the list of alternative energy sources. All present considerable problems apart from the expense. For each source point out some of the problems.

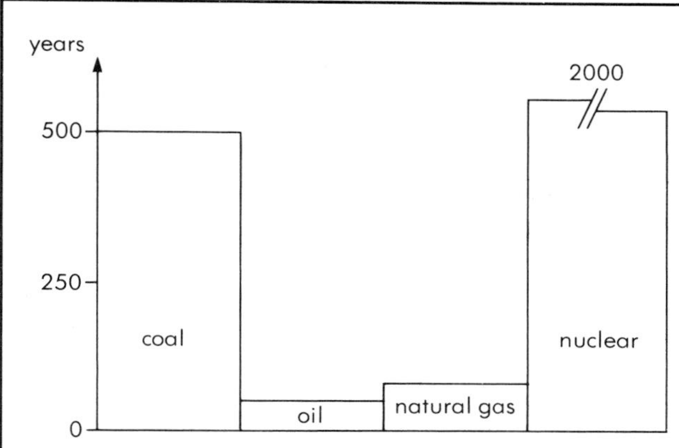

Figure 1 Estimates of how long our present energy sources will last

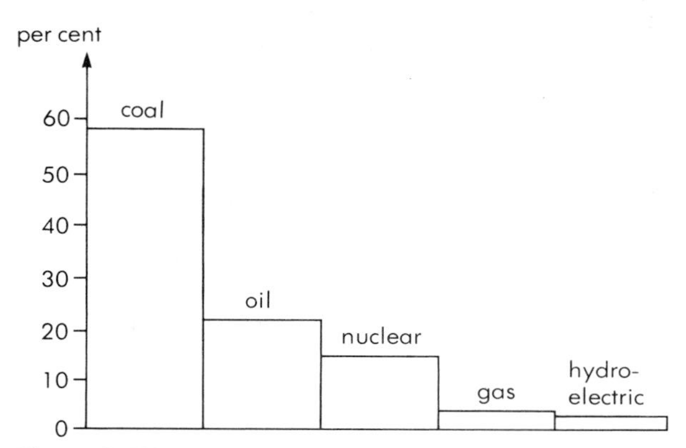

Figure 2 Where our electricity comes from (1981)

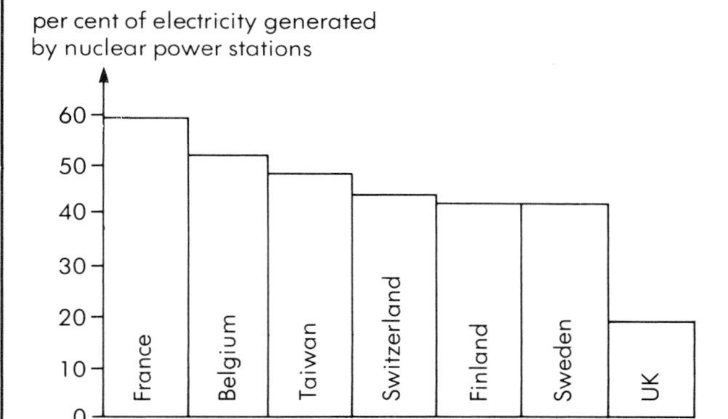

Figure 3 Who uses nuclear power (1984)

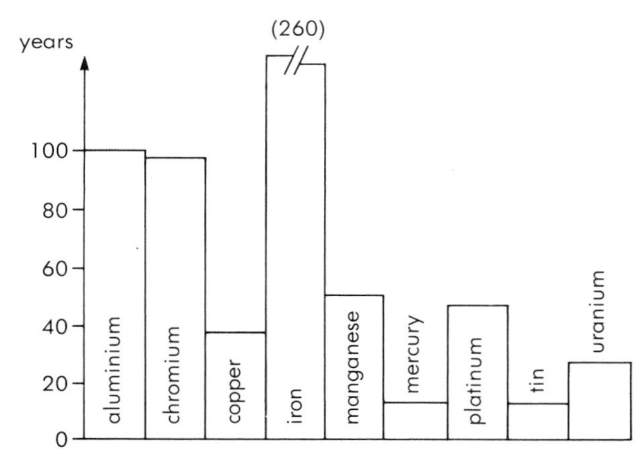

Figure 4 Existing mineral reserves (estimated)

Figure 5 The location of Britain's nuclear power stations

Figure 6 The World's largest nuclear reactor in Canada

Figure 7 A geothermal power station in New Zealand

14.3 Pollution

How is pollution created?

Pollution is a big problem in the world today. It is created when potentially harmful substances are added to the environment at a rate faster than the environment can deal with. These substances, called '**pollutants**', build up in the atmosphere, rivers, lakes and soil and interfere with natural ecological systems.

We now more than ever need to think about waste products especially when we decide what source of energy to use.

How are we polluting the land?

Our activities which pollute the land are:

- **The dumping of waste materials:**
  1. **Refuse** In the UK about 15 million tons of refuse are produced every year. If this were allowed to build up, it could create a serious health hazard. This happened during the refuse collectors' strike in 1975 (figure 1).

     Rotting organic matter provides an ideal breeding ground for disease-causing organisms and their vectors. Often epidemics of typhoid and cholera are started in this way. How refuse is dealt with is outlined on page 210.
  2. **Industrial wastes** from manufacturing and processing industries. These range from scrap metal to wood shavings (figure 2). Coal slag is particularly difficult to get rid of. This is the unusable material also brought out when coal is mined. From some mines, nearly as much slag is produced as coal. Slag can be used as hardcore in road building or to fill holes, but a lot of it is just dumped, forming great mounds. One such slag heap collapsed and destroyed a school in Aberfan in Wales in 1966. Many children lost their lives in the disaster. (See page 232).
  3. **Radioactive waste** The present type of nuclear reactors used in the UK only make use of about 0.7 per cent of the **uranium** they are supplied with. Of the residue, about 1 per cent is useful **plutonium** and the rest is considered to be waste. After **reprocessing** to extract the plutonium, this waste must be stored until it becomes harmless or a use can be found. The safe storage of this potentially dangerous material has been the subject of much research. Most of the UK's most radioactive waste has so far been set in glass and stored in concrete lined pits (figure 3). As the amount of radioactive waste increases, new and more permanent storage areas must be found. One idea is to bury it very deep in stable rock. Some people say there is no safe way of storing radioactive waste (figure 4).
- **Using fertilisers and pesticides.** Pesticides and fertilisers are used to improve food production:
  1. Fertilisers are produced from fossil fuels, a process that in itself creates pollutants. They are put on the land to increase the amount of nutrients in the soil. If applied carelessly, or in the wrong amounts they can seriously imbalance the ecosystem.
  2. Pesticides are a particular type of chemical sprayed onto crops or land to reduce competition (kill unwanted species). These are usually very persistent (figure 5), and over a short period of time can build up, becoming very poisonous to other organisms they are not supposed to kill.

  Pesticides and fertilisers can both do a lot more damage if they enter the waterways.
- **By creating dereliction** If houses, factories and land are left untended they can become breeding grounds for vermin which often spread disease.

Questions

1. Pollution started getting worse about the same time as the industrial revolution. What is the link?
2. List the ways in which we pollute the land.
3. A farmer sprays his crops with a pesticide. Later the same year many rare insect eating birds die. A post mortem reveals the cause of death as being a large dose of the same pesticide. The farmer insists that he cannot be held responsible as he only used a very dilute concentration of the pesticide. Write a letter to him explaining why he is responsible.
4. Everybody at some time adds to the pollution of the land. This might be simply by throwing away a sweet wrapping paper. Think carefully and then write down a list of all the ways you have helped to create pollution during the last few weeks.

Figure 1 When refuse builds up it becomes a health hazard

Figure 2 Abandoned cars. Often not thought of as pollution. What do you think?

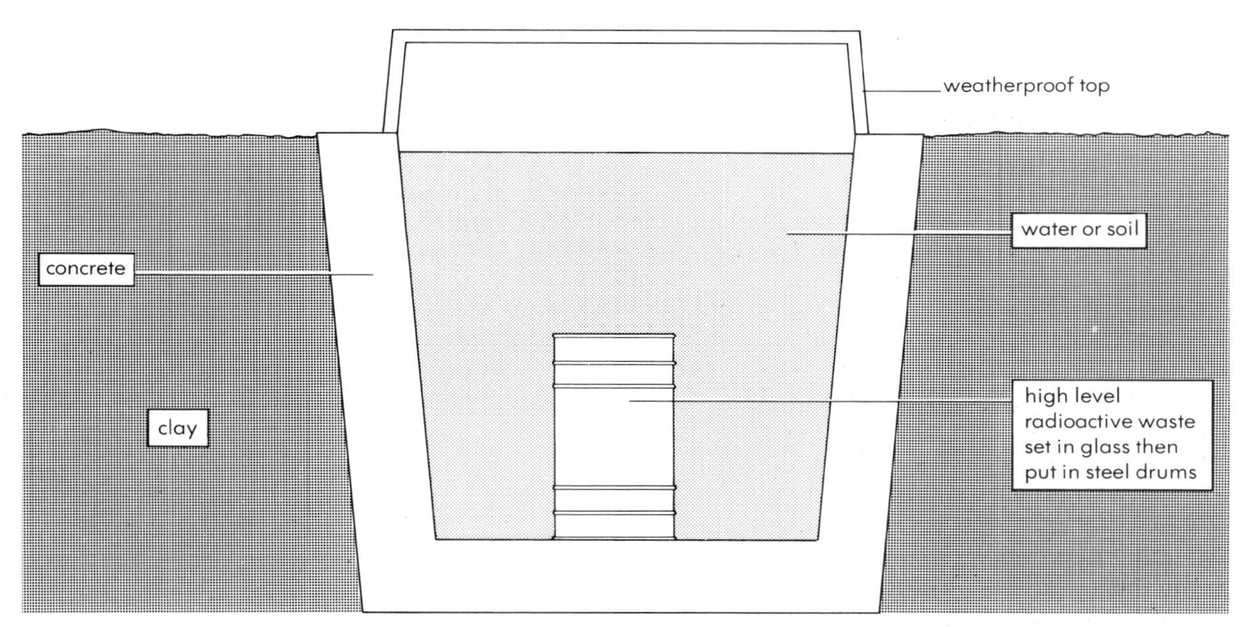

Figure 3 How Britain's nuclear waste is stored

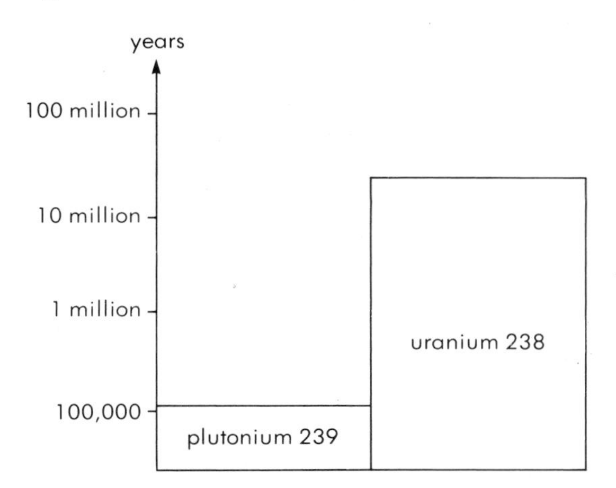

Figure 4 How long does nuclear waste stay radioactive?

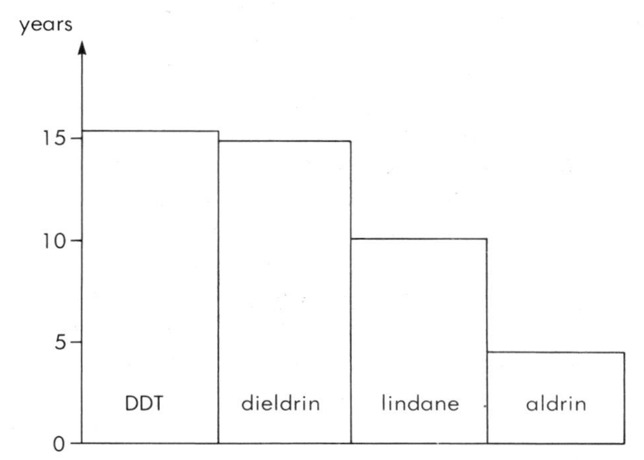

Figure 5 How long do some common pesticides stay active for?

14.4 Polluting the air

Spreading pollution

A lot of harmful waste chemicals and other substances are released straight into the atmosphere. This is very worrying because it means they can be carried anywhere, especially when dissolved in rain, causing harm in all parts of the world.

How are we polluting the air?

When coal, coke, paraffin and oil are burnt in their pure forms they release carbon dioxide, water, oxides of nitrogen, soot and heat into the atmosphere. But when impure forms (as they nearly always are) are burnt, sulphur dioxide (figure 1), carbon monoxide, some metal vapours and metal oxides can also be produced, and discharged into the air.

All these pollutants can do a great deal of damage to life forms in the environment, but the one which is the biggest worry at present is **sulphur dioxide**.

What harm does sulphur dioxide do?

- It irritates the delicate mucous membranes lining the respiratory system and causes lung diseases.
- It dissolves in rain, forming **sulphuric acid** which can corrode stonework on buildings. You may have read of the problems many countries are facing in trying to protect ancient buildings, like the Acropolis in Athens.
- It can be carried many miles away from its source before falling as **acid rain**. This acid rain can damage plants, acidify water in rivers and lakes and cause unwanted chemical reactions in soil. Trees are particularly badly affected by acid rain (figure 2), and it has been estimated that one third of West Germany's forests have been damaged, and one fifth completely destroyed. Twenty per cent of Sweden's lakes are also damaged, one fifth of these so badly that no fish can live in them. A similar situation exists in Norway. Much of the acid rain causing this was produced with sulphur dioxide made in the UK (figure 1).
- Many plants absorb sulphur dioxide directly and this kills them. **Lichens** in particular absorb it well, some species more than others. These are often used as indicators of the level of sulphur dioxide pollution in an area.

What about pollution from petrol fumes?

The exhaust fumes from petrol powered motor vehicles contain large quantities of carbon monoxide, oxides of nitrogen, soot and hydrocarbons. The major concern here is the carbon monoxide. This is an odourless gas which will combine irreversibly with haemoglobin in blood, thereby reducing its oxygen carrying capacity.

Of equal concern is the lead content of exhaust fumes. Lead is added to petrol to improve the performance of a car engine. High concentrations in the body of a child can cause brain damage and then death.

How else do we pollute the air?

- **Radioactive materials** are now used extensively in medicine, power stations and bombs. It is very difficult to handle them without at sometime releasing some radiation into the atmosphere. Figure 4 shows the main sources of radiation in the atmosphere.

  The big worry with radiation is that it can cause mutations that can be passed on to future children (see page 178). It has also been linked with some forms of cancer.
- The mining and processing of asbestos, coal and stone produces dust containing fibres and particles. Inhaling the dust produced can lead to the **pneumoconiosis** diseases, such as **asbestosis**.
- Aerosol cans produce chlorofluorocarbons which destroy the **ozone** layer in the upper atmosphere. This ozone usually filters out most of the sun's harmful radiations such as ultraviolet rays. A large dose of these rays can cause cancers.

  Heat produced from crowded communities such as towns can interfere with the climate, causing flash floods, lightning and the build up of smogs. The city of Los Angeles suffers from particularly bad petrochemical smogs.

Questions

1. Design a table to summarise air pollution.
2. Write a list of all the aerosol sprays you and your family use in a week.
3. How can you personally help to prevent air pollution?
4. List some reasons to convince your parents to change to lead-free petrol.

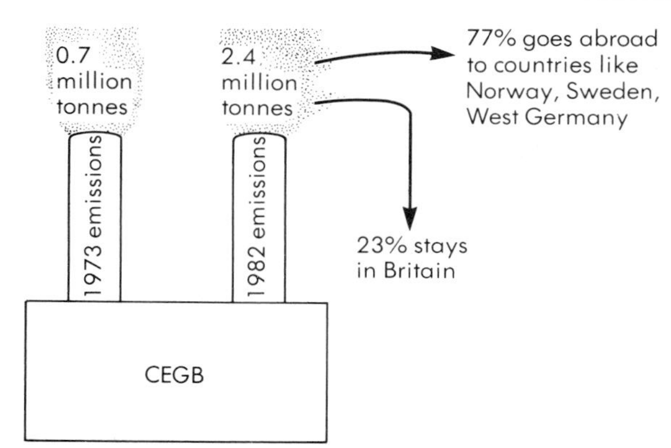

Figure 1 Most sulphur dioxide in the UK is produced by the Central Electricity Generating Board

Figure 2 Damage to a forest caused by acid rain

Figure 3 Testing atomic bombs produces harmful radiation

Radiation source	Percentage of total
1 background	
– cosmic rays	26
– terrestrial	35
– within your body	22
2 medical	
– X-rays	10
– others	4
3 man made	
– bomb tests	2
– nuclear power wastes	less than 1

Figure 4 The sources of radiation pollution (UK 1974)

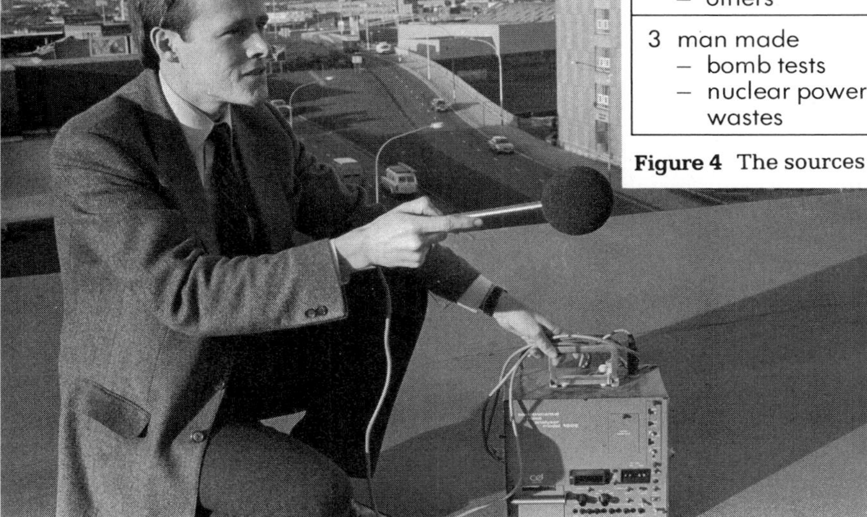

Figure 5 The level of noise pollution is continually monitored

14.5 Polluting the water

The importance of water

Water is essential to us and all other living organisms. By polluting the water we are doing immediate harm to the fish and organisms that live in it and long term harm to the balance of nature and to ourselves.

How are we polluting the water?

Dumping waste materials such as domestic sewage, industrial wastes and heat, into rivers, lakes and seas (figure 1):

- **Domestic sewage** This contains two main ingredients: **organic matter** in the form of human excrement and **detergents**. If it is discharged raw into the water, the organic matter acts like fertiliser, feeding the bacteria and enabling them to grow rapidly. This large increase in the bacteria population uses up much of the oxygen in the water and other organisms have to move on or suffocate.

  The decomposition of the organic matter also releases **nitrates** into the water. The decomposition of detergents releases **phosphates**. These minerals enable the **algae** to grow more rapidly producing an algal bloom which can choke the waterway. When the algae die, they fall to the bottom where they become food for the bacteria. These bacteria grow and remove oxygen from the water and so on. This whole process is called **eutrophication** and can lead to vast areas of water becoming lifeless.
- **Industrial wastes** These often contain chemicals such as **lead**, **mercury**, **arsenic**, **cyanide** and **ammonia**. Some of these are poisonous, even in small amounts, whereas others first need to be concentrated along a food chain (figure 2).
- **Heat** This is produced when electricity is generated. This heat can be disposed of in two ways, either in a cooling tower or in a nearby body of water. Heat discharged into water can have two main effects:
  1. Warm water holds less oxygen and therefore will not support the more active animals.
  2. Growth and development of organisms proceeds more rapidly in warmer water. Many insects for example, may reach adulthood before their food is present and may not have enough food to sustain life. Algae will bloom, again starting off the process of eutrophication.

What about fertilisers and pesticides?

Fertilisers and pesticides often soak through the soil into nearby rivers and lakes. Pesticides build up in the food chains and may even enter our bodies when we eat fish, prawns etc. Fertilisers can cause eutrophication.

What happens when oil is spilled?

About six million tonnes of oil are lost every year into the rivers and seas. This comes from industrial discharges, from leaks during drilling on rigs and from ships washing out their tanks (figure 3). Serious oil spillages also occur when large tankers carrying oil are involved in accidents (figure 4).

Oil lost in these ways has several harmful effects:

- It causes a lot of economic damage by spoiling beaches. Many take up to five years to recover.
- It kills and damages seabirds.
- It poisons marine invertebrates, imbalancing important food webs.
- It poisons other plants and animals.

Questions

1. If the amount of pollution in water is low, then the organisms in the water can cope and even clean it up, provided there is enough oxygen. The level of pollution in water can be indicated by the amount of oxygen needed to clean it up. This is called the Biological Oxygen Demand (BOD). Explain why the dumping of sewage into a river increases the BOD of that river.
2. The Americans operate a system whereby if someone wants to discharge waste into a river, they must draw their drinking water from that river from a point nearer the estuary. Is this a good or bad idea? State your reasons.
3. State a scientific reason for each of the following:
   a) Fertiliser entering a river will kill the fish.
   b) Discharging heat into a lake will affect the food webs.
   c) Discharging lead into a river at a concentration of one part per million can still result in brain damage in humans.

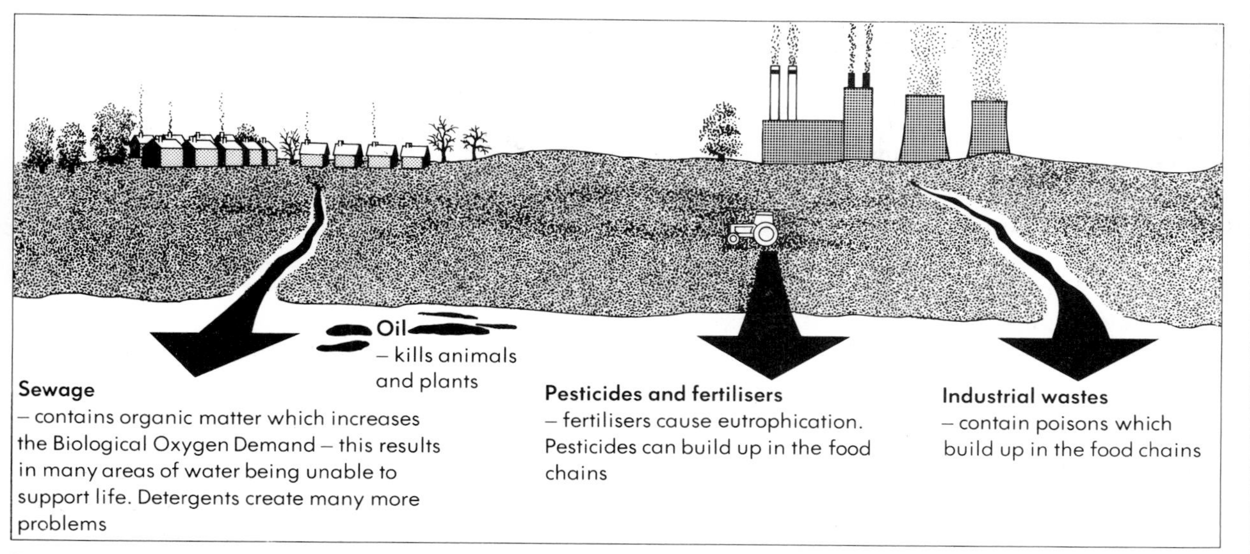

Figure 1 Some of the many ways we pollute waters

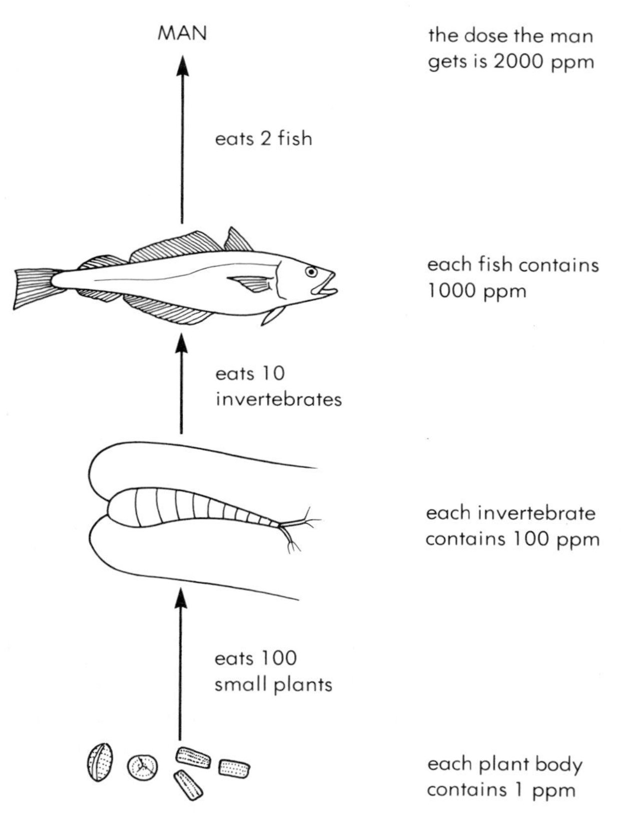

Figure 2 Some chemicals are concentrated along a food chain

Figure 3 Oil spillages over the last 20 years have killed at least 20 000 birds around the coast of the UK

Source	Percentage
Factories and refineries	50%
Discharges from ships	34%
Accidents	6%
Natural sources	10%

Figure 4 Sources of oil in water

Figure 5 Detergent has caused foaming on this river

14.6 Conservation

What is conservation?

Deforestation, over-exploitation of resources and excessive pollution are all serious threats to the world we live in. They are also all products of human interference. Fortunately, we now recognise this and, to some extent, have even started trying to do something about it. This is what **conservation** is concerned with – the use of the earth's natural resources *wisely*, not only for our benefit but for all other organisms too.

What is being done?

The problem is being tackled from four different directions:

- **By not repeating past mistakes**. Tragedies have highlighted past errors and research has told us why they happened. In the light of this Governments have acted to avoid further disasters. For example, after the dreadful London smog in 1952–53, which killed about 4000 people, the **Clean Air Acts** were introduced. Among other things these brought in smokeless zones and restricted the sale of coal. Smogs are now a thing of the past in England. Since then, many further regulations have been introduced. The **Environmental Health Service** was set up in part to enforce these.

  There are also many voluntary bodies, such as **Greenpeace**, which act as watchdogs. The WHO and other UN agencies are influential in bringing about more widespread changes based on their own information and research.

- **By better management of existing resources** The fishing industry offers a good example. Over-fishing for many years has been a problem and many people thought it could not be solved. However, detailed studies of fish populations have shown that they do need to be fished, but not quite so extensively. Too much fishing results in smaller fish, too little can result in larger fish of poorer quality because of old age and disease. The ideal amount to fish is somewhere in between and this can be obtained by careful management.

  There are also now several bodies in the UK which are concerned with protecting organisms from exploitation and extinction (killing so that none remain).

  Examples of such organisations are:

  1. **The Forestry Commission** which protects and creates forests.
  2. **The National Trust** and **Nature Conservancy Council** which oversee vast areas of the countryside and protect the natural ecosystems (figure 1).
  3. **The RSPB and RSPCA** which protect animals.

- **By finding and exploiting new and better resources**, preferably renewable ones. There is a lot of research going on into new energy and food sources. This is dealt with in the relevant sections.
- **By correcting problems created in the past** In the UK, the most notable success is the cleaning up of the River Thames. In the 1950s the portion of the Thames from the sea to Chelsea was so badly polluted that few, if any, fish could live there. The introduction of new sewage and waterworks has changed this. Over 80 species of fish now live there, and even salmon, which will only live in highly oxygenated water, can survive.

Can we put everything right?

Not all problems can be put right. The cancers, deaths and mutations caused by radiation leaks will never go away. Land made useless by the overuse of pesticides will only heal by itself and the organisms which are now extinct will never return.

Questions

1. It has been suggested that the success of intelligent life forms often leads to their destruction. What do you think this means?
2. Conservation is really about careful management. Explain how careful management can help protect a species yet still enable us to use it as food.
3. So far this century about 100 species of animal have become extinct. Part of the reason must be because we exploit them. Make a list of all the ways in which we make use of animals.
4. In 1981 the Wildlife and Countryside Act became law. This law provides protection for many native plants and animals in the UK. What else have past governments done to protect wildlife? What else should they do?

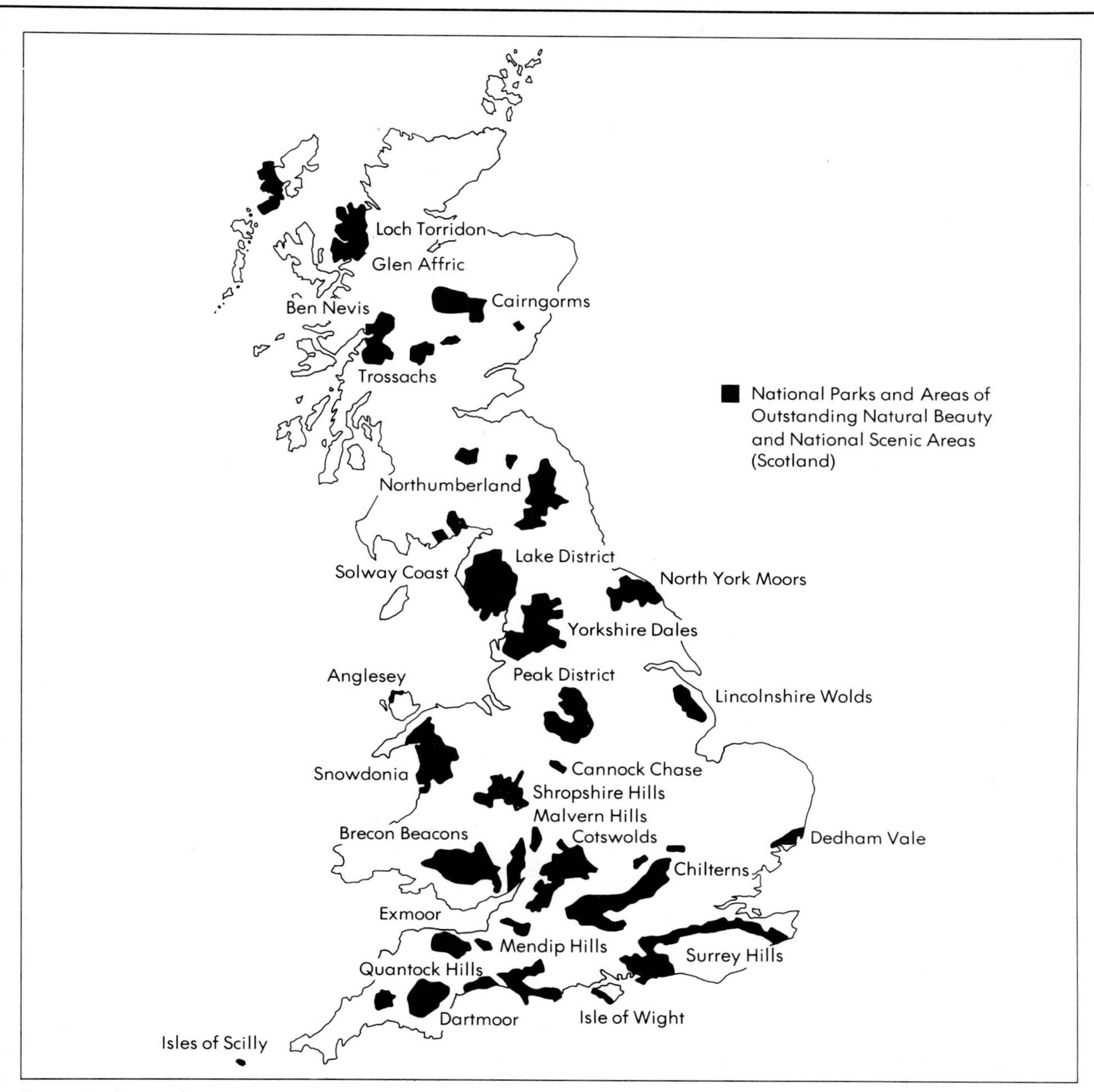

Figure 1 Some of the protected areas in the UK

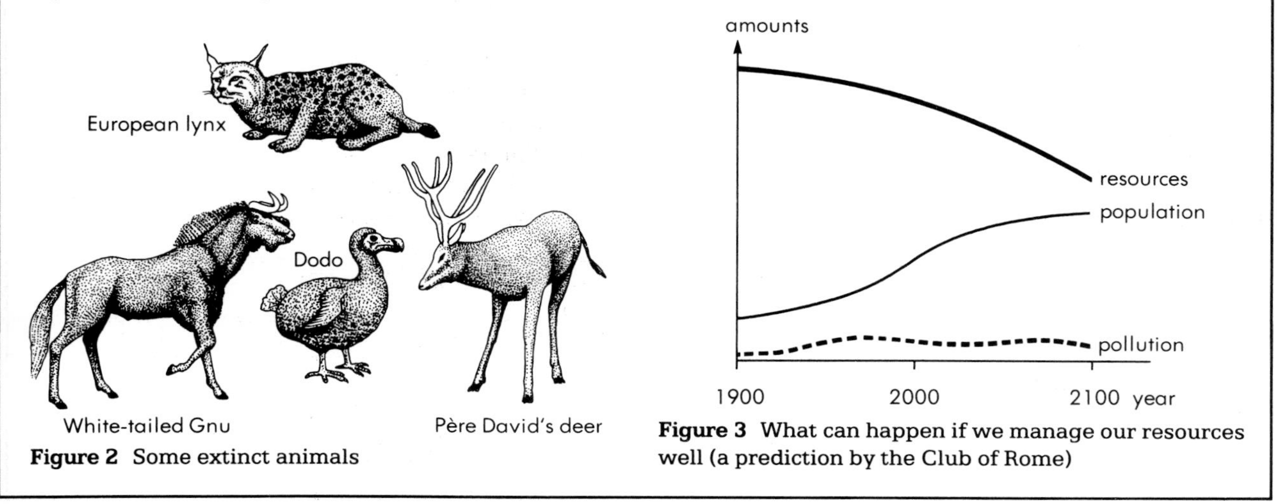

Figure 2 Some extinct animals

Figure 3 What can happen if we manage our resources well (a prediction by the Club of Rome)

Questions

Recall and understanding

1 Deforestation does not result in:
A infertile land
B more rain
C more available land
D more atmospheric carbon dioxide

2 Most of our energy at present comes from:
A oil C nuclear
B coal D gas

3 Lichens are good indicators of:
A radioactivity levels
B rainfall
C carbon monoxide pollution
D sulphur dioxide pollution

4 Acid rain is formed when:
A acid escapes from industry
B oxides of sulphur and nitrogen dissolve in rain
C radioactivity is high
D acid spills into waterways

5 Most atmospheric radiation is due to:
A natural events
B use of medical equipment
C atom bomb testing
D nuclear waste reprocessing

6 Eutrophication is due to:
A oil spillage into water
B discharging industrial chemicals into water
C oxygen depletion of water
D pesticides building up along a food chain

7 An increase in population will not result in:
A more pollution
B more resources
C less resources
D less food per person

8 The acceptable level of noise pollution is:
A 100 dB C 120 dB
B 0 dB D 80 dB

9 Acid rain does not:
A destroy trees
B damage stone buildings
C kill fish
D damage cars

10 Which of the following rivers is the most seriously polluted?
A BOD of 200 C BOD of 147
B BOD of 10 D BOD of 1000

Interpretation and application of data

11 The data below refers to the UK:

All figures are in million tonnes

Year	Sulphur dioxide: emissions from fuel combustion			
	Domestic	Power stations	Other industry	All sources
1951	0.87	1.02	2.88	4.77
1956	0.78	1.29	3.06	5.13
1961	0.79	1.98	2.93	5.70
1966	0.68	2.29	2.96	5.93
1971	0.46	2.80	2.57	5.83
1976	0.28	2.69	2.01	4.98
1977	0.29	2.74	1.95	4.98
1978	0.26	2.81	1.95	5.02
1979	0.26	3.10	1.97	5.34
1980	0.22	2.87	1.58	4.67
1981	0.21	2.71	1.31	4.22
1982	0.20	2.62	1.20	4.01
1983	0.20	2.53	0.97	3.69
1984	0.16	2.54	0.84	3.53

a) Where did most sulphur dioxide come from in 1984?
b) Has this source always produced the most?
c) Is the total amount of sulphur dioxide emitted rising or falling?
d) Why do you think the amount of sulphur dioxide emitted from power stations has increased?
e) The table below shows the percentage sulphur dioxide (of total) produced by the two main types of fuel.

Type of fuel used by power station	Coal	Oil	Other
Percent sulphur dioxide	57	9	34

i) Which type of power station produces most sulphur dioxide?
ii) Name some of the power stations in the 'other' category.

12

Radiation exposure of the population: by source, 1984 United Kingdom

Source of radiation	Percentage of total dose
Natural sources	
Cosmic rays	14.0
Terrestrial gamma rays	18.6
Internal activity	17.2
Radon decay products	32.4
Thoron decay products	4.7
Total natural sources	86.9
Man-made sources	
Medical irradiation	11.6
Fallout from weapon testing	0.5
Occupational exposure	0.4
Disposal of radioactive waste	0.1
Miscellaneous	0.5
Total man-made sources	13.1

a) Based solely on the above figures should we be more concerned about the disposal of radioactive waste or the use of medical equipment?
b) What sort of medical equipment is involved?
c) The table below shows the risks of providing energy.

	Coal	Gas	Oil	Nuclear
Death rate per million workers	230	70	90	15

i) Which industry is the safest to work in?
ii) Which industry is the most dangerous?

3 The figures below relate to the use of pesticides.

Amount of pesticide used (ppm)	Resistant organisms (%)
1	0
10	1
100	4
1000	17
10 000	23

a) What does the above data tell you about the dangers of using pesticides?
b) One danger is that pesticides build up in food chains.

Organisms	Amount of last organism eaten	Concentration of pesticide in body (ppm)
Phytoplankton	-	0.1 ppm
Zooplankton	10	
Herring	400	
Human	2	

i) Copy and complete this table to show how the pesticide builds up along the food chain.
ii) The chances of building up will depend on how persistent the pesticide is. What does this mean?

c) What is the alternative to using pesticides?

14 The graph below shows how the consumption of various minerals has increased with time.

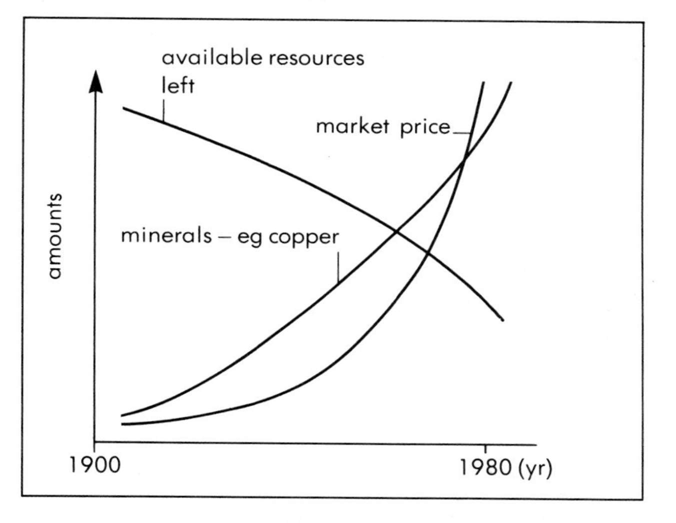

a) If present trends continue, what will happen in the near future to our mineral reserves?
b) As a substance becomes more scarce, its price goes up. Why might this be a good thing?
c) Another solution is to recycle more resources. Consider the following:

	Percentage which could be recycled	Available resources left (yr)
Aluminium	25	100
Zinc	40	40

i) How much would recycling extend the normal life of aluminium and zinc?
ii) Recycling is expensive. Suggest why it will always become economically viable sooner or later.

Index

ABO blood system 49
absorption 40, 41
acclimatisation 63
accommodation 80, 81
acid rain 240
active transport 12
adenosine tri-phosphate (ATP) 64
adolescence 150
adrenalin 96
ageing 152, 153
agriculture 26, 27, 28, 29
AIDS 157, 228, 229
air pressure 62
alcohol 92, 93
alimentary canal 36, 37
allele 174
allergy 224
alternative food sources 16, 17
amino acids 18, 31
amniocentesis 136, 137
anabolism 10, 11
analgesics 90
angina 56
anorexia nervosa 32, 34
antagonistic muscles 108, 109
antenatal care 136
antenatal clinic 136
antibiotics 222, 223
antibody 196, 224, 225
antigen 224, 225
antiseptic 222
arteries 52, 53
arthritis 106
artificial insemination 127
asbestosis 240
asexual reproduction 128
astigmatism 80, 81
athletes foot 202, 203
autonomic nervous system (ANS) 84

bacteria
  as food 17
  diseases caused by 199
  structure of 196, 197
balance 82
barbiturates 90, 91
basal metabolism 64
bends 63
bile 38
binocular vision 78
biogas 26, 236
biological control 196, 220
biological washing powders 2, 3
biomass 26
biotechnology 194, 195
birth 134,135
birth control 166, 167
bleeding 58, 59
blinking 78
blood 50, 51
  bank 48
  clot 59
  groups 48, 49, 50, 51
    inheritance of 176, 177
  sugar 122, 123
  transfusions 48, 49
  vessels 52, 53
bones 102, 104, 105
  growth of 105
booster 226
brain 86, 87
breast feeding 145
breathing 66, 68, 69
  at altitude 62, 63
  and exercise 68
  rate 68, 69
  under water 63
breeds 160
bronchitis 70

caesarian birth 134
cancer 6, 7
capillaries 52, 53
carbohydrates 18
carbon cycle 24, 25
cardiac cycle 54, 55
cardiac muscle 54
cartilage 104, 105
catabolism 10, 11
cell
  division 6, 7, 128, 129, 173
  membrane 4
  structure 4, 5
census 162
central nervous system (CNS) 84, 86
cerebrospinal fluid 86
child
  development 146, 147
  growth 146, 147
chlorophyll 22
chloroplasts 4
chromosomes 5, 172
circulatory system of the newborn baby 145
Club of Rome 164, 245
co-dominance 176, 177
cochlea 76, 83
conditioned reflex 88
conservation 244, 245
constipation 38
consumers 22
continuous variation 178, 179
contraception 166, 167
coronary vessels 54, 57
cramp 64
cross matching of blood 49
cultivators and herders 28
cytoplasm 4

decomposers 22
defaecation 118
deforestation 234, 235
dementia 188
Department of Health and Social Security (DHSS) 208, 216, 217
depressant drugs 90, 91
depression 94, 95
diabetes 113, 122
dialysis 120
diarrhoea 38
diastole 54
diet 30, 35
diffusion 12, 13
digestion 36, 37
discontinuous variation 178, 179
disinfectant 222
division of labour 28
Down's Syndrome 136
drugs 90, 91, 101, 223
  clinical trials 75
  dependence on 92
  screening of 75
  sources of 74
dysentery 200

ear 82, 83
ectothermic 114
Ehrlich, Paul 222
ejaculation 130
embryo 130, 132
emotional development of a child 148, 149
emphysema 70, 71
endocrine glands 96, 97
endocrine system 96, 97
endothermic 114
energy 64
Environmental Health Service (EHS) 218, 219
enzymes 2, 3, 10, 36, 38
  and genes 173
  in industry 2, 3
  properties of 10, 11
epidural 134
episiotomy 134
eugenics 161
excretion 118, 120, 121
exercise 154, 155
eye 79

fats 18
fatty acids 18, 31
faeces 38
fermentation 202
fermenter 3, 195
fertilisation 130, 131
fertiliser 26, 27, 28, 29, 242
fertility drugs 127
fibre 30, 31
fitness 154, 155
flatworms 204, 205
Fleming, Alexander 222
fluoride 30, 44
focusing 79, 80, 81
foetus 132, 133
food 18
  additives 20
  chain 22, 26, 27
  energy 26, 27, 28, 65
  handling 212, 213
  mountains 35
  poisoning 198, 199